Mechatronic Systems 1

Mechatronic Systems 1
Applications in Transport, Logistics, Diagnostics, and Control

Edited by

Waldemar Wójcik
Department of Electronics and Information Technology, Lublin University of Technology, Lublin, Poland

Sergii Pavlov
Vinnytsia National Technical University, Vinnytsia, Ukraine

Maksat Kalimoldayev
Institute of Information and Computational Technologies of the Science Committee of the Ministry of Education and Science of Republic of Kazakhstan, Kazakhstan

LONDON AND NEW YORK

MATLAB® and Simulink® are trademarks of The MathWorks, Inc. and are used with permission. The MathWorks does not warrant the accuracy of the text or exercises in this book. This book's use or discussion of MATLAB® and Simulink® software or related products does not constitute endorsement or sponsorship by The MathWorks of a particular pedagogical approach or particular use of the MATLAB® and Simulink® software.

First published 2021
by Routledge/Balkema
Schipholweg 107C, 2316 XC Leiden, The Netherlands
e-mail: enquiries@taylorandfrancis.com
www.routledge.com – www.taylorandfrancis.com

Routledge/Balkema is an imprint of the Taylor & Francis Group, an informa business

© 2021 selection and editorial matter, Waldemar Wójcik, Sergii Pavlov, and Maksat Kalimoldayev; individual chapters, the contributors

The right of Waldemar Wójcik, Sergii Pavlov, and Maksat Kalimoldayev to be identified as the authors of the editorial material, and of the authors for their individual chapters, has been asserted in accordance with sections 77 and 78 of the Copyright, Designs and Patents Act 1988.

All rights reserved. No part of this book may be reprinted or reproduced or utilised in any form or by any electronic, mechanical, or other means, now known or hereafter invented, including photocopying and recording, or in any information storage or retrieval system, without permission in writing from the publishers.

Although all care is taken to ensure integrity and the quality of this publication and the information herein, no responsibility is assumed by the publishers nor the author for any damage to the property or persons as a result of operation or use of this publication and/ or the information contained herein.

Library of Congress Cataloging-in-Publication Data
A catalog record has been requested for this book

ISBN: 978-1-032-10583-3 (hbk)
ISBN: 978-1-032-12335-6 (pbk)
ISBN: 978-1-003-22413-6 (ebk)

DOI: 10.1201/9781003224136

Typeset in Times New Roman
by codeMantra

Contents

List of contributors ix

1 **Method of project calculation of hydroimpulsive device for vibroturning with an incorporated cycle spring pressure pulse generator** 1
   Roman Obertyukh, Andrii Slabkyi, Leonid K. Polishchuk, Serhii Andrukhov, Larysa E. Nykiforova, Andrzej Smolarz, Aimurat Burlibayev, and Zhanar Azeshova

2 **Determination of the optimal parameters of the driver of a resonance vibratory stand for diagnostics of dampers** 17
   Volodymyr Gursky, Igor Kuzio, Olena Lanets, Yaroslav Zinko, Pavlo Nosko, Konrad Gromaszek, Paweł Droździel, Maksat Kalimoldayev, and Amandyk Tuleshov

3 **Algorithm to calculate work tools of machines for performance in extreme working conditions** 29
   Leonid A. Khmara, Sergej V. Shatov, Leonid K. Polishchuk, Valerii O. Kravchuk, Piotr Kisala, Yedilkhan Amirgaliyev, Amandyk Tuleshov, and Mukhtar Junisbekov

4 **Vibration diagnostic of wear for cylinder-piston couples of pumps of a radial piston hydromachine** 39
   Vladimir Shatokhin, Boris Granko, Vladimir Sobol, Leonid K. Polishchuk, Olexander Manzhilevskyy, Konrad Gromaszek, Mukhtar Junisbekov, Nataliya Denissova, and Kuanysh Muslimov

5 **Mathematical modeling of dynamic processes in the turning mechanism of the tracked machine with hydrovolume transmission** 53
   Vladimir Shatokhin, Boris Granko, Vladimir Sobol, Victor Stasiuk, Maksat Kalimoldayev, Waldemar Wójcik, Aigul Iskakova, and Kuanysh Muslimov

6 **Forecasting reliability of use of gas-transmitting units on gas transport systems**   65
Volodumyr Grudz, Yaroslav Grudz, Vasyl Zapuklyak, Lubomyr Poberezhny, Ruslan Tereshchenko, Waldemar Wójcik, Paweł Droździel, Maksat Kalimoldayev, and Aliya Kalizhanova

7 **Development of main gas pipeline deepening method for prevention of external effects**   75
Vasyl Zapuklyak, Yura Melnichenko, Lubomyr Poberezhny, Yaroslava Kyzymyshyn, Halyna Grytsuliak, Paweł Komada, Yedilkhan Amirgaliyev, and Ainur Kozbakova

8 **Increasing surface wear resistance of engines by nanosized carbohydrate clusters when using ethanol motor fuels**   89
Olga O. Haiday, Volodymir S. Pyliavsky, Yevgen V. Polunkin, Yaroslav O. Bereznytsky, Olexandr B. Yanchenko, Andrzej Smolarz, Paweł Droździel, Saltanat Amirgaliyeva, and Saule Rakhmetullina

9 **Method of experimental research of steering control unit of hydrostatic steering control systems and stands for their realization**   101
Mykola Ivanov, Oksana Motorna, Oleksiy Pereyaslavskyy, Serhiy Shargorodskyi, Konrad Gromaszek, Mukhtar Junisbekov, Aliya Kalizhanova, and Saule Smailova

10 **Possibility of improving the dynamic characteristics of an adaptive mechatronic hydraulic drive**   113
Leonid Kozlov, Yurii Buriennikov, Volodymyr Pyliavets, Vadym Kovalchuk, Leonid Polonskyi, Andrzej Smolarz, Paweł Droździel, Yedilkhan Amirgaliyev, Ainur Kozbakova, and Kanat Mussabekov

11 **Application of feedback elements in proportional electrohydraulic directional control valve with independent flows control**   127
Dmytro O. Lozinskyi, Oleksandr V. Petrov, Natalia S. Semichasnova, Konrad Gromaszek, Maksat Kalimoldayev, and Gauhar Borankulova

12 Optimization of design parameters of a counterbalance valve for a hydraulic drive invariant to reversal loads   137
Leonid Kozlov, Leonid Polishchuk, Oleh Piontkevych, Viktor Purdyk, Oleksandr V. Petrov, Volodymyr M. Tverdomed, Piotr Kisala, Saltanat Amirgaliyeva, Bakhyt Yeraliyeva, and Aigul Tungatarova

13 Calculations of unsteady processes in channels of a hydraulic drive   149
Alexander Gubarev, Alona Murashchenko, Oleg Yakhno, Alexander Tyzhnov, Konrad Gromaszek, Aliya Kalizhanova, and Orken Mamyrbaev

14 Analysis, development, and modeling of new automation system for production of permeable materials from machining waste   161
Oleksandr Povstyanoy, Oleg Zabolotnyi, Olena Kovalchuk, Dmytro Somov, Taras Chetverzhuk, Konrad Gromaszek, Saltanat Amirgaliyeva, and Nataliya Denissova

15 Study of effect of motor vehicle braking system design on emergency braking efficiency   173
Andrii Kashkanov, Victor Bilichenko, Tamara Makarova, Olexii Saraiev, Serhii Reiko, Andrzej Kotyra, Mukhtar Junisbekov, Orken Mamyrbaev, and Mergul Kozhamberdiyeva

16 Essential aspects of regional motor transport system development   185
Volodymyr Makarov, Tamara Makarova, Sergey Korobov, Valentine Kontseva, Piotr Kisala, Paweł Droździel, Saule Smailova, Kanat Mussabekov, and Yelena Kulakova

17 Improvement of logistics of agricultural machinery transportation technologies   197
Ievgen Medvediev, Iryna Lebid, Nataliia Luzhanska, Volodymyr Pasichnyk, Zbigniew Omiotek, Paweł Droździel, Aisha Mussabekova, and Doszhon Baitussupov

18 Optimization of public transport schedule on duplicating stretches   209
Iryna Lebid, Nataliia Luzhanska, Irina Kravchenya, Ievgen Medvediev, Andrzej Kotyra, Paweł Droździel, Aisha Mussabekova, and Gali Duskazaev

19 **Dynamic properties of symmetric and asymmetric layered materials in a high-speed engine** 221
Orest Horbay, Bohdan Diveyev, Andriy Poljakov, Oleksandr Tereschenko, Piotr Kisala, Mukhtar Junisbekov, Samat Sundetov, and Aisha Mussabekova

20 **Selection and reasoning of the bus rapid transit component scheme of huge capacity** 233
Volodymyr Sakhno, Victor Poliakov, Victor Bilichenko, Igor Murovany, Andrzej Kotyra, Gali Duskazaev, and Doszhon Baitussupov

21 **Physical-mathematical modelling of the process of infrared drying of rape with vibration transport of products** 243
Igor P. Palamarchuk, Mikhailo M. Mushtruk, Vladislav I. Palamarchuk, Olena S. Deviatko, Waldemar Wójcik, Aliya Kalizhanova, and Ainur Kozbakova

22 **Energy efficiency of gear differentials in devices for speed change control through a carrier** 255
Volodymyr O. Malashchenko, Oleh R. Strilets, Volodymyr M. Strilets, Vladyslav L. Lutsyk, Andrzej Smolarz, Paweł Droździel, A. Torgesizova, and Aigul Shortanbayeva

23 **Influence of electrohydraulic controller parameters on the dynamic characteristics of a hydrosystem with adjustable pump** 267
Volodymyr V. Bogachuk, Leonid H. Kozlov, Artem O. Tovkach, Valerii M. Badakh, Taras V. Tarasenko, Yevhenii O. Kobylianskyi, Zbigniew Omiotek, Gauhar Borankulova, and Aigul Tungatarova

24 **High-precision ultrasonic method for determining the distance between garbage truck and waste bin** 279
Oleh V. Bereziuk, Mykhailo S. Lemeshev, Volodymyr V. Bogachuk, Piotr Kisala, Aigul Tungatarova, and Bakhyt Yeraliyeva

List of contributors

Saltanat Amirgaliyeva Institute of Information and Computational Technologies CS MES RK, Almaty, Kazakhstan and IT department, Academy of Logistics & Transport, Almaty, Kazakhstan

Yedilkhan Amirgaliyev Institute of Information and Computational Technologies CS MES RK, Almaty, Kazakhstan

Serhii Andrukhov Industrial Engineering Department, Vinnytsia National Technical University Vinnytsia, Ukraine

Zhanar Azeshova Institute of Automation and Information Technologies, Satbayev University, Almaty, Kazakhstan

Valerii M. Badakh Hydro Gas Systems Department, National Aviation University, Kyiv, Ukraine

Doszhon Baitussupov IT department, Academy of Logistics & Transport, Almaty, Kazakhstan

Oleh V. Bereziuk Safety of Life and Security Pedagogy Department, Vinnytsia National Technical University, Vinnytsia, Ukraine

Yaroslav O. Bereznytsky Homogeneous Catalysis and Additives for Petroleum Products Department, V.P Kukhar Institute of Bioorganic Chemistry and Petrochemistry NAS of Ukraine, National University of "Kyiv-Mohyla Academy", Kyiv, Ukraine

Victor Bilichenko Automobiles and Transport Management Department,Vinnytsia National Technical University, Vinnytsia, Ukraine

Volodymyr V. Bogachuk Scientific and Research Department, Vinnytsya National Technical University, Vinnytsya, Ukraine

Gauhar Borankulova Faculty of Information Technology, Automation and Telecommunications, M.Kh.Dulaty Taraz Regional University, Taraz, Kazakhstan

Yurii Burienikov Machine-building technology and Automation Department, Vinnytsia National Technical University, Vinnytsia, Ukraine

Aimurat Burlibayev Faculty of Information Technology, Al-Farabi Kazakh National University, Almaty, Kazakhstan

Taras Chetverzhuk Applied Mechanics and Mechatronics Department, Lutsk National Technical University, Lutsk, Ukraine

Nataliya Denissova East Kazakhstan State Technical University named after D.Serikbayev, Ust-Kamenogorsk, Kazakhstan

Olena S. Deviatko Technology Production and Processing of Livestock and Veterinary Products Faculty, Farm Animals and Aquatic Bioresources Departments, Vinnytsia National Agrarian University, Vinnytsia, Ukraine

Bohdan Diveyev Mechanical Engineering and Transport Institute Automotive Engineering Department, National University «LvivskaPolytechnika», Lviv, Ukraine

Paweł Droździel Faculty of Mechanical Engineering, Lublin University of Technology, Lublin, Poland

Gali Duskazaev IT department, Academy of Logistics & Transport, Almaty, Kazakhstan

Boris Granko Theoretical Mechanics Department, Kharkiv National University of Civil Engineering and Architecture, Kharkiv, Ukraine

Konrad Gromaszek Faculty of Electrical Engineering and Computer Science, Lublin University of Technology, Lublin, Poland

Volodumyr Grudz Oil and Gas Pipelines and Storage Facilities Department, Ivano-Frankivsk National Technical University of Oil and Gas, Ivano-Frankivsk, Ukraine

Yaroslav Grudz Oil and Gas Pipelines and Storage Facilities Department, Ivano-Frankivsk National Technical University of Oil and Gas, Ivano-Frankivsk, Ukraine

Halyna Grytsuliak Oil and Gas Pipelines and Storage Facilities Department, Ivano-Frankivsk National Technical University of Oil and Gas, Ivano-Frankivsk, Ukraine

Alexander Gubarev Applied Hydro-Aeromechanics and Mechatronics Department, Igor Sikorsky Kyiv Polytechnic Institute, Kyiv, Ukraine

Volodymyr Gursky Department of Mechanics and Automation of Mechanical Engineering, Institute of Engineering Mechanics and Transport, Lviv Polytechnic National University, Lviv, Ukraine

Olga O. Haiday Homogeneous Catalysis and Additives for Petroleum Products Department, V.P Kukhar Institute of Bioorganic Chemistry and Petrochemistry NAS of Ukraine, Kyiv, Ukraine

Orest Horbay Mechanical Engineering and Transport Institute Automotive Engineering Department, National University «LvivskaPolytechnika», Lviv, Ukraine

Aigul Iskakova Institute of Automation and Information Technologies, Satbayev University, Almaty, Kazakhstan

Mykola Ivanov Machinery and Equipment of Agricultural Production Department, Vinnytsia National Agrarian University, Vinnytsia, Ukraine

Mukhtar Junisbekov Faculty of Information Technology, Automation and Telecommunications, M.Kh.Dulaty Taraz Regional University, Taraz, Kazakhstan

Maksat Kalimoldayev Institute of Information and Computational Technologies CS MES RK, Almaty, Kazakhstan

Aliya Kalizhanova Institute of Information and Computational Technologies CS MES RK and Faculty of Information Technology, Al-Farabi Kazakh National University, Almaty, Kazakhstan

Andrii Kashkanov Automobiles and Transport Management Department, Vinnytsia National Technical University, Vinnytsia, Ukraine

Leonid A. Khmara Construction and Road Machinery Department, Pridneprovsk State Academy of Civil Engineering and Architecture, Dnipro, Ukraine

Piotr Kisala Faculty of Electrical Engineering and Computer Science, Lublin University of Technology, Lublin, Poland

Yevhenii O. Kobylianskyi Industrial Engineering Department, National Aviation University, Kyiv, Ukraine

Paweł Komada Faculty of Electrical Engineering and Computer Science, Lublin University of Technology, Lublin, Poland

Valentine Kontseva Department of Finance, Accounting and Auditing, National Transport University, Kiev, Ukraine

Sergey Korobov Automobiles and Transport Management Department, National Technical University of Vinnytsia, Vinnytsia, Ukraine

Andrzej Kotyra Faculty of Electrical Engineering and Computer Science, Lublin University of Technology, Lublin, Poland

Olena Kovalchuk Applied Mechanics and Mechatronics Department, Lutsk National Technical University, Lutsk, Ukraine

Vadym Kovalchuk Machine-building technology and Automation Department, Vinnytsia National Technical University, Vinnytsia, Ukraine

Ainur Kozbakova Institute of Information and Computational Technologies CS MES RK, Almaty, Kazakhstan and Almaty Technological University, Almaty, Kazakhstan

Mergul Kozhamberdiyeva Faculty of Information Technology, Al-Farabi Kazakh National University, Almaty, Kazakhstan

Leonid H. Kozlov Machine-building technology and Automation Department, Vinnytsia National Technical University, Vinnytsia, Ukraine

Irina Kravchenya Road Traffic Management Department, Belarusian State University of Transport, Gomel, Belarus

Valerii O. Kravchuk Industrial Engineering Department, Vinnitsia National Technical University, Vinnitsia, Ukraine

Yelena Kulakova Institute of Automation and Information Technologies, Satbayev University, Almaty, Kazakhstan

Igor Kuzio Department of Mechanics and Automation of Mechanical Engineering, Institute of Engineering Mechanics and Transport, Lviv Polytechnic National University, Lviv, Ukraine

Yaroslava Kyzymyshyn Oil and Gas Pipelines and Storage Facilities Department, Ivano-Frankivsk National Technical University of Oil and Gas, Ivano-Frankivsk, Ukraine

Olena Lanets Department of Mechanics and Automation of Mechanical Engineering, Institute of Engineering Mechanics and Transport, Lviv Polytechnic National University, Lviv, Ukraine

Iryna Lebid Transport Technologies Department, National Transport University, Kiev, Ukraine

Mykhailo S. Lemeshev Safety of Life and Security Pedagogy Department, Vinnytsia National Technical University, Vinnytsia, Ukraine

Dmytro O. Lozinskyi Machine-building technology and Automation Department, Vinnytsia National Technical University, Vinnytsia, Ukraine

Vladyslav L. Lutsyk Industrial Engineering Department, Vinnytsia National Technical University, Vinnytsia, Ukraine

Nataliia Luzhanska Transport Technologies Department, National Transport University, Kiev, Ukraine

Tamara Makarova Automobiles and Transport Management Department, Vinnytsia National Technical University, Vinnytsia, Ukraine

Volodymyr Makarov Automobiles and Transport Management Department, National Technical University of Vinnytsia, Vinnytsia, Ukraine

Volodymyr O. Malashchenko Machine Parts Department, Lviv Polytechnic National University, Lviv, Ukraine

Orken Mamyrbaev Institute of Information and Computational Technologies CS MES RK, Almaty, Kazakhstan

Olexander Manzhilevskyy Industrial Engineering Department, Vinnytsia National Technical University, Vinnytsia, Ukraine

Ievgen Medvediev Logistics Management and Traffic Safety Department, Volodymyr Dahl East Ukrainian National University, Severodonetsk, Ukraine

Yura Melnichenko Oil and Gas Pipelines and Storage Facilities Department, Ivano-Frankivsk National Technical University of Oil and Gas, Ivano-Frankivsk, Ukraine

Oksana Motorna Machinery and Equipment of Agricultural Production Department, Vinnytsia National Agrarian University, Vinnytsia, Ukraine

Alona Murashchenko Applied Hydro-Aeromechanics and Mechatronics Department, Igor Sikorsky Kyiv Polytechnic Institute, Kyiv, Ukraine

Igor Murovany Automobiles and Transport Management Department, Vinnitsia National Technical University, Vinnitsia, Ukraine

Mikhailo M. Mushtruk Food Technologies and Quality Management of Agricultural Products Faculty, Processes and Equipment for Processing of Agricultural Products Department, National University Life and Environmental Sciences of Ukraine, Vinnytsia, Ukraine

Kuanysh Muslimov Institute of Automation and Information Technologies, Satbayev University, Almaty, Kazakhstan

Aisha Mussabekova IT department, Academy of Logistics & Transport, Almaty, Kazakhstan

Kanat Mussabekov IT department, Academy of Logistics & Transport, Almaty, Kazakhstan

Pavlo Nosko Department of Mechanical Engineering, National Aviation University, Kyiv, Ukraine

Larysa E. Nykiforova Automation and Robotic Systems named acad. I.I. Martynenko Department, National University of Life and Environmental Sciences of Ukraine, Kiev, Ukraine

Roman Obertyukh Industrial Engineering Department, Vinnytsia National Technical University, Vinnytsia, Ukraine

Zbigniew Omiotek Faculty of Electrical Engineering and Computer Science, Lublin University of Technology, Lublin, Poland

Igor P. Palamarchuk Food Technologies and Quality Management of Agricultural Products Faculty, Processes and Equipment for Processing of Agricultural Products Department, National University Life and Environmental Sciences of Ukraine, Vinnytsia, Ukraine

Vladislav I. Palamarchuk Technology Production and Processing of Livestock and Veterinary Products Faculty, Farm Animals and Aquatic Bioresources Departments, Vinnytsia National Agrarian University, Vinnytsia, Ukraine

Volodymyr Pasichnyk Theoretical and Computer Mechanics Department, Oles Honchar Dnipropetrovsk National University, Dnipro, Ukraine

Sergii Pavlov Vinnytsia National Technical University, Vinnytsia, Ukraine

Oleksiy Pereyaslavskyy Machinery and Equipment of Agricultural Production Department, Vinnytsia National Agrarian University, Vinnytsia, Ukraine

Oleksandr V. Petrov Machine-building technology and Automation Department, Vinnytsia National Technical University, Vinnytsia, Ukraine

Oleh Piontkevych Machine-building technology and Automation Department, Vinnytsia National Technical University, Vinnytsia, Ukraine

Lubomyr Poberezhny Oil and Gas Pipelines and Storage Facilities Department, Ivano-Frankivsk National Technical University of Oil and Gas, Ivano-Frankivsk, Ukraine

Victor Poliakov Automobiles Department, National Transport University, Kyiv, Ukraine

Leonid Polishchuk Industrial Engineering Department, Vinnytsia National Technical University, Vinnytsia, Ukraine

Andriy Poljakov Automobiles and Transport Management Department, Vinnitsia National Technical University, Vinnitsia, Ukraine

Leonid Polonskyi Mechanical Engineering Department, State University "Zhytomyr Polytechnic", Zhytomyr, Ukraine

Yevgen V. Polunkin Homogeneous Catalysis and Additives for Petroleum Products Department, V.P Kukhar Institute of Bioorganic Chemistry and Petrochemistry NAS of Ukraine, Kyiv, Ukraine

Oleksandr Povstyanoy Applied Mechanics and Mechatronics Department, Lutsk National Technical University, Lutsk, Ukraine

Viktor Purdyk Machine-building technology and Automation Department, Vinnytsia National Technical University, Vinnytsia, Ukraine

Volodymyr Pyliavets Machine-building technology and Automation Department, Vinnytsia National Technical University, Vinnytsia, Ukraine

Volodymir S. Pyliavsky Homogeneous Catalysis and Additives for Petroleum Products Department, V.P Kukhar Institute of Bioorganic Chemistry and Petrochemistry NAS of Ukraine, Kyiv, Ukraine

Saule Rakhmetullina East Kazakhstan State Technical University named after D.Serikbayev, Ust-Kamenogorsk, Kazakhstan

Serhii Reiko Engineering, Economic, Commodity Research Department, Zhitomir Scientific Research Forensic Center of the Ministry of Internal Affairs of Ukraine, Zhytomyr, Ukraine

Volodymyr Sakhno Automobiles Department, National Transport University, Kyiv, Ukraine

Olexii Saraiev Automobiles Department, Kharkiv National Automobile and Highway University, Kharkiv, Ukraine

Natalia S. Semichasnova Machine-building technology and Automation Department, Vinnytsia National Technical University, Vinnytsia, Ukraine

Serhiy Shargorodskyi Machinery and Equipment of Agricultural Production Department, Vinnytsia National Agrarian University, Vinnytsia, Ukraine

Vladimir Shatokhin Theoretical Mechanics Department, Kharkiv National University of Civil Engineering and Architecture, Kharkiv, Ukraine

Sergej V. Shatov Construction and Road Machinery Department, Pridneprovsk State Academy of Civil Engineering and Architecture, Dnipro, Ukraine

Aigul Shortanbayeva Faculty of Information Technology, Al-Farabi Kazakh National University, Almaty, Kazakhstan

Andrii Slabkyi Industrial Engineering Department, Vinnytsia National Technical University, Vinnytsia, Ukraine

Saule Smailova East Kazakhstan State Technical University named after D.Serikbayev, Ust-Kamenogorsk, Kazakhstan

Andrzej Smolarz Faculty of Electrical Engineering and Computer Science, Lublin University of Technology, Lublin, Poland

Vladimir Sobol Theoretical Mechanics Department, Kharkiv National University of Civil Engineering and Architecture, Kharkiv, Ukraine

Dmytro Somov Applied Mechanics and Mechatronics Department, Lutsk National Technical University, Lutsk, Ukraine

Victor Stasiuk Civil Safety Department, Lutsk National Technical University, Lutsk, Ukraine

Oleh R. Strilets Theoretical Mechanics, Engineering Graphics and Mechanical Engineering Department, Mechanical Institute, National University of Water and Environmental Engineering, Rivne, Ukraine

Volodymyr M. Strilets Theoretical Mechanics, Engineering Graphics and Mechanical Engineering Department, Mechanical Institute, National University of Water and Environmental Engineering, Rivne, Ukraine

Samat Sundetov Academy of Logistics & Transport, Almaty, Kazakhstan

Taras V. Tarasenko Hydro Gas Systems Department, National Aviation University, Kyiv, Ukraine

Oleksandr Tereschenko Automobiles and Transport Management Department, Vinnitsia National Technical University, Vinnitsia, Ukraine

Ruslan Tereshchenko VRTP "UKRGAZENERGOSERVIS", Kyiv, Ukraine

Ainur Torgesizova IT department, Academy of Logistics and Transport, Almaty, Kazakhstan

Artem O. Tovkach Machine-building technology and Automation Department, Vinnytsya National Technical University, Vinnytsya, Ukraine

Amandyk Tuleshov Institute of Mechanics and Engineering Science CS MES RK, Almaty, Kazakhstan

Aigul Tungatarova Faculty of Information Technology, Automation and Telecommunications, M.Kh.Dulaty Taraz Regional University, Taraz, Kazakhstan

Volodymyr M. Tverdomed Faculty of Infrastructure and Railway Rolling Stock, State University of Infrastructure and Technology, Kyiv, Ukraine

Alexander Tyzhnov Antonov State Enterprise, Kyiv, Ukraine

Waldemar Wójcik Faculty of Electrical Engineering and Computer Science, Lublin University of Technology, Lublin, Poland

Oleg Yakhno Applied Hydro-Aeromechanics and Mechatronics Department, Igor Sikorsky Kyiv Polytechnic Institute, Kyiv, Ukraine

Olexandr B. Yanchenko Industrial Engineering Department, Vinnytsia National Technical University, Vinnytsia, Ukraine

Bakhyt Yeraliyeva Faculty of Information Technology, Automation and Telecommunications, M.Kh.Dulaty Taraz Regional University, Taraz, Kazakhstan

Oleg Zabolotnyi Applied Mechanics and Mechatronics Department, Lutsk National Technical University, Lutsk, Ukraine

Vasyl Zapuklyak Oil and Gas Pipelines and Storage Facilities Department, Ivano-Frankivsk National Technical University of Oil and Gas, Ivano-Frankivsk, Ukraine

Yaroslav Zinko Department of Mechanics and Automation of Mechanical Engineering, Institute of Engineering Mechanics and Transport, Lviv Polytechnic National University, Lviv, Ukraine

Chapter 1

Method of project calculation of hydroimpulsive device for vibroturning with an incorporated cycle spring pressure pulse generator

Roman Obertyukh, Andrii Slabkyi, Leonid K. Polishchuk, Serhii Andrukhov, Larysa E. Nykiforova, Andrzej Smolarz, Aimurat Burlibayev, and Zhanar Azeshova

CONTENTS

1.1 Introduction 1
1.2 Materials and results of research 2
1.3 Conclusion 14
References 15

1.1 INTRODUCTION

On the basis of an analysis of publications (Danilchik 2018, Altintas et al. 2008, Śniegulska-Grądzka et al. 2017), an increasing trend of interest among researchers and manufacturers was found in vibrating cutting processes (turning, drilling, etc.).Vibration cutting is cost effective in processing materials with high strength, wear resistance, heat resistance, hardness, and so on. Depending on the mode of the vibrating load of the cutter (the main parameters being frequency and amplitudes of vibrations), vibration turning is used to improve the accuracy of machining (Obertyukh et al. 2015) [five to six qualities of precision (more than 15 kHz) and continuous chopping (up to 200 Hz)] (Altintas et al. 2008, Obertyukh et al. 2015). It has been established that hydropulse drive, whose advantages over other types of vibration actuators are known, is the most promising for the construction of low-frequency devices for vibration cutting, in particular vibroturning (Obertyukh et al. 2015). A promising direction in the development of small-sized hydropulse vibration devices is the use of elastic elements of high rigidity such as slotted, plate, and ring springs (RSs). Hydroimpulse devices for vibration cutting (HDVC), power (hydraulic cylinders), and distribution (pressure pulse generator, PPG) links are constructed on elastic elements of high rigidity (or combined with them) such as slit (SS), ring (RS), or plate (PS) springs. In comparison with devices for vibration cutting with other types of drives (hydraulic, pneumatic, etc.), these have small dimensions, a wide range of adjustment of the vibro-loading parameters of the object under treatment, and do not require a structural and kinematic change of metal cutting machines (Obertyukh et al. 2018, 2019, Polishchuk 2016).

DOI: 10.1201/9781003224136-1

1.2 MATERIALS AND RESULTS OF RESEARCH

A constructive scheme for a HDVC, for example, for radial vibroturning, is shown in Figure 1.1. The device consists of two hydraulically connected blocks: PPG and hydrocylinder of the drive in a vibratory movement, for example, a lathe cutter for radial vibroturning. The PPG contains a shut-off element in the form of a valve spool of 1 mass m_{vs}, the right-hand face (on the shaft) which is a support ring of a circular spring PPG (RS1) with rigidity k_1 and mass m_{k_1}, and a set of outer 2 and internal 3 rings, which interact with each other by the mediation of the inner and outer conical surfaces. The preliminary deformation, y_{01}, of the PPG RS is controlled by screw 4, which moves the pressure plunger 5, and the left-hand side (on the drawing line) is designed as the second bearing ring RS1. Plunger 5 is sealed with a rubber ring of 6 round sections. The body of the PPG and the device, in general, is not conventionally shown in Figure 1.1.

With the block of the hydraulic cylinder of the drive of the cutter (located in the same housing as the PPG), the PPG is connected through a common pressure cavity A (hole diameter a_1). The hydraulic cylinder consists of a plunger 7 with a mass m_{dp} sealed by a rubber ring 8. On the right (behind the drawing), the ends of the plunger 7 are forming a protuberance, which is the supporting and guiding surface of the bearing ring 9 of the cylinder ring spring (RS2), with rigidity k_2 and mass m_{k_2}.

RS2 consists of outer 10 and inner 11 rings and support rings 9 and 12. The RS2 rings are in contact with each other through the inner and outer conical surfaces. The pre-deformation y_{02} of the RS2 is regulated by means of a collar nut 13 which is screwed onto the cutting device 14 of the device body, which is conventionally shown by the icon "x". The collar nut 13 clicks onto the stepped cap 15 in the inner dull groove

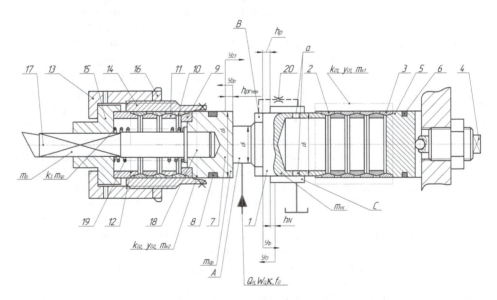

Figure 1.1 Constructive scheme hydroimpulse devices for vibration cutting (HDVC) for vibroturning with a built-in ring spring pressure pulse generator (PPG).

which is placed at the left (behind the drawing) and adjusts the support ring 12 RS2. Nut 13 is hinged with a spline nut 16.

In the rectangular aperture of step cap 15, a cutter 17 is provided, equipped with a cylindrical rod 18 with a supporting protrusion. The right part of the rod 18 (behind the supporting protrusion) enters the approaching landing in the central non-through hole of the plunger 7, and a spring 19 is installed on the left part of rod 18 (in front of the borehole): one end rests on the bust of the rod and the other on the cover 15. Spring 19 is installed during the assembly of the hydraulic cylinder with the estimated pre-deformation and carries the axial pre-fixing of the cutter 17.

To provide a stable mode of landing for the valve spool 1 at the end of its reverse, a throttle 20 is provided. The function of the throttle 20 is achieved by the gap in the conjugation of the spool part of the valve spool 1 by the diameter d_2 of the mortar in the body (or the sleeve in the real construction) of the device or by the experimentally selected flat plane $l_{fp} \geq h_p$ (here h_p is a positive overlap of the spool valve part, spool 1), which connects the intermediate B and drain C cavity of the device. Cavity RS1 is connected to the drainage cavity C with radial openings "a" in the guide part of the valve spool 1.

The device is powered by a working fluid (energy carrier) from a compact water pump station (not shown in Figure 1.1 conventionally), which connects to the device with two flexible hoses of high pressure. The sleeve of the energy supply joins the pressure cavity A, and the sleeve of the drain cavity C combines the latter with the hydrobox of the hydraulic pump station of the HDVC drive. Given the radial stretching of the outer rings 2 and 10, RS1 and RS2, during their working deformation, the cylindrical surfaces of the outer rings 2 and 10, RS1 and RS2, are directed to the holes of the device body and are tied to the surfaces of the holes of the hull for runways not exceeding 9–10 qualifications. Since, in absolute value, the radial deformations of the rings are small, the gaps that provide the named qualities are guaranteed to exclude the possibility of wedging of the rings.

The method of design calculation for this HDVC is based on the results of theoretical and experimental studies of hydropulse drives and devices of various technical and technological purposes (Obertyukh et al. 2018, 2019), according to the initial data, the content and composition of which is specified in the technical task (TT) for the development of the device.

The content and composition of the basic input data during the design calculation of the developed hydropulse device are determined by such data:

- Range of adjustment of vibration load parameters of the object of treatment (radial turning of detail).
- Vibration frequency: 10–100 Hz; amplitudes of vibrations: $0.2–2 \times 10^{-3}$ m.
- Maximum component of the cutting force F_y (depending on the mechanical characteristics of the treated material (metal) and cutting modes).
- Tentative masses m_1 and device m_2 (see Figure 1.1).
- Maximum pressure of energy p_1 "open" of PPG depending on the type of hydraulic pump of the hydropump station drive device. From the works of Obertyukh et al. (2018, 2019), it was established that the most reliable and stable in hydropulse drives are volumetric hydraulic pumps of gear type for which the average nominal pressure of the energy carriers is 16 MPa.

- Range regulation of the previous deformations RS1 and RS2. Taking into account the design features of the precipitating RSs, $y_{01} = (4.0...6.0)\cdot 10^{-3}$ m for RS1 and $y_{02} = (0.1...4.0)\cdot 10^{-3}$ (see Figure 1.1).
- Permanent pre-deformation $y_{02\max} = 10\cdot 10^{-3}$.
- The qualities of the accuracy of the couplings of the guiding surfaces of the valve spool 1, the plunger 7, the cutter 17 (rectangular mobile coupling with a rectangular aperture in the cover 15), the guides RS1 and RS2 (cylindrical surfaces of the outer rings RS1 and RS2), and other moving parts of the device (see Figure 1.1). According to the experience of designing and operating the hydropulse drives and their PPG, it has been established that the accuracy of the spool pairs should be no less than 6–7 qualifications, the accuracy of the coupling of the plunger 7 with its guiding surface not less than 7–8 qualifications, the rectangular conjugation of the cutter 17 not lower than 8–9 qualifications, and the same accuracy of the conjugation of the pressure plunger 5. The precision of the conjugation of the auxiliary links of the device, for example, the rod 18 of the cutter 17 in the guide aperture of the plunger 7 (see Figure 1.1), is intended for constructive and technological reasons at the level of 8–10 qualities of accuracy.
- The minimum positive blocking $h_p = 2\cdot 10^{-3}$ m of the valve spool 1, designed for the given accuracy of the conjugation of the guiding surface of the valve spool 1 and the conditions for ensuring the required sealing of the second degree of sealing PPG.
- Marks of materials of the main parts of the device and the types of their thermal or chemical-thermal treatment (the priority this item acquires during the direct development of the design of the device).
- The permissible speeds $[V]$ of the energy carrier in the pressure and drainage lines of the device and through the open gaps of the first and second stages of the PPG sealing.
- The given initial data necessary for the design calculation of the device may be supplemented in the calculation process with additional data to specify the design features of individual units and elements of the device, for example, to obtain the maximum frequency of the vibrations $v_{\max}$ of the cutter 17, it is expedient to preset the maximum possible volume $V_{0\max}$ of the pressure cavity in hydrosystems drive device. Also, it can be assumed previously that the negative overlap of the valve spool 1 is $h_N = h_k - h_p \leq h_p$, which can be refined during theoretical and experimental studies of the device's dynamics.

According to the proposed cycle diagram of the device's working cycle (see Figure 1.2), in the first approximation, the energy balance of the direct stroke of the valve spool 1 and plunger 7 (cutter 17) (see Figure 1.1) can be given as the equation

$$A_{pw1} + A_{pw2} = \Delta E_{pk1} + \Delta E_{pk2} + \Delta E_{pk3} + A_{f\Sigma} + \Delta E_{HL} \tag{1.1}$$

where $A_{pw2} = p_{m1}\cdot h_{pr\max}\cdot A_3$ is the average work of forces of pressure of energy during t_{dp}, the direct travel of the plunger 7 (cutter 17); $A_{pw1} = p_1\cdot h_k\cdot A_2$ is the average work of forces of pressure of energy during time t_{ds} of a direct passage of a valve spool; $\Delta E_{pk1} = 0.5 k_1 h_k^2$ is the growth of potential energy of deformation RS1; $\Delta E_{pk2} = 0.5 k_2 h_{pr\max}^2$ is the growth of potential energy of deformation RS2; $\Delta E_{pk3} = 0.5 k_3 h_{pr\max}^2$ is the increase in the potential energy of strain of the spring 19; $\Delta_{f\Sigma} = F_{f\Sigma}(h_k + h_{pr\max})$ is the total conditional work of frictional forces during the direct passage of the valve spool 1 and plunger 7 (cutter 17); and $\Delta E_{HL} = 0.5 k_{or} h_{or}^2$ is the increase in the potential energy of deformation HL. The total work of the friction forces $\Delta D_{f\Sigma}$ can be divided into two parts: the work of friction forces during the displacement of the valve spool 1 $A_{fk1} = F_{f1} p_{m1} \cdot h_k$ and the work of friction forces during the displacement of the plunger 7 (cutter 17) $A_{fPK} = F_{f2} h_{prmax}$, wherein forces F_{f1} and F_{f2} should also include the friction forces between the rings RS1 and RS2 and the elastic seal of the plunger 7. It is obvious that the coefficients of friction in the details of the hydraulic cylinder drive are larger than in the conjugations of the PPG, and hence $F_{f2} > F_{f21} \cdot y \cdot A_{fPK}$, and it is also advisable to include the work of the force F_y: $A_{fy} = F_y h_{Pmax}$.

Equation (1.1) can be divided into two relatively independent equations:

$$A_{pw2} = \Delta E_{pk2} + A_{fPK} + \Delta E_{pk3} = p_{m1} h_{Pr\,max} \cdot A_3 = 0.5 k_2 h_{Pr\,max}^2 + F_2 \cdot h_{Pr\,max} + 0.5 k_3 h_{Pr\,max}^2 \quad (1.2)$$

$$A_{pw1} = \Delta E_{PK1} + A_{fk1} + \Delta E_{HL} = p_1 h_k A_2 = 0.5 k_1 h_k^2 + F_{f1} \cdot h_k + 0.5 k_{or} x_{or}^2 \quad (1.3)$$

where $p_{m1} = 0.5(p_c + p_1)$ is the average energy pressure in the time interval $t_{dp} = (t_h - t_c) + t_N = (t_h - t_c) + t_{ds}$ (see Figure 1.2).

The values of the energy pressures p_c (see (1.2)) and p_2 can be determined with given accuracy only for known values of the cross-sectional area of the plunger 7 A_3 and the valve spool 1 A_1 and A_2, as well as the stiffnesses RS1 and RS2 (see Figure 1.1). At the stage of indicative calculations A_1, A_2, and A_3, the assumption that the main operation of the device under study will be carried out at vibration frequencies close to the maximum v_{max}, it is possible to base the calculations on the studies of hydropulse drives p_c and $p_c/p_1 = 0.5...0.8$ (Obertyukh et al. 2018, 2019). Let us take average values from these recommended ranges and assign them: $p_2 = 0.3 p_1$ and $p_c = 0.6 p_1$.

From the structural scheme of the device, it is obvious that the throughput of the PPG is determined by the area of the cross-section A_1 and the diameter d_1 (the first degree of sealing PPG) is essentially the diameter d_{yh} of the conventional passage PPG. To determine d_{yh}, it is necessary to find the supply of the hydraulic pump Q_H of the hydropump station of the device by the formula (Obertyukh et al. 2019, Polishchuk 2019, Abramov et al. 1997):

$$Q_H = K_{P\Delta} v_{max} \cdot p_{1max} \cdot W_0 k^{-1} \cdot \eta_{vh}^{-1}, \quad (1.4)$$

where $K_{cw} = 1 + (t_{hT} + t_{VT} + t_{ZT})/t_h$ is the coefficient conventionally named in Obertyukh et al. (2015) by the cyclic coefficient of the energy pressure pulse and η_{vh} is the volumetric efficiency of the hydraulic pump. According to the graph (see Figure 1.2), it is obvious that $K_{cw} > 1$. With the required given accuracy, the value of the coefficient K_{cw} can be found from the results of the analysis of the mathematical model of the hydropulse drive device, the adequacy of which is confirmed by experimental research of the device sample. Estimate of the level K_{cw} can be set using the cycle diagram of the device's working cycle (see Figure 1.2), introduced for curve 1 $p_r = f(t)$. The relative values of the individual components of the period T_T of the energy pressure pulse on the basis of the concept of the scale of the pulse of pressure:

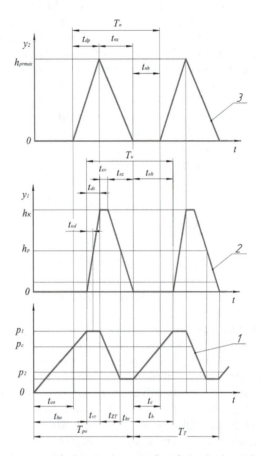

Figure 1.2 Tentative cyclogram of the working cycle of the hydropulse device for vibration cutting: 1, curve of pressure change in the pressure cavity A of the device; 2, displacement curve of valve spool 1; and 3, displacement curve of plunger 7 (cutter 17).

$$\mu_{tp} = T_T / ac' = oc' \cdot v_{\max} \tag{1.5}$$

where oc' is the length of the segment on the graph (Figure 1.3) (see Figure 1.2, curve 1) in mm

$$t_h = \mu_{tp} \cdot ac'; t_c = \mu_{tp} \cdot ab'; t_c = \mu_{tp} \cdot ab'; t_{VT} = \mu_{tp} \cdot c'd'; t_{VT} = \mu_{tp} \cdot c'd'; t_{hT}$$
$$= \mu_t \cdot oa; t_{ZT} = \mu_t \cdot d'o \tag{1.6}$$

Considering (1.6) in K_{cw}, we will find $K_{cw} = 1 + (oa + c'd' + d'o)/ac'$.

As the cyclogram of the working cycle of the device is marked as conditional (indicative), the coefficient K_c is also indicative and estimated, therefore it is expedient to enter the stock factor in formula (1.4) (Iskovich-Lototsky et al. 1982) $K_2 = 1.1...1.25$:

$$Q_{Hy} = K_2 \cdot K_{P\Delta} v_{max} p_{1max} k^{-1} \cdot \eta_{vh}^{-1} \qquad (1.7)$$

where Q_{Hy} is the specified value of the calculated delivery of the hydraulic pump of the device drive. Time t_{ZT} can be estimated by dependence, similarly (Obertyukh et al. 2019, Abramov et al. 1997):

$$t_{1T} = (p_1 - p_2) W_0 (Q_{TH} \cdot k)^{-1} \qquad (1.8)$$

where Q_{TH} is the average energy consumption through the open slit $A_{CH} = \pi d_2 h_N$ of the PPG device and level can be found in the cyclogram of the device's operating cycle. See Figure 1.2 for the assumptions about the linearity of the function $p_r = f(t)$ (Obertyukh et al. 2015, 2018):

$$t_{1T} / t_h = Q_\mathcal{H} / Q_{TH} = d'o / ac' = \tau_{ZT} \qquad (1.9)$$

whence

$$Q_{TH} = Q_\mathcal{H} \cdot \tau_{ZT}^{-1} \qquad (1.10)$$

where τ_{ZT} is the relative time of reduction of energy in the pressure device from level p_1 to level p_2. According to the graph, it is obvious that $t_{ZT} < t_h$ and $t_{ZT} < 1$, from which it follows that $Q_{TH} = Q_H$.

According to the design of hydraulic drives (Iskovich-Lototsky et al. 1982), the average speed of energy V_H through the gap A_{CH} PPG should not exceed the permissible $[V_H]$:

$$V_H = Q_{m\tau} / (\pi d_2 h_N) \leq [V_H] \qquad (1.11)$$

whence

$$Q_{TH} \leq (\pi d_2 h_N) \cdot [V_H] = \pi d_2 h_N \cdot [V_H] \qquad (1.12)$$

Conditional passage of PPG can be found by known formula (Iskovich-Lototsky et al. 1982):

$$d_{yh} = \sqrt{4 Q_{Hy} / ([V_H] \cdot \pi)} \approx 1.13 \sqrt{Q_{Hy} / [V_H]} \qquad (1.13)$$

The width of the sealing chamfer b_k (Figure 1.4) of the first degree of PPG sealing by analogy with the distribution valves of internal combustion engines operating in such devices with hydropulse devices can be recommended within $b_k = 2...4$ mm (Obertyukh et al. 2015, 2019). To simplify the process of grasping the valve spool 1 (see Figure 1.1), it is desirable for $b_k \leq 2.0...2.5$ mm. The angle of the cone α_k for preventing the jamming of the valve spool 1 in the saddle is assigned within $\alpha_k = 60°...90°$ (Obertyukh et al. 2015). Choosing, $\alpha_k = 60°$ for Figure 1.2, we will find

$$d_1' = d_1 + 2 b_k tg(0.5 \alpha_k) = d_1 + 2 tg 30° = d_1 + 1.15 \cdot b_k \qquad (1.14)$$

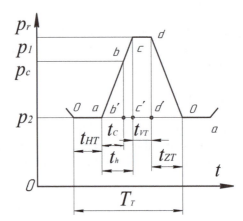

Figure 1.3 To the concept of the scale of the impulse of energy pressure.

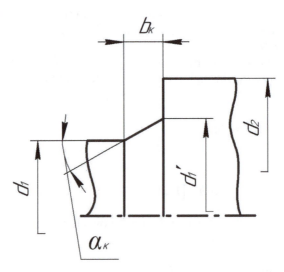

Figure 1.4 The shape of the sealing chamber of the first degree of pressure pulse generator (PPG) sealing.

The level of pressure of the energy p_1 "open" PPG is inversely proportional to the square (d_1^2) of the diameter of the first degree of PPG sealing, provided that the main sealing is carried out at the minimum diameter of the chamfer b_k (see Figure 1.4). Technically, this condition cannot be realized. It is most likely that sealing occurs at the diameter of the chamfer b_k, close to the average

$$d_{1T} = 0.5(d_1 + d_1') = d_1 + 0.575 b_k \qquad (1.15)$$

Receiving $b_k = 2$ mm, we get

$$d_{1T} = d_1 + 1.15 \text{ mm}, \tag{1.16}$$

where d_k is in millimeters.

Taking into account (1.15), we find

$$p_1 \geq 0.785 k_1 y_{o1} \cdot d_{1T}^{-2} \tag{1.17}$$

According to the observations made earlier, we take $d_{yu} = d_{1T}$ and introduce into the formula (1.17) the stock factor $K_{SH} = 1.05\ldots1.1$ to transform the inequality (1.17) into equality:

$$p_1 = K_{SH} \cdot 0.785 k_1 y_{o1} \cdot d_{1T}^{-2} = (0.82\ldots0.86) k_1 y_{o1} \cdot d_{1T}^{-2} \tag{1.18}$$

where according to the technical requirements (input data) of the maximum – the pressure $p_{1\max}$ of the energy carrier and the previous deformation $y_{o1\max}$ – we find the required value of the rigidity of the RS1

$$k_1 = p_{1\max} \cdot d_{1T}^2 / \left[(0.82\ldots0.86) y_{o1\max} \right] = (1.16\ldots1.22) p_{1\max} d_{1T}^2 \cdot y_{o1\max}^{-1} \tag{1.19}$$

Knowing (Obertyukh et al. 2019) the relationship between energy pressures "closing" p_2 and "opening" the PPG p_1:

$$p_2 \leq p_1 \cdot A_1 \cdot A_2^{-1} + k_1 h_k \cdot A_2^{-1} = p_1 d_1^2 \cdot d_2^{-2} + 4 k_1 h_k \cdot \pi^{-1} d_2^{-2} \cong p_1 d_1^2 \cdot d_2^{-2} + 1.274 k_1 h_k \cdot d_2^{-2} = (p_1 d_1 + 1.274 k_1 h_k) d_2^{-2} \tag{1.20}$$

Let us transform the dependence (1.20) to the form

$$p_2 / p_1 \leq d_{1T}^2 \cdot d_2^{-2} + 1.274 k_1 h_k \cdot d_2^{-2} p_1^{-1} = d_{1T} \cdot d_2^{-2} + K'_{CH}, \tag{1.21}$$

where $K'_{CH} = 1.274 k_1 h_k \cdot d_2^{-2} p_1^{-1} = $ const. is a peculiar coefficient of increasing the pressure of "closing" the PPG at the end of the direct stroke of the valve spool 1 (see Figure 1.1). From (1.20), we establish that:

$$d_2 \leq \left[d_{1T}^2 / (0.3 - K'_{CH}) \right]^{1/2} \tag{1.22}$$

where $p_2/p_1 = 0.3$ is accepted with the remarks outlined above. It is obvious that formula (1.21) will have physical content if $0 < K'_{CH} < 0.3$, that is, it lies in a narrow range of values. Taking this interval $= 0.1\ldots0.2$, we will find:

$$d_2 \leq (2.22\ldots3.13) d_{1T}. \tag{1.23}$$

Let us take the value of the diameter of the valve spool 1 (Figure 1.1) (the second – the spool sealing degree of PPG) at the lower interval:

$$d_2 = 2 d_{1T}. \tag{1.24}$$

Taking into account (1.23) in (1.20) and previously accepted $h_k = 2h_p$, we write the expression (1.20) in the form:

$$p_2 \leq 0.25 p_1 + 0.637 k_1 h_p \cdot d_{1T}^{-2} \tag{1.25}$$

Obviously, a condition must be provided to ensure the vibration cutting process

$$p_c A_3 3 F_y + k_2 (y_{o2} + y_p) + F_{f2}, \tag{1.26}$$

where y_p is the radial feed of the cutter 17 (as shown in Figure 1.1).

From (1.26), the required cross-sectional area of the plunger 7 (Figure 1.1)

$$A_3 \left[F_y + k_2 (y_{o2} + y_p) + F_{f2} \right] p_c^{-1} \Rightarrow \left[F_y + k_2 (y_{o2} + y_p) + F_{f2} \right] (0.6 p_1)^{-1} \tag{1.27}$$

where $p_c = 0.6 p_1$. Since the level p_c is determined by the level p_1 being adjusted (see Figure 1.1), the device must be adjusted for each given vibration cutting mode that determines the size of the cutting force F_y component. According to the made note, the process of vibroturning with the maximum possible force $F_{y\max}$ may be for adjusting the PPG device to the level of "opening" pressure $p_{1\max}$.

To analyze the effect on the magnitude of the cross-sectional area of the spring 7 of the constituent formulas (1.26), we introduce the relative values of forces in comparison with the forces arising from the previous deformation of the RS2 $F_{ORS2} = k_2 y_{o2}$:

$$k_{Fy} = F_y / F_{ORS2} = F_y / (k_2 y_{o2}); \tag{1.28}$$

$$k_{f2} = F_{f2} / F_{ORS2} = F_{f2} / (k_2 y_{o2}); \tag{1.29}$$

$$S_p = k_2 y_p / k_2 y_{o2} = y_p / y_{o2}. \tag{1.30}$$

Assuming that $y_{o2} \neq 0$ (see input data for calculation) and taking into account the significant value of the rigidity k_2 of the RS2, the coefficients $k_{Fy} < 1$ and $k_{f2} < 1$, and the coefficient S_p, which by its nature is a relative radial feed of the cutter 17 (see Figure 1.1), depending on the size y_{o2} and y_p may have a limit of $S_p \leq 1$. Since, as a rule, the component of the cutting force F_y and the total force F_{f2} in absolute magnitude are small but preliminarily taking approximate values $k_{Fy} = 0.2$, $k_{f2} = 0.1$, and $S_p = 0.1$, let us write dependence (1.28) in the form (Iskovich-Lototsky et al. 1982):

$$A_3^3 2.3 \cdot k_2 y_{o2} (0.6 p_1)^{-1} = 3.8 k_2 y_{o2} \cdot p_1^{-1} \tag{1.31}$$

According to the considerations of compactness and technological design of a hydropulse device for vibration cutting and the possibility of direct mounting it in a tool holder, for example, a lathe-screw machine, it is advisable to have the diameter of the plunger 7 d_3 equal to the diameter d_2 of the valve spool 1 (see Figure 1.1), then $A_3 = A_2$.

A condition must be fulfilled to ensure the vibration cutting process

$$p_{1\min} \cdot A_2 \geq k_2 y_{o2\min} (k_{Fy} + k_{f2} + S_p + 1) = 2.3 k_2 y_{o2\min}, \tag{1.32}$$

where $p_{1min} \geq p_c$. By introducing into (1.32) the stock factor $K_{SC}=2.0...2.2$, we transform the inequality (1.32) into equality

$$p_{1min} \cdot A_3 \equiv p_{1min} \cdot A_2 = K_{SC} \cdot 2.3 k_2 y_{o2min} = (4.6...5.1) k_2 y_{o2min}, \quad (1.33)$$

whence

$$k_2 = p_{1min} \cdot A_3 / y_{o2min}, \quad (1.34)$$

The construction of RS1 and RS2 (see Figure 1.1) can be realized in different ways and conditions; for example, it is advisable to increase the diameter of the springs by minimizing the number of rings of springs, and on the condition of minimizing the transverse dimensions of the device it seems that the best approach will be when the diameters of the RS1 rings and RS2, especially internal ones, are equal, especially, d_2 of the valve spool and d_3 plunger 7 (see Figure 1.1), but the stiffness of the rings determined by known dependencies is required. Obviously, in our case, the second approach will be more acceptable, which, moreover, increases the production capacity of the device body.

Since it has been previously assumed that $A_3 = A_2$, from the point of equality $d_3 = d_2$, the basic dimensions of the inner rings RS1 and RS2 will be taken (Birger et al. 1993):

$$D_{i3} = d_2; d_{i9} = d_2 \cdots d_{i11} = d_2; d_{i12} = d_2, \quad (1.35)$$

where d_{i3}, d_{i9}, d_{i11}, and d_{i12} are the outer diameters of the inner rings RS1 and RS2 (see Figure 1.1).

It is estimated that the inner and outer rings RS1 and RS2 have the same calculated average square cross-section $A_v = A_z = A_t$, which results in the equality of modulus of stresses stretching σ_p for outer rings and grips σ_c for internal rings RS1 and RS2. Since for materials, including metals and their alloys, the tensile stress is more dangerous, then the strength of the outer rings determines its strength RS2 and RS2 in general.

Based on the comments made and taking additional input data – $\beta = 12°$ (Birger et al. 1993), the angle of the taper of the rings; material of rings, steel 60C2A GOST 15959-79, $[\sigma_p] = 314$ MPa (Sorokin et al. 1989); $H_k = (3...5) b_{tk}$ (here b_{tk} is the average thickness of the rings) is the height of the rings – general geometric ratios are found for calculating the size of the rings RS1 and RS2 (Figure 1.5a and b).

According to Figure 1.5a, the calculated section $p_e - p_e$ is located at a distance from the middle of the inner ring (as well as from its end) $H_k/4$.

The outer diameter of the inner ring in the section is easy to determine by simple geometric calculations based on Figure 1.4 and 1.5a:

$$d_T = d_{ipi} = d_2 - 0.5 H_k tg\beta \quad (1.36)$$

The average area of the cross-section in the section $p_i - p_i$ of the inner and outer rings of the RS can be found by the formula:

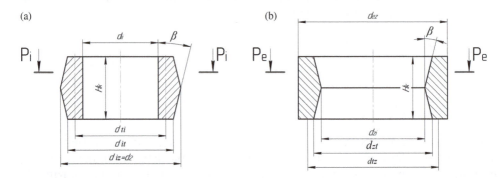

Figure 1.5 Sketches and geometric characteristics of inner (a) and external (b) RS rings d_{iz}, d_{ez}, d_{ti}, d_{tz}, d_{ii}, and d_{zi} – accordingly, external, average, and internal diameters of internal (b) rings of RS; d_{it}, d_{zt} – respectively, the outer and inner diameters of the ends of the inner (a) and outer (b) rings of the RS; $\beta = 12°$ – the angle of the taper of the rings RS; H_k – rings height; $p_i - p_i$, $p_e - p_e$ – calculated cross sections of rings.

$$A_m = \pi\left(d_{ipi}^2 - d_{ii}^2\right)/4 = 0.785[4b_{k\max}(d_2 - b_{k\max}) + H_k tg\beta(H_k - d_2 \cdot tg\beta)], \quad (1.37)$$

where $d_{ii} = d_2 - 2b_{k\max}$ and $b_{k\max}$ is the thickness of the inner ring of the RS in the middle (largest) section. The thickness of the rings $b_{k\max}$ and the inner diameters d_{ii} and d_{zi} determine the rigidity of the inner and outer rings and the RS in general. To obtain an acceptable rigidity (compliance) of the RS, we take the thickness of the inner ring in the middle section to be $b_{k\max} = 4$ mm, then $d_{ii} = d_2 - b_{k\max}$ and $d_2 - 8$ mm.

The average thickness of the inner ring of the RS is calculated by the arithmetic mean:

$$b_{mk} = [0.5(d_2 - d_{ii}) + 2 \cdot 0.5(d_{it} - d_{ii})]/3 = (0.5d_2 + d_{it} - 1.5d_{ii})/3$$
$$= [1.5(d_2 - d_{ii}) - H_k tg\beta]/3, \quad (1.38)$$

where $d_b = d_2 - H_k tg\beta$.

Given (1.38) values d_{ii}, we find:

$$b_{mk} = b_{k\max} - (H_k tg\beta/3), \quad (1.39)$$

From the recommended range of ratios H_k and b_{mk}, let u's take:

$$H_k = 4b_{mk} = 4(b_{k\max} - H_k tg\beta)/3 \quad (1.40)$$

where

$$H_k = 12b_{k\max}/(3 + 4 \cdot tg\beta) \quad (1.41)$$

Substituting in (1.41) and (1.39) the numerical values, we calculate:

$$H_k = 12 \cdot 4 / (3 + 4 \cdot tg12°) = 12.5 \text{ mm} \tag{1.42}$$

$$b_{mk} = 4 - (12.5 \cdot tg12° / 3) = 3.125 \text{ mm} \tag{1.43}$$

Taking into account (1.37) numerical values H_k and $b_{k\max}$, we obtain

$$A_m = 12.12 \cdot d_2 - 22.48 \text{ mm}^2 \tag{1.44}$$

where d_2 is in mm.

The diameter of the inner ring of the RS and the outer rings in the calculated cross-section $p_i - p_i$ and $p_e - p_e$ is equal to each other: $d_{ipi} = d_{zpb} = d_m = (d_{ti} - d_{tz})/2$.

Proceeding from the equality of the cross-section of the internal and external rings, we determine using Figure 1.4b and geometric calculations, the main dimensions of the outer ring of the RS:

$$d_{ez} = \left(1.27 A_m + d_m^2\right)^{1/2} \tag{1.45}$$

$$d_{zi} = d_m - 0.5 H_k \cdot tg\beta \tag{1.46}$$

$$d_{zt} = d_m + 0.5 H_k \cdot tg\beta \tag{1.47}$$

$$b'_{mk} = [0.5(d_{ez} - d_{zi}) + 2 \cdot 0.5(d_{ez} - d_m) = 1.5(d_{ez} - d_m) - 0.25 \cdot H_k \cdot tg\beta \tag{1.48}$$

Here b'_{mk} is the average thickness of outer ring RS.

Maximum thickness of outer ring RS:

$$b'_{mk} = 0.5(d_{ez} - d_{zi}) = 0.5(d_{ez} - d_m) = 1.5(d_{ez} - d_m) + 0.25 \cdot H_k \cdot tg\beta \tag{1.49}$$

Taking into account (1.49) in (1.48), we find:

$$b'_{mk} = 2(d_{ez} - d_{zi}) - b'_{k\max} \tag{1.50}$$

Substituting in (1.46) and (1.47), we obtain:

$$d_{zi} = d_2 - H_k \cdot tg\beta = d_{it} \tag{1.51}$$

$$d_{zt} = d_2 \tag{1.52}$$

The maximum axial displacement $\delta'_{\max}$ of the outer ring of the RS relative to the internal one, taking into account (1.36) depending on (1.49), can be calculated by the formula

$$\delta'_{\max} = [\sigma_p] (d_2 - 0.5 \cdot H_k \cdot tg\beta) / (E \cdot tg\beta) = [\sigma_p][d_2 \cdot (tg\beta)^{-1} - 0.5 \cdot H_k] \cdot E^{-1} \tag{1.53}$$

or taking into account numerical values $[\sigma_p]$ = 314 MPa, $\beta = 12°$, H_k =12.5 mm, and $E = 2 \cdot 10^5$ steel (Iskovich-Lototsky et al. 1982).

$$\delta'_{max} = 1.46 \cdot 10^3 (4.26 \cdot d_2 - 6.25) \text{ mm.} \qquad (1.54)$$

where d_2 is in millimeters.

The number n_{k1} of working rings is calculated according to the formula obtained on the basis of works by Birger et al. (1993), Ponomarev (1980), and Roganov and Karnaukh (2000), where $h_k = 2h_p$ and $\delta_{max1} = \delta'_{max}$.

$$n_{k1} \geq 2.25 \cdot h_p (\delta'_{max})^{-1} + 3 \qquad (1.55)$$

The number of working n_{k2} RS2 we find from (1.54) for $\delta_{max2} = \delta'_{max}$ and the following conditions:

$$\lambda_{max2} \geq h_{n\,max2} + y_{o2max} = y_{pmax} + y_{o2max}, \qquad (1.56)$$

where y_{pmax} is the maximum possible radial feed of the cutter 17 (see Figure 1.1), which is set during the design process vibration cutting, for example, radial vibroturning:

$$n_{k2} \geq (y_{pmax} + y_{o2max}) \cdot (\delta'_{max})^{-1} + 2, \qquad (1.57)$$

The purpose of the spring 19 (see Figure 1.1) is the axial fixing of the cutter 17 relative to the spring 7. Best of its function, the spring 19 will perform in the mode of resonance work, characterized by the condition (Birger et al. 1993)

$$\omega'_{03} = (k_3 / m_p) \cdot (\delta'_{max})^{1/2} \geq 2\pi v_{max} \qquad (1.58)$$

where ω'_{03} is the own circular frequency of the cutter 17 relative to the spring 19. Under condition (1.58), the stiffness of the spring 19 required is

$$k_3 \geq 8\pi^2 \cdot m_p \cdot v_{max}^2 = 787.88 \cdot m_p \cdot v_{max}^2 \qquad (1.59)$$

The initial F_{03} and labor force F_{03max} of the spring 19 can be calculated from simple dependencies:

$$F_{03} = k_3 \cdot y_{03max}; \; F_{03max} = k_3 \left(y_{03max} + h_{prmax} \right) \qquad (1.60)$$

The geometric parameters of the spring 19 should be selected from the standard row, determined constructively by placing the spring 19 on the rod 18 (see Figure 1.1) of the cutter in the central opening of the RS2 with outer D_{ez} and inner D_{zi} diameters and compressed pre-force F_{03} at height H_{03}. All other parameters of the spring 19 are determined by the standard.

1.3 CONCLUSION

The method of design calculation of the hydropulse device for vibroturning with integrated circular spring PPG allows for simple dependencies to determine all the main power and geometric parameters of the PPG and the hydraulic cylinder of the drive in the vibrating motion of the cutter, as well as to find all the power and design parameters of the RSs PPG and the hydraulic cylinder.

In order to increase the accuracy of the calculation of the device parameters in the calculation formulae of the proposed method, some coefficients may be adjusted (refined) according to the results of the theoretical study of the mathematical model of the device and the experimental verification of the adequacy of this model with a prototype device.

The approaches and principles of calculating the parameters of the hydropulse device for vibroturning (vibration cutting) described in the method can be the basis for constructing methods of design calculations for other similar hydropulse devices.

REFERENCES

Abramov, E.I., Kolisnichenko, K.A. & Maslov, V.T. 1977. *Elements of the Hydraulic Drive: Reference Book*. Kiev: Tehnika.
Altintas, Y., Eynian, M & Onozuka, H. 2008. Identification of dynamic cutting force coefficients and chatter stability with process damping. *CIRP Annals - Manufacturing Technology* 57(1): 371–374.
Birger, I.A., Shorr, B.F. & Iosilevich, G.B. 1993. *Calculation of the Strength of Machine Parts*. Moscow: Mechanical Engineering.
Danilchik, S.S. 2018. *Vibration Turning of Structural Steels*. Minsk: BNTU.
Iskovich-Lototsky, R.D. Matveev, I.B. & Krat V.A. 1982. *Vibration and Vibro-Impact Action Machines*. Kiev: Technics.
Obertyukh, R.R., Andrii, V. Slabkyi, A.V., et al., 2019. Method of design calculation of a hydropulse device for strain hardening of materials. *Przegląd Elektrotechniczny* R95(4): 65–73.
Obertyukh, R.R. & Slabkyy, A.V. 2015. *Devices for Vibroturning on the Basis of a Hydropulse Drive*. Vinnitsa: VNTU.
Obertyukh, R.R., Slabkyi, A.V., Marushchak, M.V. et al., 2018. Dynamic and mathematical models of the hydraulic-pulse device for deformation strengthening of materials. *Proceedings of SPIE Photonics Applications in Astronomy, Communications, Industry, and High-Energy Physics Experiments* 10808: 108084Y.
Polishchuk, L., Kharchenko, Y., Piontkevych, O. & Koval, O. 2016. The research of the dynamic processes of control system of hydraulic drive of belt conveyors with variable cargo flows. *Eastern-European Journal of Enterprise Technologies* 2(8): 22–29.
Polishchuk, L.K., Kozlov, L.G., Piontkevych, O.V. et al. 2019. Study of the dynamic stability of the belt conveyor adaptive drive. *Przeglad Elektrotechniczny* 95(4): 98–103.
Ponomarev, S D. & Andreeva, L.E. 1980. *Calculation of Elastic Elements of Machines and Instruments*. Moscow: Machine-building.
Roganov, L.L. & Karnaukh, S.G. 2000. *Calculation of Springs and Elastic Shock Absorbers*. Kramatorsk: DSEA.
Śniegulska-Grądzka, D., Nejman, M. & Jemielniak, K. 2017. Cutting force coefficients determination using vibratory cutting. *Procedia CIRP* 62: 205–208.
Sorokin, V.G. et al 1989. *Database of Steels and Alloys*. Moscow: Engineering.

Chapter 2

Determination of the optimal parameters of the driver of a resonance vibratory stand for diagnostics of dampers

Volodymyr Gursky, Igor Kuzio, Olena Lanets, Yaroslav Zinko, Pavlo Nosko, Konrad Gromaszek, Paweł Droździel, Maksat Kalimoldayev, and Amandyk Tuleshov

CONTENTS

2.1 Introduction 17
2.2 Analysis of literary sources and statement of the problem 18
2.3 Purpose and tasks of the research 19
2.4 Materials and methods 20
2.5 Results and discussion 23
2.6 Conclusion 27
References 27

2.1 INTRODUCTION

Vibration systems are widely used in various technological processes as well as in diagnostic and testing equipment (Shpachuk et al. 2018). The effectiveness of the functioning of vibration devices is determined by implementing energy-efficient principles when adjusting their operation. For vibration equipment, the energy-efficient mode is the mode which is close to the main resonance. Concerning the drive, contactless drives are optimal (Nosko 2010), in particular the inertial unbalanced (Michalczyk 2012, Yaroshevich et al. 2015, Yatsun et al. 2018) and electromagnetic types (Gursky et al. 2018). Inertial unbalanced drives are easy to control by the frequency and nominal value of the perturbation force. The accessibility of facilities and control systems for these types of drive ensures their high prevalence of use. Well-known objects requiring qualitative diagnostics are dampers and shock absorbers. Standard procedures for testing and diagnostics of shock absorbers and dampers have already been established, as well as the determination of their elastic-dissipative characteristics (Zhao et al. 2016).

DOI: 10.1201/9781003224136-2

2.2 ANALYSIS OF LITERARY SOURCES AND STATEMENT OF THE PROBLEM

In the general case, the elastic-dissipative characteristics of the damping systems are designed to ensure effective quenching of vibrations in transport systems and buildings (Özcan et al. 2013). It is well-known that the defining characteristics of dampers are the maximum amplitude h_d and the nonlinear damping characteristic, which depend on the speed of the motion. The latter can be presented as a group of linear: piecewise linear and piecewise nonlinear dependencies in accordance with the basic function (Chen and Sun 2017, Fujita et al. 2014):

$$\bar{F}_d = b_d |\dot{y}|^\beta \text{sign}(\dot{y}) \tag{2.1}$$

where b_d is a coefficient that depends on the physical and mechanical characteristics of the energy carrier (Sikora 2018, Elliott et al. 2015) and can be determined on the basis of experimental research, and β is an index that determines the nature of the dependence of the damping force on the speed of motion.

At present, the most effective means for reducing vibrations is the use of asymmetric piecewise linear dependences of the damping behavior of the dampers (Silveira et al. 2014).

The principles of the choice of parameters b_d and β for typical characteristics of the dampers, which can be used as a vibration damper in various industries, are established in Chen and Sun (2017). The technical characteristics of the test stand for shock absorbers are presented in research by Cui et al. (2010), where an experimental dependence on the damping force is obtained. The principal problem of minimizing the total quadratic error between the given and simulative functions for the shock absorber resistance force is considered. The use of a polynomial characteristic for the representation of the damping force dependence on the speed was substantiated, and an estimation of the influence of the frequency of oscillations on the discrepancy in the results was carried out. These researches are fundamental; they can be extended to different types of shock absorbers and dampers.

The possibility of linearization of dynamical systems with nonlinear viscous friction was investigated in Elliott et al. (2015). It is established that such systems, if they are determined by the cubic dependence for the friction force, can be determined analytically, in particular by the harmonic balance method. However, this approach has not been tested for systems with piecewise linear and piecewise nonlinear dependencies.

A nonlinear dissipative characteristic of the damper is applied as the initial data, which should be ensured during its vibration test (Figure 2.1). According to (2.1), the loading characteristic of the stand will be:

$$\bar{F}_d = \begin{cases} b_{1d} |\dot{y}|^{0.5} \text{sign}(\dot{y}), & \dot{y} \geq 0 \\ b_{2d} |\dot{y}| \text{sign}(\dot{y}), & \dot{y} < 0 \end{cases} \tag{2.2}$$

where the coefficients are $b_{1d} = 300$ N·(s/m)$^{0.5}$, and $b_{1d} = 50$ N·s/m.

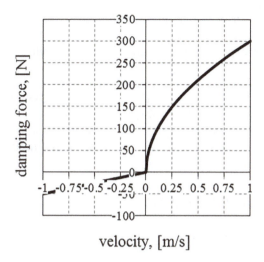

Figure 2.1 Damping force dependence on velocity.

The decisive assumption for using model (2.2) is the acceptance of the condition that the characteristic and the physical and mechanical properties of the damper are not dependent on the frequency of oscillations and are quite constant.

In the fundamentals of the research of vibrational processes of vibration systems and machines, the dynamics of the drive is taken into account, in particular, as presented in the works (Yaroshevich et al. 2015, Yatsun et al. 2018). However, to simplify the procedure for the synthesis and optimization of elastic force and structural parameters, it is expedient to consider only the mechanical component of the system (Gursky et al. 2018). The obtained parameters, which are refined and analyzed at the next stage of simulation of dynamic processes, take into account the characteristics of the selected drive type (Lanets 2013).

The hydraulic drive represented in the works (Surace et al. 1992, Obertyukh et al. 2018, 2019) is used for the stands, which are used for testing shock absorbers and dampers of various types in the automotive industry. For the development of a stand for a particular type, in particular, short-stroke dampers, such a necessity is not required, but it is more expedient to use standard industrial engines or motor vibrators, which significantly reduce the cost of their production (Polishchuk et al. 2016a, 2016b, 2018).

At present, the existing research does not contain sufficient information on the calculation of optimal parameters and conditions of perturbation of vibration stands with an inertia drive for the diagnosis and testing of objects with nonlinear damping characteristics. This leads to the need for the solution of such a problem, which will provide the opportunity to obtain sufficient information for further design and practical implementation of vibration stands.

2.3 PURPOSE AND TASKS OF THE RESEARCH

The purpose of the presented research is to ensure the effective operation of the vibration stand for the testing of short-stroke dampers with viscous nonlinear friction on the basis of the implementation of the minimum drive force and the specified nonlinear damping characteristic of the damper (Polishchuk et al. 2019, Kozlov et al. 2019, Kukharchuk et al. 2017b).

To achieve this purpose, it is necessary to solve the following problems:

1. the formation of mathematical dependencies for the estimation of the general characteristics of force, energy, and kinematics of the vibration stand;
2. characterization of the optimal force and structural parameters and rotational frequency of the inertial unbalanced drive of the vibration stand, which is based on ensuring the minimum difference between the required and realized damping characteristic.

2.4 MATERIALS AND METHODS

A two-mass oscillatory system with vertical motion of masses was chosen for the research of the dampers (Figure 2.2). Operational 1 (m_1) and reactive 2 (m_2) masses are joined by the element 3 with linear elastic-dissipative characteristics (c_{12}, b_{12}). The insulators 5 (c_{is}) are provided to install the system on the frame 4, and an inertia vibrator with unbalanced mass 6 (m_d), which rotates with frequency ω, is used for perturbation. The damper 7 is rigidly mounted vertically, and its rod is connected to the operational mass 1. To ensure rational modes of testing the damper, it is necessary to establish such disturbance conditions that provide the minimum deviation of the effective resistance of the damper and the minimum power consumption of the drive.

The motion of a system, loaded by the effective resistance force $\bar{F}_d(\dot{y}_1)$ in accordance with (2.2), is described by three generalized coordinates (Lanets 2013):

$$m_1\ddot{y}_1 + c_{12}(y_1 - y_2) + c_{is}y_1 + b_{12}(\dot{y}_1 - \dot{y}_2) + b_{is}\dot{y}_1 + \bar{F}_d(\dot{y}_1) = 0, \quad (a)$$

$$m_2\ddot{y}_2 - c_{12}(y_1 - y_2) - b_{12}(\dot{y}_1 - \dot{y}_2) = m_d r(\ddot{\varphi}\sin\varphi + \dot{\varphi}^2\cos\varphi), \quad (b) \qquad (2.3)$$

$$J\ddot{\varphi} - m_d r\ddot{y}_2\sin\varphi = M(\dot{\varphi}) - b_\varphi\dot{\varphi}, \quad (c)$$

where y_1 and y_2 are vertical amplitudes of the damper rod and the reactive mass, respectively; φ is the twist angle of the unbalanced mass; $J = J_d + m_d r^2$ is the summarized moment of inertia of the motor rotor J_d and unbalanced mass m_d; $M(\dot{\varphi})$ is the mechanical characteristic of the drive satisfying the value M_0 at the calculated frequency of perturbation f; b_φ is the coefficient of the resistance to the rotation of the unbalanced mass; $\omega = 2\pi f$ is the rotational frequency of the unbalanced mass; and b_{12} and b_{is} are coefficients of viscous resistance of the main spring and vibration isolator, respectively (Kukharchuk et al. 2016a, 2016b, 2017a, Vedmitskyi et al. 2017).

At the stage of parameter synthesis in the system (2.3), rejecting the equation (c), we consider only the motion according to generalized coordinates y_1 and y_2, which are described by the system of differential equations:

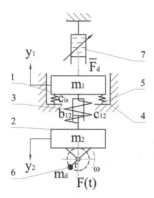

Figure 2.2 Structural scheme of vibratory stand for research of a damper: 1 and 2, working and reactive masses; 3, spring; 4, frame; 5, isolators; 6, unbalanced mass; and 7, damper.

$$\left.\begin{array}{l} m_1\ddot{y}_1 + c_{12}(y_1 - y_2) + c_{is}y_1 + b_{12}(\dot{y}_1 - \dot{y}_2) + b_{is}\dot{y}_1 + \overline{F}_d(\dot{y}_1) = 0, \\ m_2\ddot{y}_2 - c_{12}(y_1 - y_2) - b_{12}(\dot{y}_1 - \dot{y}_2) = F(t), \end{array}\right\} \quad (2.4)$$

where $F(t) = F_0 \sin\omega t = m_d\omega^2 r \sin\omega t$ is the equation which represents the changing perturbation force in the vertical direction.

The functional dependence for the coefficient of rigidity c_{12} and the coefficient of viscous resistance b_{12} on the cyclic frequency of perturbation f is in the system of equations (2.5):

$$c_{12}(f) = \frac{m_1 m_2}{m_1 + m_2}\left(\frac{2\pi f}{z}\right)^2, \quad b_{12}(f) = 2\frac{m_1 m_2}{m_1 + m_2}\xi(2\pi f) \quad (2.5)$$

where $z = f/f_0$ is a resonant adjustment, defined as the ratio of the frequency of forced oscillations f to free oscillations f_0, and ξ is the dimensionless damping parameter.

The root mean square error value (RMSE) of the damping force can be determined in a known order:

$$\text{RMSE} = \sqrt{\frac{1}{n}\sum_{i=1}^{n}\left(F_{d_i} - \overline{F}_{d_i}\right)^2} \quad (2.6)$$

where n is the number of points for the estimation of the error; F_{d_i} and $\overline{F}_{d_i}$ are the values of the damping force at the i-th point, calculated according to (2.4), and are valid in accordance with (2.2).

The task of minimizing the error, which depends on the desired characteristics of the drive, is typically:

$$\text{RMSE}(f, F_0) \to \min \quad (2.7)$$

However, the minimization of the error of the implementation of the damping characteristics must be solved taking into account the dynamic and kinematic characteristics of the vibration stand, based on its design features and drive type. Therefore, it is necessary to establish such a group of restrictions that guarantee the solution of the problem and that the obtained solutions can be practically implemented. The parameters that are determined from the simplified model (2.4) are the restrictions on the calculation stage. In particular, this is the maximum value of accelerating the operating mass $\ddot{y}_{1\max}$, as well as the range in which the frequency of the perturbation is searched $f \in (f_{\min} \dots f_{\max})$. The determining factor for choosing the frequency is the maximum value of the acceleration of the operating mass $\ddot{y}_{1\max}$. Its value increases with increasing perturbation frequency, and based on the characteristic of damping, the velocity value must be the same for different frequencies. Therefore, the choice of a lower frequency value in this case is desirable. From a practical point of view, the value of the frequency of oscillations can limit the maximum value of acceleration of the operating mass, which depends on the mass and dimensional parameters of the system and the capabilities of the drive.

Thus, the general view of the optimization problem (2.7) is supplemented by the limitation:

$$\left. \begin{array}{l} \ddot{y}_{1\max}(f, F_0) \leq [\ddot{y}_1], \\ f_{\min} < f \leq f_{\max}, \end{array} \right\} \qquad (2.8)$$

where $[\ddot{y}_1]$ is the accepted value of accelerating the working mass.

The nominal values of the perturbation force F_0 and the frequency of the perturbation f, obtained by the solution of the optimization problem, enable determination of the static moment of the unbalanced mass

$$m_d r = F_0 / (2\pi f)^2 \qquad (2.9)$$

as well as the required elastic-dissipative coefficients (2.5) of the vibration system.

The power of the drive of the vibration stand is calculated as the sum of the two components:

$$p(t) = p_v(t) + p_f(t) \qquad (2.10)$$

where $p_v(t) = F(t)v_2(t)$ is the power to be expended in actuating the vibration stand; $p_f(t) = \mu F(t)(2\pi f) d/2$ is the power dissipated by friction in the bearings; μ is the coefficient of rolling friction in the bearings, $\mu = 0.004–0.008$; and d is diameter of the shaft.

The nominal value of the engine torque at the calculated frequency of perturbation is established and is dependent on the power consumption of the drive:

$$M_0 = \mathrm{stdev}\bigl[p(t)\bigr] / 2\pi f \qquad (2.11)$$

Having chosen the type of drive, it is expedient to perform additional simulations confirming the calculated parameters, the adequacy of the characteristics of the selected drive type, and the appropriate frequency of perturbation. The main functional dependencies for checking the drive efficiency and providing the calculated technological parameters of the vibration stand, obtained on the simplified model, are determined from the solutions of the system of equations (2.3).

2.5 RESULTS AND DISCUSSION

The research was carried out on the basis of the numerical solution of the system of nonlinear differential equations (2.3) for output characteristics of the damper in accordance with (2.2) and the following parameters of the vibration stand: $m_1 = 30$ kg, $m_2 = 15$ kg, $z = 1$, $\xi = 2{,}000$ N/m, and $d = 0.035$ m.

The primary result of the study is the analysis of the dependence of the RMSE of the damping force (2.6) on the main characteristics of the disturbing force. Its graphical representation (Figure 2.3) indicates the optimization character of the error relative to the value of the nominal perturbation force F_0.

Solutions for (2.4) and (2.7) can be presented as a group of values for a specific value of the frequency of perturbation f: $Z_f = \begin{bmatrix} \text{RMSE}^{\min} & F_0 & \ddot{y}_{1\max} \end{bmatrix}$. The values of the global minima of the root mean square deviation of the damping effort (Figure 2.4) are found by the numerical Levenberg–Marquardt method for the frequencies of perturbation 25 and 50 Hz, which are mostly applied for the vibration mechanisms: $Z_{[25\text{ Hz}]} = [21.5\text{ N} \; 1.857\text{ kN} \; 111.16\text{ m/s}^2]$ and $Z_{[50\text{ Hz}]} = [16\text{ N} \; 3.763\text{ kN} \; 225.13\text{ m/s}^2]$.

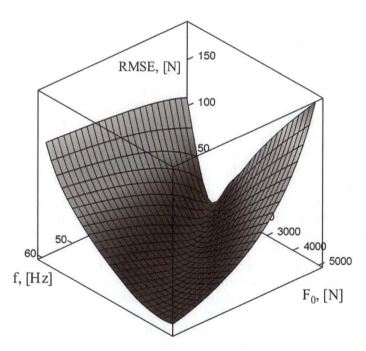

Figure 2.3 Dependence of root mean square error value (RMSE) of damping force from frequency and nominal force value.

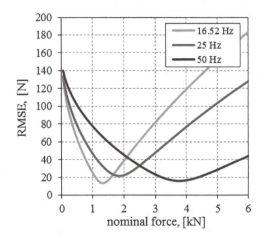

Figure 2.4 Dependence of root mean square error value (RMSE) of damping force from nominal force value.

However, the maximum obtained value for the operating mass acceleration $\ddot{y}_{1\max}$ for both frequencies is larger than 1,000 m/s², and therefore the values of these frequencies are unacceptable (2.8): $\ddot{y}_{1\max}(f, F_0) \leq 100$ m/s², 12 Hz $< f <$ 25 Hz. This obtains the value of the optimal frequency of perturbation $f^{\mathrm{opt}} = 16.52$ Hz, for which the calculated parameters acquired more optimal values: $Z_{[25\ \mathrm{Hz}]} = [14.98\ \mathrm{N}\ 1.22\ \mathrm{kN}\ 73\ \mathrm{m/s}^2]$.

After obtaining optimal power parameters, the static moment of the unbalanced mass will be $m_d r = 0.113$ kg/m in accordance with (2.9). For an unbalanced mass $m_d = 2$ kg, the radius of its center of mass should be $r = 0.057$ m.

The power consumed in a vibration stand according to its components determined by (2.10) gives the numerical values: stdev $[p_v(t)] = 554.54$ W and stdev$[p_f(t)] = 8.92$ W. The nominal value of the engine torque at the determined perturbation frequency is also calculated according to (2.11) as $M_0 \approx 5.5$ Nm. Thus, the determined structural and power parameters of the vibrator are $m_d r = 0.113$ kg/m, $n = 991.2$ rpm, and $F_0 = 1.22$ kN. In modern industrial vibrators, the parameters $m_d r$, f, and F_0 can be adjusted, which significantly increase the functionality and limits of optimal modes of operation of vibration stands.

For the numerical calculation of the system of nonlinear differential equations (2.3), the next values of the parameters are taken: $M(\dot{\phi}) = M_0 = 5.5$ Nm, $J_d = 0.001$ kg/m, and $b\varphi = 0.0001$ N/m/s/rad. The main kinematic and technological characteristics (Figure 2.5) indicate the efficiency of the vibrating stand and its operational stability. The actual displacement of the operating mass (Figure 2.5b) occurs relative to the point of its dynamic equilibrium, with acceleration which satisfies the imposed restrictions (Figure 2.5c). The implemented damping force is asymmetric in the time domain (Figure 2.5d) due to its piecewise nonlinear characteristic.

The determined results of this dynamic model are the main time dependences, which allow a parametric form of their performance. A valid damping characteristic was obtained (Figure 2.6).

Optimal parameters of the driver 25

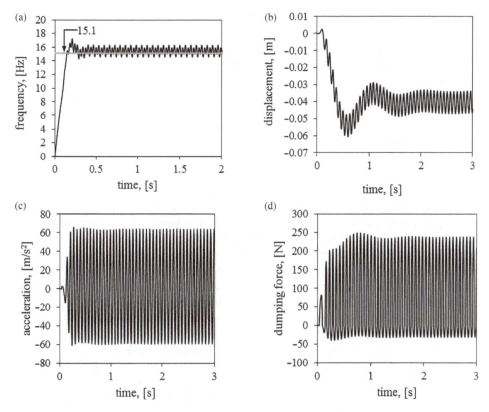

Figure 2.5 Time dependences of the main characteristics of vibratory stand: (a) frequency of oscillations, (b) displacement, (c) acceleration, and (d) damping force.

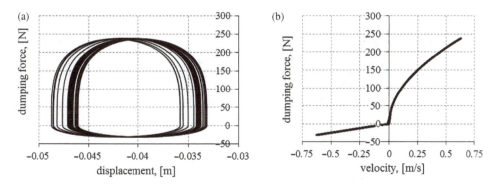

Figure 2.6 Parametric dependences of damping force on (a) displacement and (b) velocity.

Providing the resonant mode of operation of the vibration test stand is proven by the maximum velocity of the operating mass at the frequency of the disturbance of $f = 16.52$ Hz (Figure 2.7a). According to the results of simulation of the system of equations (2.3), taking into account the motion of the unbalanced mass, the average value of the rotational frequency of the unbalanced shaft (Figures 2.5a and 2.8) is in the range $f = (15.1...15.5)$ Hz.

The obtained results are useful for the design of low-frequency resonant vibration stands, which are intended for the analysis of objects with given nonlinear characteristics, presented in the form of a piecewise nonlinear function (2.2). An optimization problem with constraints on the main kinematic and dynamic characteristics of the designed device is established for this, with the possibility of its numerical solution, in particular on the basis of a simplified mechanical model (2.4). As a result, the frequency of oscillations and the nominal perturbation force are obtained, which guarantees a minimum difference between the actual and the calculated characteristics of the damper. Under certain conditions, structural and power parameters and energy characteristics of the inertial drive of the vibration stand are calculated.

Having executed the refined modeling taking into account the dynamics of the drive, according to the equations (2.3), the time dependence of the oscillation frequency of the vibration stand is obtained (Figure 2.6a), the average value of which is $f = 15.1$ Hz, which differs from the calculated value by 8.65%. As a result, there is a discrepancy in providing a damping characteristic $F_d = (-31...239)$N, the perfect form of which varies within $F_d = (-31...239)$N. The obtained dependence $f(M_0)$ (Figure 2.7b) is a calculation to specify the actual value of the nominal torque M_0, which will provide the estimated value of the rotational frequency of the drive. Therefore, solving the nonlinear equation $f(M_0) = f^{opt} = 16.52$ Hz, the value of the nominal torque is obtained, being $M_0 = 7.3$ Nm (1.3 times more than calculated), which will improve the technological characteristics of the vibration stand and provide a more acceptable range $F_d = (-34...246)$N. In this case, during the start of the vibration system, attention must be paid to the condition $M_0 < M_{0max} = 8$ Nm, in the case of an excess of which there is a breakdown of the perturbation frequency due to the phenomenon of Sommerfeld

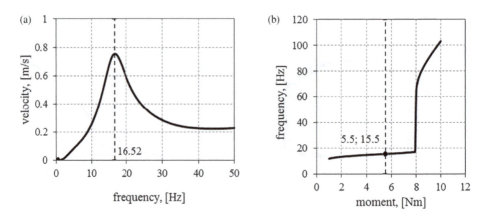

Figure 2.7 (a) Frequency dependence of velocity of vibratory booth and (b) dependence of frequency of oscillations of vibratory booth from nominal moment.

effect to the magnitude of $f = 65.2$ Hz, at which the work of the vibration stand will be disturbed (Figure 2.7b).

2.6 CONCLUSION

Based on the proposed approach, testing and diagnostics of a nonlinear damper can be provided on vibration stands with a controlled inertia drive. Mathematical dependencies, which are within the range of the permissible limits for engineering calculations, provide an opportunity to get closer to the real characteristics of the damper. The obtained optimization models allow for the calculation of the rational structural and power parameters of vibration stands, and the main results are necessary for the design process. This research can be improved by the use of more advanced models of dampers, taking into account their structural and physical-mechanical peculiarities, and also by describing their rheological behavior.

REFERENCES

Chen, L. & Sun, L. 2017. Steady-state analysis of cable with nonlinear damper via harmonic balance method for maximizing damping. *Journal of Structural Engineering* 143(2): 04016172.

Cui, Y., Kurfess, T.R. & Messman, M. 2010. Testing and modeling of nonlinear properties of shock absorbers for vehicle dynamics studies. *Proceedings of the World Congress on Engineering and Computer Science* October 20–22, 2010, San Francisco II: 2187.

Elliott, S.J., Tehrani, M.G. & Langley, R.S. 2015. Nonlinear damping and quasi-linear modelling. *Philosophical Transactions of the Royal Society A: Mathematical, Physical and Engineering Sciences* 373(2051): 1–30.

Fujita, K. et al. 2014. Optimal placement and design of nonlinear dampers for building structures in the frequency domain. *Earthquakes and Structures* 7(6): 1025–1044.

Gursky, V., Kuzio, I. & Korendiy, V. 2018. Optimal synthesis and implementation of resonant vibratory systems. *Universal Journal of Mechanical Engineering* 6(2): 38–46.

Kozlov, L.G. et al. 2019. Experimental research characteristics of counter balance valve for hydraulic drive control system of mobile machine. *Przeglad Elektrotechniczny* 95(4): 104–109.

Kukharchuk, V.V., Bogachuk, V.V., Hraniak, V.F. et al. 2017a. Method of magneto-elastic control of mechanic rigidity in assemblies of hydropower units. *Proc. SPIE, Photonics Applications in Astronomy, Communications, Industry, and High Energy Physics Experiments* 10445: 104456A.

Kukharchuk, V.V., Kazyv, S.S. & Bykovsky, S.A. 2017b. Discrete wavelet transformation in spectral analysis of vibration processes at hydropower units. *Przeglad Elektrotechniczny* R93(5): 65–68.

Kukharchuk, V.V. et al. 2016. Noncontact method of temperature measurement based on the phenomenon of the luminophor temperature decreasing. *Proc. SPIE, Photonics Applications in Astronomy, Communications, Industry, and High-Energy Physics Experiments* 2016 10031: 100312F.

Lanets, O.V. et al. 2013. Realizatsiia efektu Zommerfelda u vibratsiinomu maidanchyku z inertsiinym pryvodom [Implementation of Sommerfeld effect in vibratory table with inertial drive]. *Avtomatyzatsiia vyrobnychykh protsesiv u mashynobuduvanni ta pryladobuduvanni [Industrial Process Automation in Engineering and Instrumentation]* 47: 12–28.

Michalczyk, J. 2012. Inaccuracy in self-synchronisation of vibrators of two-drive vibratory machines caused by insufficient stiffness of vibrators mounting. *Archives of Metallurgy and Materials* 57(3): 823–828.

Nosko, P., Breshev, V., Fil, P. & Boyko, G. et al. 2010. Structural synthesis and design variants for non-contact machine drives. *TEKA Commission of Motorization and Power Industry in Agriculture. OL RAN, IOB*: 77–86.

Obertyukh, R.R. et al. 2018. Dynamic and mathematical models of the hydraulic-pulse device for deformation strengthening of materials. *Proceedings of SPIE, Photonics Applications in Astronomy, Communications, Industry, and High-Energy Physics Experiments 2018. International Society for Optics and Photonics* 10808: 108084Y.

Obertyukh, R.R. et al. 2019. Method of design calculation of a hydropulse device for strain hardening of materials. *Przeglad Elektrotechniczny* R95(4): 65–73.1

Özcan, D., Sönmez, Ü. & Güvenç, L. 2013. Optimisation of the nonlinear suspension characteristics of a light commercial vehicle. *International Journal of Vehicular Technology* 1–2 Doi: 10.1155/2013/562424.

Polishchuk, L. Bilyy, O. & Kharchenko, Y. 2016a. Prediction of the propagation of crack-like defects in profile elements of the boom of stack discharge conveyor *Eastern-European Journal of Enterprise Technologies* 6(1): 44–52.

Polishchuk, L., Kharchenko, Y., Piontkevych, O. & Koval, O. 2016b. The research of the dynamic processes of control system of hydraulic drive of belt conveyors with variable cargo flows. *Eastern-European Journal of Enterprise Technologies* 2(8): 22–29.

Polishchuk, L.K. et al. 2018. Study of the dynamic stability of the conveyor belt adaptive drive. *Proc. SPIE, Photonics Applications in Astronomy, Communications, Industry, and High-Energy Physics Experiments 2018* 10808: 1080862.

Polishchuk, L.K. et al. 2019. Study of the dynamic stability of the belt conveyor adaptive drive. *Przeglad Elektrotechniczny* 95(4): 98–103.

Shpachuk, V.P., Zasiadko, M.A. & Dudko, V.V. 2018. Investigation of stress-strain state of packet node connection in spatial vibration shakers. *Naukovyi Visnyk Natsionalnoho Hirnychoho Universytetu* 3: 74–79.

Sikora, M. 2018. Modeling and operational analysis of an automotive shock absorber with a tuned mass damper. *Acta mechanica et automatica* 12(3): 234–251.

Silveira, M., Pontes, B.R. & Balthazar, J.M. 2014. Use of nonlinear asymmetrical shock absorber to improve comfort on passenger vehicles. *Journal of Sound and Vibration* 333(7): 2114–2129.

Surace, C., Worden, K. & Tomlinson G.R. 1992. On the non-linear characteristics of automotive shock absorbers. *Proceedings of the Institution of Mechanical Engineers, Part D: Journal of Automobile Engineering* 206(1): 3–16.

Vedmitskyi, Y.G., Kukharchuk, V.V. & Hraniak, V.F. 2017. New non-system physical quantities for vibration monitoring of transient processes at hydropower facilities, integral vibratory accelerations. *Przeglad Elektrotechniczny* R93(3): 69–72.

Yaroshevich, M.P., Zabrodets, I.P. & Yaroshevich, T.S. 2015. Dynamics of vibrating machines starting with unbalanced drive in case of bearing body flat vibrations. *Naukovyi Visnyk Natsionalnoho Hirnychoho Universytetu* 3: 39–45.

Yatsun, V. Filimonikhin, G., Nevdakha, A. & Pirogov, V. 2018. Experimental study into rotational-oscillatory vibrations of a vibration machine platform excited by the ball auto-balancer. *Eastern-European Journal of Enterprise Technologies* 4 7(94): 34–42.

Zhao, L., Zhou, C., Yu, Y. & Yang, F 2016. A method to evaluate stiffness and damping parameters of cabin suspension system for heavy truck. *Advances in Mechanical Engineering* 8(7): 1–9.

Chapter 3

Algorithm to calculate work tools of machines for performance in extreme working conditions

Leonid A. Khmara, Sergej V. Shatov, Leonid K. Polishchuk, Valerii O. Kravchuk, Piotr Kisala, Yedilkhan Amirgaliyev, Amandyk Tuleshov, and Mukhtar Junisbekov

CONTENTS

3.1 Introduction 29
3.2 Analysis of destroyed buildings and structures 29
3.3 Materials and methods 30
3.4 Results and discussion 32
3.5 Conclusion 34
References 36

3.1 INTRODUCTION

The impact of manmade and natural disasters (e.g. gas explosions, fires, earthquakes and landslides) results in damage to and destruction of buildings and structures. The pieces of such objects form debris, the disassembly of which is a dangerous process but must be performed as quickly as possible. This creates extreme conditions for the performance of restorations. Currently, the dismantling of blockages is carried out by a technique that does not meet the requirements of these works, which leads to an increase in the time to dismantle the blockages and to work on imperfect technological schemes (Kochetkov et al. 1995, Bakin & Batygin 1989, Belikov et al. 2014). Therefore, the actual problem is the calculation of the capacities of the working tools of machines for dismantling the rubble of destroyed buildings and structures and their improvement (Goncharenko et al. 2013, Polishchuk et al. 2016a).

3.2 ANALYSIS OF DESTROYED BUILDINGS AND STRUCTURES

An analysis of literary sources shows that the dismantling of debris is carried out in the following sequence: clearing roads for the movement of machinery, preparing a site for the execution of works, the collapse of damaged building structures that are under threat of collapsing, the destruction of damaged structures and large-scale debris and loading and removal of disassembly products of blockages (Glazkov et al. 2000, Kazakov & Chadov 2007, Vedmitskyi et al. 2017). In the case that the construction objects were only deformed and are in an unstable position, they are fastened with supports and a decision on their restoration is made – level the building with the

DOI: 10.1201/9781003224136-3

Figure 3.1 Destroyed earthquake objects in (a) Iran (2017) and (b) Nepal (2015).

help of jacks or carry out measures to provide stability and repair the damaged parts (Kukharchuk et al. 2017, Vasyl et al. 2016, 2017).

During 2015–2017, significant earthquake destruction took place in Iran, Italy, Nepal, Indonesia and Taiwan. Buildings and structures were completely or partially destroyed (Figure 3.1). The fragments were separated structural elements and structures, as well as deformed parts of these objects. The largest and most massive fragments were columns and inter-floor overlays. The volume of these components of buildings amounted to 0.8–1.2 m^3 each. The location of the fragments was chaotic. The distance from the ruins of the destroyed buildings was in the range of 20–30 m. A large number of blockages blocked the driving part of the roads, which complicated the restoration work (Vedmitskyi et al. 2017, Kukharchuk et al. 2017, Vasyl et al. 2016).

At all stages of the work, various construction machineries were used: telescopic lifts, cranes, excavators and loaders (Kudryavcev 1989, Markov & Markova 2008, Khmara et al. 2015). In the case of significant destruction and to shorten the time for cleaning up the wreckage of buildings, the organization of works was to attract a significant amount of equipment. In this case, excavators with different types of work tools were used: buckets and clamping tools. Such a technological scheme of the use of equipment for dismantling the blockages provides high productivity but can be used only during significant volumes of work and when there are no restrictions in the working areas of the machines, which is uncommon when disassembling debris. The disadvantage of the well-known working tools for dismantling rubble is the lack of solutions for extracting fragments under compressed conditions when one or two cars may be located in the area of rubble (Figure 3.2).

The purpose of the research is to develop an algorithm for calculating the most efficient use of working machines for dismantling the rubble of destroyed buildings and structures and their improvement (Livinskiy et al. 2012, Miroshnychenko 2017, Berni et al. 2008).

3.3 MATERIALS AND METHODS

The algorithm for calculating the work tools of machines for dismantling the rubble of destroyed buildings and structures implies the consistent implementation of the following steps: determining the nature of the destruction of buildings and the fractional

Figure 3.2 Use of equipment for dismantling building debris: (a) telescopic lifts, (b) truck cranes, (c) excavators and (d) loaders.

composition of the debris (Shatov 2011, 2013, Senderov 1991); available choice of types of machines and their equipment; development of technology to manage the location and movement of the machines in the area of rubbish and determine the number of appropriate types of machines. All stages of the algorithm include the use of computer programs, which allows, in the shortest time after receipt of incoming information about the nature of the destruction of objects, calculation of the quantitative composition of cars to be performed and the information received to be provided to the relevant repairers – units of the State Emergency Services (Tarakanov 1984, Khmara & Shatov 2010).

The created algorithm allows you to perform calculations of the work tools of machines at all stages of repair works – from initial operations to their completion (Khmara & Shatov 2015, Chumak 2001).

Performance of loaders (Polishchuk et al. 2019):

- when working with a bucket (disassembly of the rubble with small debris) $П_{ек}$:

$$П_{ек} = \frac{3,600}{T_{ts}} \cdot q \cdot K_H \cdot K_B \qquad (3.1)$$

where q = capacity of the bucket, m³; K_H = ladle filling factor, $K_H = 0.8...0.9$; K_B = coefficient of loader usage in time, $K_B = 0.8...0.85$ and T_{ts} = duration of the working cycle, s.

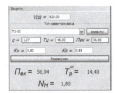

Figure 3.3 Window of the program of calculation of loaders.

- when working with grips Π_{e3}:

$$\Pi_{e3} = \frac{3{,}600}{T_{II}} \cdot Q \cdot K_{\Gamma} \cdot K_{B} \qquad (3.2)$$

where Q = load capacity, t and K_{Γ} = coefficient of use of loader by load capacity (Polishchuk et al. 2016b).

Duration of the working cycle T_{ts}:

$$T_{ts} = t_{3a\theta.} + t_{nep.} + t_{po3\theta.} + t_M + t_n \qquad (3.3)$$

where $t_{3a\theta.}$ = bucket loading time or grab time, s; $t_{nep.}$ = time to move the loader from the fall and load and back, s; $t_{po3\theta.}$ = unloading time, s; t_M = time to manoeuvre the loader, s and t_n = time for gear shifting, s.

Time to disassemble a blockage T_P^{HK}:

$$T_P^{HK} = \frac{V_{p\partial}}{\Pi_{eK}} \qquad (3.4)$$

where $V_{p\partial}$ = volume of traffic jams on the road, m³.

Required number of loaders N_H taking into account the time factor T_ϕ, the blockage needs to be disassembled in 8 hours (Polishchuk et al. 2018, Kozlov et al. 2019):

$$N_H = \frac{T_P^{HK}}{T_\phi} \qquad (3.5)$$

The 'Zaval' software is used to determine the number of means of mechanization (Figure 3.3 shows the interface with the results of calculations) (Shatov 2013, Senderov 1991, Shkinev 1976).

3.4 RESULTS AND DISCUSSION

To carry out demolition work on demolished buildings and structures under compressed conditions and with a limited number of units, it is proposed to install multipurpose equipment (MPE) based on diggers, which have (Patent UA 2006) a bucket of

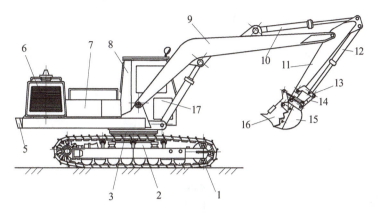

Figure 3.4 Multipurpose equipment based on a crawler excavator.

a backhoe shovel excavator and a hydraulic jaw or grip (Figure 3.4) (Xemmond 1960, Patent UA, 2006, Patent UA, 2009).

The execution of the MPE with the jaw allows the following technological operations to be performed:

- when jaw 16 is closed to bucket 15, they should be filled with small debris (Figure 3.5a);
- open then close jaw 16 towards bucket 15 to grab and hold large rubble debris (Figure 3.5b);
- collapse unstable elements of destroyed or damaged buildings (Figure 3.5c);
- carry out the restoration of utility networks: develop the soil and replace the damaged areas of pipelines (Figure 3.5d) (Patent UA, 2011a);
- perform technological operations using other work tools which are grasped by the MPE: to grind large-sized fragments with a hydraulic hammer (Figure 3.5e) and to consolidate soil or other building materials when the engineering networks are restored (Vedmitskyi et al. 2017);
- load the wreckage onto vehicles or unload them into warehouses (Figure 3.5f);
- when there is a technological need to manage soils and other media or construction materials (Figure 3.5a).

Various structures (Figure 3.6) of this equipment have been developed, depending on the specific conditions of the production of the work: swivel buckets with a jaw (Figure 3.6a and d) and with several grapples (Figure 3.6f), for changing the position of the bucket at the free location of fragments in space; a bucket with a jaw and lateral ribbed overlays (Figure 3.6b), for reliable seizure of different shapes of debris and buckets with a controlled tooth of lobed-capture (Figure 3.6c and e), for grinding large-sized debris and their grabbing. All designs are protected by Ukrainian patents (Patent UA, 2009, Patent UA, 2011b).

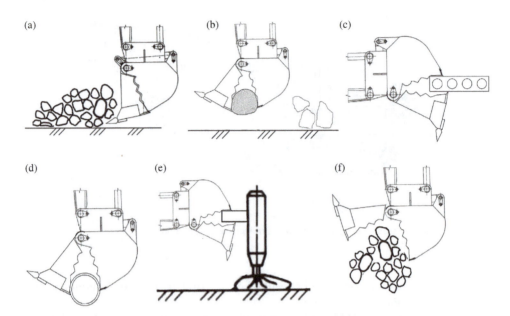

Figure 3.5 Technological operations performed by MPE: (a) removal of small debris; (b) grabbing large debris; (c) collapse of unstable elements of buildings; (d) renovation of engineering networks; (e) – grasping and holding work tools and (f) loading of fragments for transport.

In the case of use of excavators with MPE for dismantling blockages, a technological scheme is developed that takes into account the specific conditions for the elimination of the emergency.

3.5 CONCLUSION

To disassemble the debris of destroyed buildings from transport networks, machinery is used that does not meet the requirements of these processes, which leads to the implementation of restorations on imperfect technological schemes and increases the time and complexity of their maintenance. Designs of working tools of construction machines in the form of grippers, which are installed on loaders and bulldozers and developed to improve the efficiency of disassembling the destruction of objects, are shown. Software for the determination of the types and number of loaders for disassembly of building waste from transport networks has been developed. An algorithm for calculating the working tools of machines for dismantling the rubble of destroyed buildings and structures is enumerated. Designs of working tools of machines for the effective performance of restorations in extreme conditions have been developed.

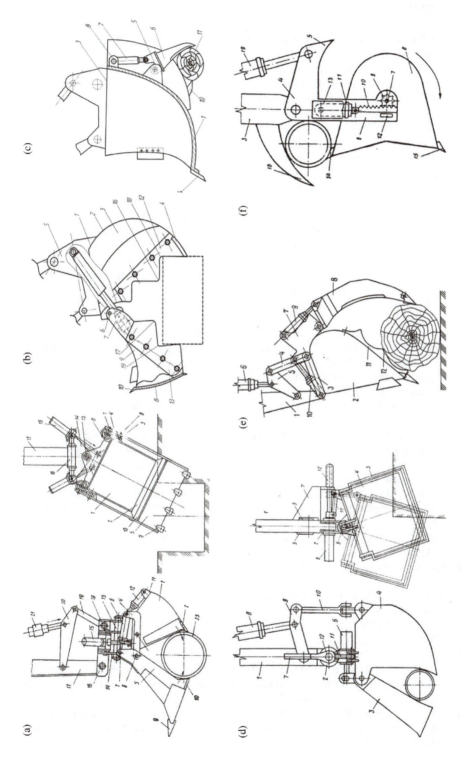

Figure 3.6 Types of working equipment for different purposes: (a, d) rotary bucket with a jaw; (b) bucket with a jaw and lateral ribbed overlays; (c, e) bucket with a controlled tooth loosening-capture and (f) swivel bucket with several teeth hooks.

REFERENCES

Bakin, V. P. & Batygin, N. S. 1989. Demolition of damaged buildings by earthquakes. *Construction Mechanization* 6: 10–11.
Belikov, A.S. et al. 2014. *Work Safety in Construction. Prydniprovs'ka State Academy of Civil Engineering and Architecture*. Kiev: Osnova.
Berni, D., Gilpin, D., Kojn, S. & Simons, P. 2008. *Unrestrained Planet. When Nature Goes Mad*. Germaniya: Dom Riderz Dajdzhest.
Caterpillar, Inc. 1999. *Technic and Operating Description of Machines of Firm Caterpillar*. Reference book. Peoria: Caterpillar Inc.
Chumak, S. P. 2001. Scope evaluation method of rescue operations certain types in their planning and preparation. Security concerns in emergencies. *VINITI. All-Russian Institute for Scientific and Technical Information* 3: 176–184.
Glazkov, A. A., Manakov, N. A. & Pankratov, A. V. 2000. *Build, Travelling and Special Technique of Domestic Production. Short Reference Book*. Moscow: Business-Arsenal.
Goncharenko, D. F., Melencov, N. A. & Konstantinov, A. S. 2013. Technology of demolition, construction and installation work by recovering of partially destroyed building. *Industrial Construction and Civil Engineering Constructions* 1: 42–44.
Kazakov, B. & Chadov, E. 2007. Organization and carrying of rescue works on residential buildings and structures. *Emergency* 6: 44–49.
Khmara, L. A., Kravets, S. V. & Skobluk, N. P. 2014. *Machines Are for Earthworks: Textbook*. Kharkov: KhNARU.
Khmara, L. A. & Shatov, S. V. 2010. Technological features of rubble demolition of destroyed buildings. *Bulletin of Prydnniprovs'ka Academy of Civil Engineering and Architecture* 7: 42–52.
Khmara, L. A. & Shatov, S. V. 2015. Improvement of technological processes of sorting out of the blasted buildings and constructions. *Bulletin of Prydnniprovs'ka Academy of Civil Engineering and Architecture* 6: 45–52.
Kochetkov, K. E., Kotlyarevskij, V. A. & Zabegaeva, A. V. 1995. *Accidents and Disasters. Prevention and Mitigation*. Moscow: ASV.
Kozlov, L. G. et al. 2019. Experimental research characteristics of counterbalance valve for hydraulic drive control system of mobile machine. *Przegląd Elektrotechniczny* 95(4): 104–109.
Kudryavcev, E. M. 1989. *Construction Complex Mechanization, Automation and Mechanical Equipment*. Moscow: Strojizdat.
Kukharchuk, V. V., Kazyv, S. S. & Bykovsky, S. A. 2017. Discrete wavelet transformation in spectral analysis of vibration processes at hydropower units. *Przegląd Elektrotechniczny* 93(5): 65–68.
Livinskiy, O.M., Dorofeev, V.S. & Ushackiy, S.A. 2012. *Technology of a Build Production*. Kyiv: UAN, MP Lesya.
Markov, A.I. & Markova, M.A. 2008. *Accidents Buildings and Constructions*. Zaporozh'e: Nastroj.
Miroshnychenko, M. 2007. This lesson should learn the state. *Gas Explosion* 10: 8–15.
Patent UA, 2006. no. 20025.
Patent UA, 2009. no. 39338.
Patent UA, 2011a. no. 63515.
Patent UA, 2011b. no. 63556.
Polishchuk, L., Bilyy, O. & Kharchenko, Y. 2016a. Prediction of the propagation of crack-like defects in profile elements of the boom of stack discharge conveyor. *Eastern-European Journal of Enterprise Technologies* 6(1): 44–52.
Polishchuk, L. et al. 2016b. The research of the dynamic processes of control system of hydraulic drive of belt conveyors with variable cargo flows. *Eastern-European Journal of Enterprise Technologies* 2(8): 22–29.

Polishchuk, L. et al. 2018. Study of the dynamic stability of the conveyor belt adaptive drive. *Proc. SPIE 10808, Photonics Applications in Astronomy, Communications, Industry, and High-Energy Physics Experiments* 1080862.

Polishchuk, L. et al. 2019. Study of the dynamic stability of the belt conveyor adaptive drive. *Przegląd Elektrotechniczny* 95(4): 98–103.

Senderov, B.V. 1991. *Residential Buildings Accidents.* Moscow: Strojizdat.

Shatov, S.V. 2011. Determination of scantling parameters of destroyed structures and building elements, which are reconstructed. *Bulletin of the Prydniprovs'ka State Academy of Civil Engineering and Architecture* 3: 8–14.

Shatov, S.V. 2013. Organizational and technological solutions of damaged and reconstructed constructions and buildings dismantling. *Bulletin of Prydnniprovs'ka Academy of Civil Engineering and Architecture* 4: 12–17.

Shkinev, A.N. 1976. *Accidents at the Building Sites, Their Causes and Methods of Warning.* Moscow: Strojizdat.

Tarakanov, N.D. 1984. *Complex Mechanization of Rescue and Emergency Restoration Works.* Moscow: Energoatomizdat.

Vasyl, V. et al. 2016. Noncontact method of temperature measurement based on the phenomenon of the luminophor temperature decreasing. *Proc. SPIE 10031, Photonics Applications in Astronomy, Communications, Industry, and High-Energy Physics Experiments* 100312F.

Vasyl V. et al. 2017. Method of magneto-elastic control of mechanic rigidity in assemblies of hydropower units. *Proc. SPIE 10445, Photonics Applications in Astronomy, Communications, Industry, and High Energy Physics Experiments* 104456A.

Vedmitskyi, Y.G., Kukharchuk, V.V. & Hraniak, V.F. 2017. New non-system physical quantities for vibration monitoring of transient processes at hydropower facilities, integral vibratory accelerations. *Przegląd Elektrotechniczny* 93(3): 69–72.

Xemmond, R. 1960. *Accidents of Buildings and Structures. Causes and Lessons of Modern Structures Accidents of Various Types.* Moscow: Gosstrojizdat.

Chapter 4

Vibration diagnostic of wear for cylinder-piston couples of pumps of a radial piston hydromachine

Vladimir Shatokhin, Boris Granko, Vladimir Sobol,
Leonid K. Polishchuk, Olexander Manzhilevskyy,
Konrad Gromaszek, Mukhtar Junisbekov, Nataliya Denissova,
and Kuanysh Muslimov

CONTENTS

4.1 Introduction 39
4.2 Analysis of dynamical processes in RPH with spherical ball pistons 40
4.3 Dynamical model of radial piston pump with HVT 40
4.4 Diagnostics algorithm for pressure surges of cylinder and piston pump
    couples of RPH 41
4.5 Diagnostics algorithm for a statistical vibration signal 45
4.6 Calculation and experimental research 47
4.7 Conclusion 49
References 49

4.1 INTRODUCTION

One of the most advanced and effective directions for designing continuously variable transmissions for mobile transport and special tracked vehicles is associated with the use of hydrovolumetric transmissions (HVT), which allow parallel flows of hydraulic and mechanical power. In the most modern cases of such transmission, radial piston hydromachines (RPH) and ball pistons are usually used. Using these, it is possible to create quite compact transmission designs. It has been established that radial-type hydraulic machines have the ability to withstand significantly greater stresses than axial-type hydraulic machines. Such devices also achieve a significantly higher rotational speed. However, such transmissions have a significant level of loading of the most critical elements, in particular the cylinder-piston couples. Their excessive wear leads to decreasing levels in the transmitted power, decreasing reliability of devices and leading even to failures. One of the structural techniques to improve the HVT reliability is a technical diagnostic. Research devoted to the study of dynamic processes in power transmissions with radial piston HVT and devoted to the development of algorithms for calculating the diagnostic parameters of RPH with ball pistons have not yet been sufficiently presented in the scientific literature.

DOI: 10.1201/9781003224136-4

4.2 ANALYSIS OF DYNAMICAL PROCESSES IN RPH WITH SPHERICAL BALL PISTONS

Interest in the use of RPH with ball pistons for transport and special tracked vehicles was increased significantly after Martin-Marietta (USA) began mass production with this type of hydromachine (Figure 4.1a). These devices are beginning to be used in domestic engineering (Borisyuk et al. 2008, Avrunin et al. 2004). In Babaev et al. (1987) and Petrov (1988), it was shown that the power losses in hydromachines with significant wear of cylinder-piston couples can be very large. Avramov et al. (1981) describe the study of the loads acting in the piston of a radial piston hydraulic pump. A number of articles are devoted to the study of various losses in hydraulic machines. Mechanical losses are analyzed in Aleksandrov and Samorodov (1984), volumetric in Avramov and Samorodov (1986), and hydraulic in Lovtsov (1967). Stress-strain state research of the elements of hydrovolumetric transmissions are presented in Tkachuk et al. (2017) and Tkachuk and Bibik (2018). In Jiang (2009) and Cai and Tian (2012), pressure pulsations in a piston chamber of a radial piston pump are investigated, and an optimization problem is solved to minimize such pulsations. In Zhao et al. (2015), a new two-row radial piston pump was proposed. It is shown that an improved design can significantly reduce the load on the main shaft of the pump.

In Shatokhin (2008) and Shatokhin et al. (2014), a sufficiently complete mathematical model of dynamic processes in an RPH pump was developed, taking into account the inertial properties of the piston and stator balls, the elastic properties of the fluid column in the control cylinder, eccentricity, charge pressure in the pressure area, and the rotor rotation frequency.

The aim of this chapter is to develop an algorithm based on this model for wear diagnostics of cylinder-piston couples of an RPH pump. It is necessary to solve the following problems: choose the diagnostic model; propose parameters characterizing the wear of the cylinder-piston couples; choose the diagnostic data which are highly informative and easy to acquire and process; perform the method generalization for vibration signals of a statistical character; and perform computational and experimental studies of the dynamic processes.

4.3 DYNAMICAL MODEL OF RADIAL PISTON PUMP WITH HVT

The kinematics diagram (Figure 4.1b) shows the following: 1, cylinder block (rotor); 2, piston ball; 3, distribution pintle; 4, stator ring; 5, HVT casing; 6, hydraulic cylinder of the control system; 7, control ring of stator ring; and 8, rotation axis of the stator ring. The rotation axis of the rotor passes through the axis of the stator ring and the axis of its rotation (perpendicular to the figure plane).

The mathematical model of the stator vibrations can be presented as (Shatokhin 2008):

$$I_{O_3z}^{(e)}\ddot{\vartheta} + b_e\dot{\vartheta} + c_e\vartheta = M_{O_3z}^{(0)} + M_{O_3z}(t) \qquad (4.1)$$

where ϑ is the stator deflection angle; $I_{O_3z}^{(e)} = \dfrac{7}{10}m_b n h^2 + I_{O_3z}^{(s)}$ is the equivalent moment of inertia of the model; m_b is the ball mass; $n = 9$ is the number of cylinders; h is the

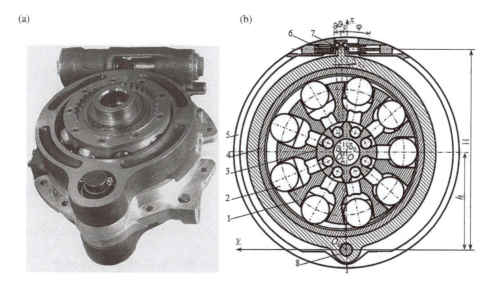

Figure 4.1 Radial piston hydromachine with ball pistons: (a) general view of a device from Martin-Marietta and (b) kinematics scheme of radial piston pump with HVT-900.

distance from the stator rotation axis to the rotor axis; $I_{O_3z}^{(s)} = I_{O_1z}^{(s)} + m_s \cdot O_3O_1^2$ is the inertia moment of the stator about its rotation axis; $I_{O_1z}^{(s)}$ is the inertia moment of the stator about the stator rotation axis; m_s is the stator mass; O_3O_1 is the distance from the stator ring axis to its mass center; $c_e = cH^2 - \frac{7}{5}m_b nh^2\omega^2$ is the equivalent stiffness; c is the elasticity equivalent coefficient of fluid in the control hydraulic cylinder; H is the distance from the stator rotation axis to the axis of the control hydraulic cylinder; ω is the rotor angular velocity; $M_{O_3z}^{(0)} = \frac{7}{5}m_b nh^2\vartheta_0\omega^2$ is the constant torque caused by the movement of the balls; ϑ_0 is the initial stator deflection angle; b_e is the equivalent drag coefficient; and $M_{O_3z}(t)$ is the disturbing torque caused by fluid pressure of the cylinder area under the piston.

4.4 DIAGNOSTICS ALGORITHM FOR PRESSURE SURGES OF CYLINDER AND PISTON PUMP COUPLES OF RPH

In the law of pressure change of the cylinder area under the piston of a ball piston in the interval of one revolution of the pump rotor, which forms the disturbing moment M_{O_3z} of the model (4.1), a two-parameter dependence is established for the pressure swing curve (Shatokhin et al. 2014). The impulse interval of the angle of rotor rotation $\Delta\varphi$ and its maximum value Δp are considered (Figure 4.2). For a particular pump, the specified interval is determined by its geometry. The level of pressure surges, which depend on the rotor rotation frequency, and the load on the HVT can significantly influence the leaks. Such leaks are caused by excessive wear of the balls and cylinder walls. The developed algorithm is based on an experimentally confirmed fact about

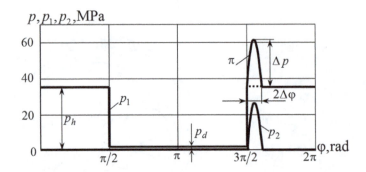

Figure 4.2 Dependence of pressure in cylinder area under piston on rotor rotation angle.

the relation between pressure surges and pump stator vibrations about the rotation axis (Polishchuk et al. 2019, Kozlov et al. 2019).

The pressure $p(\varphi)$ for first cylinder can be presented as the following sum:

$$p(\varphi) = p_1(\varphi) + \eta p_2(\varphi) \tag{4.2}$$

where $p_1(\varphi)$ is the pressure without surge; $p_2(\varphi)$ is the pressure surge without wear; φ is the rotor rotation angle ($0 \leq \varphi \leq 2\pi$); η is the parameter characterizing the level of pressure surge $0 \leq \eta \leq 1$; p_h is the pressure in the pressure area; p_d is the charge pressure; and Δp is the value of pressure surge. The value $\eta = 1$ corresponds to zero wear and the value $\eta = 0$ to limiting wear.

Now, the disturbing torque from the first cylinder, taking to account one revolution of a rotor, can be presented as:

$$M_{O3z}(\varphi) = q_1(\varphi) + \eta q_2(\varphi) \tag{4.3}$$

where $q_1(\varphi) = \dfrac{\pi r^2}{\cos \psi}(x \sin(\varphi + \psi) - y \cos(\varphi + \psi)) p_1(\varphi)$; $q_2(\varphi) = \dfrac{\pi r^2}{\cos \psi}(-x \sin(\varphi + \psi) + y \cos(\varphi + \psi)) p_2(\varphi)$; r is the ball radius; $x = h + \rho \cos(\varphi) + r \cos(\varphi + \psi)$; $y = -(\rho \sin(\varphi)) + r \sin(\varphi + \psi)$; $\rho = R_0 - \delta_1 \sin \varphi$; R_0 is the radius of the circle around which the centers of the balls move; and $\psi = \dfrac{\delta_1}{R_0} \cos \varphi$.

For ρ and ψ, we give approximate expressions with an accuracy up to the terms of the first order of smallness.

Let us present an approximately disturbing torque from the first cylinder by a finite trigonometric series, expanding expression (4.3) in a Fourier series:

$$M_1(\varphi) = M_{O3z}(\varphi) = \sum_{i=1}^{m} \left(b_{ci}^{(1)} \cos i\varphi + b_{si}^{(1)} \sin i\varphi \right) + \eta_1 \sum_{i=1}^{m} \left(a_{ci}^{(1)} \cos i\varphi + a_{si}^{(1)} \sin i\varphi \right) \tag{4.4}$$

where $b_{ci}^{(1)}$, $b_{si}^{(1)}$ is the cosine and sine amplitude of the i-th harmonic in Fourier series expansion $p_1(\varphi)$; $a_{ci}^{(1)}$ $a_{si}^{(1)}$ is the cosine and sine amplitude of the i-th harmonic in Fourier series expansion $p_2(\varphi)$; and m is the number of harmonics.

Then, for the torque from the k-th cylinder, we will have the following formula:

$$M_k(\varphi) = \sum_{i=1}^{m}\left(b_{ci}^{(1)}\cos i\tilde{\varphi}_k + b_{si}^{(1)}\sin i\tilde{\varphi}_k\right) + \eta_k \sum_{i=1}^{m}\left(a_{ci}^{(1)}\cos i\tilde{\varphi}_k + a_{si}^{(1)}\sin i\tilde{\varphi}_k\right) \quad (4.5)$$

where $\tilde{\varphi}_k = \varphi - \gamma_k$ and $\gamma_k = (k-1)\dfrac{2\pi}{\omega n}$ is the phase shift of the k-th disturbing torque ($k = \overline{1,n}$).

Taking into account the expression for $\tilde{\varphi}_k$, the last formula can be presented finally:

$$M_k(\varphi) = \sum_{i=1}^{m}\left(b_{ci}^{(k)}\cos i\varphi + b_{si}^{(k)}\sin i\varphi\right) + \eta_k \sum_{i=1}^{m}\left(a_{ci}^{(k)}\cos i\varphi + a_{si}^{(k)}\sin i\varphi\right), \quad (4.6)$$

where $b_{ci}^{(k)} = b_{ci}^{(1)}\cos i\gamma_k - b_{si}^{(1)}\sin i\gamma_k$, $b_{si}^{(k)} = b_{ci}^{(1)}\sin i\gamma_k + b_{si}^{(1)}\cos i\gamma_k$; $a_{ci}^{(k)} = a_{ci}^{(1)}\cos i\gamma_k - a_{si}^{(1)}\sin i\gamma_k$ and $a_{si}^{(k)} = a_{ci}^{(1)}\sin i\gamma_k + a_{si}^{(1)}\cos i\gamma_k$.

The amplitudes of the cosine and sine harmonic components in the pump stator motion of the given diagnostic model, which are obtained on the basis of experimental data, can be denoted as, respectively, ϑ_{Ci} and ϑ_{Si}. Here, we assume that i takes the numbers values of the recorded harmonics of the vibration signal ($i = i_1, i_2, \ldots, i_\nu$; ν = number of harmonics used for the diagnostic process).

To form a system of linear algebraic equations by the definition of parameters η_k ($k = \overline{1,n}$), we use harmonic coefficients of influence. The cosine and sinus amplitudes of the stator vibrations from the torque of an individual amplitude $\sin i\omega t$ are denoted by $\alpha_{Ci}^{(s)}$ and $\alpha_{Si}^{(s)}$, respectively, and from the torque of an individual amplitude $\cos i\omega t$ by $\alpha_{Ci}^{(c)}$ and $\alpha_{Si}^{(c)}$.

For model (4.1), the influence coefficients can be obtained on the basis of the known formulas for the amplitude and phase of the forced vibrations of an oscillator with a linear-viscous resistance (Loytsyanskiy & Lure 1983):

$$\alpha_{ci}^{(s)} = -\alpha_{si}^{(c)} = -\frac{2\beta i\omega}{W}; \alpha_{si}^{(s)} = \alpha_{ci}^{(c)} = \frac{\omega_0^2 - (i\omega)^2}{W} \quad (4.7)$$

where $\omega_0^2 = \dfrac{c_e}{I_{O3z}^{(e)}}$, $\beta = \dfrac{b_e}{2I_{O3z}^{(e)}}$, and $W = I_{O3z}^{(e)}\left[\left(\omega_0^2 - (i\omega)^2\right)^2 + 4\beta^2(i\omega)^2\right]$.

Let us use formulas (4.6) and (4.7) to form diagnostic equations. We write the expressions for the amplitudes of the cosine and sine components of each harmonic in the stator reaction from the corresponding disturbing harmonics and equate them with the corresponding experimental harmonic values. After simple transformations, we will have the following expressions:

$$\left.\begin{aligned}&\left(\alpha_{ci}^{(c)}a_{ci}^{(1)}+\alpha_{ci}^{(s)}a_{si}^{(1)}\right)\eta_1+\left(\alpha_{ci}^{(c)}a_{ci}^{(2)}+\alpha_{ci}^{(s)}a_{si}^{(2)}\right)\eta_2+\cdots\\ &+\left(\alpha_{ci}^{(c)}a_{ci}^{(n)}+\alpha_{ci}^{(s)}a_{si}^{(n)}\right)\eta_n=\vartheta_{ci}-\alpha_{ci}^{(c)}\sum_{k=1}^{n}b_{ci}^{(k)}-\alpha_{ci}^{(s)}\sum_{k=1}^{n}b_{si}^{(k)},\\ &\left(\alpha_{si}^{(c)}a_{ci}^{(1)}+\alpha_{si}^{(s)}a_{si}^{(1)}\right)\eta_1+\left(\alpha_{si}^{(c)}a_{ci}^{(2)}+\alpha_{si}^{(s)}a_{si}^{(2)}\right)\eta_2+\cdots\\ &+\left(\alpha_{si}^{(c)}a_{ci}^{(n)}+\alpha_{si}^{(s)}a_{si}^{(n)}\right)\eta_n=\vartheta_{si}-\alpha_{si}^{(c)}\sum_{k=1}^{n}b_{ci}^{(k)}-\alpha_{si}^{(s)}\sum_{k=1}^{n}b_{si}^{(k)},\end{aligned}\right\} \quad (4.8)$$

where $i = i_1, i_2, \ldots, i_v$.

The system of linear algebraic equations (4.8) can be presented in matrix form:

$$\mathbf{A}\boldsymbol{\eta} = \mathbf{b} \quad (4.9)$$

where $\mathbf{A}$ is the matrix with dimension $2v \times n \cdot (v \geq 5)$; $\boldsymbol{\eta} = [\eta_1, \eta_2, \ldots, \eta_n]^T$ is the vector of diagnostic parameters; and $\mathbf{b} = [b_1, b_2, \ldots, b_{2v}]^T$ is the vector of right-hand sides (T is the transpose symbol).

Since the experimental data are not accurately determined, the system (4.9) can be replaced by a system of so-called conditional equations:

$$\mathbf{A}\boldsymbol{\eta} = \tilde{\mathbf{b}} \quad (4.10)$$

where $\tilde{\mathbf{b}}$ is the approximate value of the vector of right-hand sides.

The number of equations $2v$ can be more than the number of unknowns. In this case, the system (4.10) is overdetermined. The usual and common way to solve overdetermined systems of linear algebraic equations is the least squares method, in which the solution is a vector $\tilde{\boldsymbol{\eta}}$ that minimizes the residual:

$$\|\mathbf{A}\tilde{\boldsymbol{\eta}} - \tilde{\mathbf{b}}\| = \inf \|\mathbf{A}\tilde{\boldsymbol{\eta}} - \tilde{\mathbf{b}}\|$$

in the whole solution space of the system (4.10). As the vector norm can take the following magnitude:

$$\|\mathbf{b}\| = \left(\sum_{i=1}^{2v} b_i^2\right)^{1/2} \quad (4.11)$$

the solution vector can be determined as:

$$\tilde{\boldsymbol{\eta}} = \mathbf{Y}\tilde{\mathbf{b}} \quad (4.12)$$

where $\mathbf{Y} = \left(\mathbf{A}^T \mathbf{A}\right)^{-1} \mathbf{A}^T$.

4.5 DIAGNOSTICS ALGORITHM FOR A STATISTICAL VIBRATION SIGNAL

The pressure surge diagnostic algorithm of the cylinder-piston couples of the RPH pump in a deterministic formulation is mentioned above. At the same time, experimental studies of the pump stator vibrations show that the recorded initial vibration signal has a clearly statistical nature, since, in each realization of the vibration signal, there are random deviations determined by various factors. This means that when we consider a diagnostics of cylinder-piston couples wear during their operation, diagnostic parameters must be determined taking into account the statistical nature of the vibration signal. For the obtained expected value of the diagnostic parameters, the confidence interval in which these parameters lie must be determined (Loytsyanskiy & Lure 1983, Yarmak 2004, Kukharchuk et al. 2017a).

For this procedure, two methods can be used. By using the first method, the diagnostic parameters for each random realization of the vibration signal are determined in a deterministic formulation and, therefore, are obtained as random values. By doing this for various realizations of the vibration signal, we can find the expected value, variance, and confidence interval of the diagnostic parameters. When we use such an approach for each experiment, the experimental material must be processed into diagnostic material. To obtain reliable statistical characteristics of the diagnostic parameters in this case, a sufficiently large number of experiments and a lot of time for results processing are needed (Kukharchuk et al. 2016, Kukharchuk et al. 2017a, 2017b, Vedmitskyi et al. 2017).

A significant reduction in the time spent on this can be achieved by using the second method, in which, by using random realizations of the initial vibration signal, its expected value and variance can be determined. Then, we can determine the diagnostic parameters and confidence interval for such a vibration signal.

Let us write the system of equations for the detuning parameters (4.6) for the j-th realization of a vibration signal:

$$\left. \begin{array}{l} s_{ci}^{(1)}\eta_1^{(j)} + s_{ci}^{(2)}\eta_2^{(j)} + \cdots + s_{ci}^{(n)}\eta_n^{(j)} = \vartheta_{ci}^{(n)} + \psi_{ci} \\ s_{si}^{(1)}\eta_1^{(j)} + s_{si}^{(2)}\eta_2^{(j)} + \cdots + s_{si}^{(n)}\eta_n^{(j)} = \vartheta_{si}^{(n)} + \psi_{si} \end{array} \right\} \quad (4.13)$$

To simplify the expression, we can use the following designations:

$$s_{ci}^{(k)} = \alpha_{ci}^{(c)}a_{ci}^{(k)} + \alpha_{ci}^{(s)}a_{si}^{(k)}; s_{si}^{(k)} = \alpha_{si}^{(c)}a_{ci}^{(n)} + \alpha_{si}^{(s)}a_{si}^{(n)} \quad (4.14)$$

$$\psi_{ci} = -\alpha_{ci}^{(c)}\sum_{k=1}^{n}b_{ci}^{(k)} - \alpha_{ci}^{(s)}\sum_{k=1}^{n}b_{si}^{(k)}; \psi_{si} = -\alpha_{si}^{(c)}\sum_{k=1}^{n}b_{ci}^{(k)} - \alpha_{si}^{(s)}\sum_{k=1}^{n}b_{si}^{(k)} \quad (4.15)$$

By using N realizations, we will have N systems of linear algebraic equations in the form (4.13). Summing up this equation system for all j and dividing them by N, we will find the equation system for the expected value determination of the diagnostic parameters:

$$\left.\begin{array}{l}s_{ci}^{(1)}m(\eta_1)+s_{ci}^{(2)}m(\eta_2)+\cdots+s_{ci}^{(n)}m(\eta_n)=m(\vartheta)+\psi_{ci},\\ s_{si}^{(1)}m(\eta_1)+s_{si}^{(2)}m(\eta_2)+\cdots+s_{si}^{(n)}m(\eta_n)=m(\vartheta)+\psi_{si},\end{array}\right. \quad (4.16)$$

where $m(\eta_k)$ is the expected value of tuning parameter η_k of the k-th cylinder (diagnostic parameters) and $m(\vartheta_{Ci})$ and $m(\vartheta_{Si})$ are the expected value cosine and sine components for the i-th harmonics of the stator motion law, respectively.

In matrix form, the equation system (4.16) can be written:

$$\mathbf{A}m(\mathbf{\eta})=m(\mathbf{b}) \quad (4.17)$$

where $m(\mathbf{\eta})=[m(\eta_1),m(\eta_2),\ldots,m(\eta_n)]^T$ is the column vector of expected values of detuning parameters and $m(\mathbf{b})=[m(\vartheta_{ci1}),m(\vartheta_{si1}),m(\vartheta_{ci2}),m(\vartheta_{si2}),\ldots,m(\vartheta_{civ}),m(\vartheta_{siv})]$ is the column vector of expected values of right-hand sides of equations.

For the expected value vector of the diagnostic parameters, we have the same as (4.12):

$$m(\mathbf{\eta})=\mathbf{Y}m(\mathbf{b}) \quad (4.18)$$

To determine the accuracy of the diagnostic parameters, it is necessary to compose a covariance matrix of a random vector of detuning cylinders $\mathbf{\eta}$:

$$\mathbf{D}(\mathbf{\eta})=m\left\{[\mathbf{\eta}-m(\mathbf{\eta})][\mathbf{\eta}-m(\mathbf{\eta})]^T\right\}=\begin{bmatrix} D_{11} & D_{12} & \cdots & D_{1n} \\ D_{21} & D_{22} & \cdots & D_{2n} \\ \vdots & \vdots & \ddots & \vdots \\ D_{n1} & D_{n2} & \cdots & D_{nn} \end{bmatrix} \quad (4.19)$$

The diagonal elements of this matrix are the variances of the corresponding components of the vector $\mathbf{\eta}$, which can be determined by:

$$D_{ii}=m\left[(\eta_i-m(\eta_i))^2\right]=D(\eta_i) \quad (i=\overline{1,n}) \quad (4.20)$$

The off-diagonal elements of matrix $\mathbf{D}(\mathbf{\eta})$ are called covariances (correlation moments) of corresponding pairs of vector components $\mathbf{\eta}$. Such elements can be determined by using this vector and its expected value:

$$D_{ik}=m[(\eta_i-m(\eta_i))(\eta_k-m(\eta_k))]=K(\eta_i,\eta_k) \quad (i,k=\overline{1,n}) \quad (4.21)$$

It is obvious that $K(\eta_i,\eta_k)=K(\eta_k,\eta_i)$, so the covariance matrix is symmetric.

Let us express the covariance matrix of the detuning parameters (4.19) in terms of the covariance matrix of the measured signals. By using formulas (4.12) and (4.19), we have:

$$\mathbf{D}(\eta) = \mathbf{D}(\mathbf{Yb}) = m\left\{[\mathbf{Yb} - m(\mathbf{Yb})][\mathbf{Yb} - m(\mathbf{Yb})]^T\right\}$$
$$= m\left\{\mathbf{Y}[\mathbf{b} - m(\mathbf{b})][\mathbf{Y}(\mathbf{b} - m(\mathbf{b}))]^T\right\}$$
$$= \mathbf{Y}m\left\{[\mathbf{b} - m(\mathbf{b})][\mathbf{b} - m(\mathbf{b})]^T\right\}\mathbf{Y}^T = \mathbf{YD}(\mathbf{b})\mathbf{Y}^T \tag{4.17}$$

where $\mathbf{D}(\mathbf{b}) = m\left\{[\mathbf{b} - m(\mathbf{b})][\mathbf{b} - m(\mathbf{b})]^T\right\}$ is the covariance matrix of right-hand sides of equation system (4.8). As a result, we have the matrix expression for variances determination $D(\eta_k)$ of the detuning parameters of the cylinders.

4.6 CALCULATION AND EXPERIMENTAL RESEARCH

In Figure 4.3, the experimental dependences of the stator deflection angle ϑ and the pressure in the control cylinder p_u for one revolution are shown, obtained by the method developed by Yarmak (2004). In this case, the wear of the cylinder-piston couples is equal to zero ($\eta_i = 1; i = \overline{1,9}$). The graphs are given in relative units (the angular stator deviations are obtained by recalculating the experimentally measured vibration accelerations). The figure also shows the revolution mark.

The calculated dependence of the change in the stator rotation angle for one complete rotor rotation in the case of a lack of leaks is shown in Figure 4.4. An analysis of the corresponding curves in Figures 4.3 and 4.4 leads to the conclusion that they are similar. The analysis showed that the eighteenth-order harmonic with the highest amplitude in the motion law is formed by a change in pressure of the cylinder respective to the law of a rectangular sine without taking into account pressure surge (Figure 4.2). The ninth-order harmonic is caused by a pressure surge when the cylinder moves from the low-pressure zone to the high-pressure zone.

In Figures 4.5 and 4.6, the influence of the character of the deviations on the stator movement from the normal operation of the third cylinder are shown. The curve presented in Figure 4.5 corresponds to the mathematical simulations of leaks in the third cylinder ($\eta_3 = 0.5$). The curve is significantly different from the curve in Figure 4.4. At

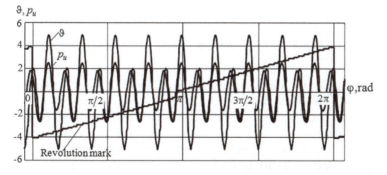

Figure 4.3 Experimental dependences of the stator deflection angle and the control cylinder pressure for one complete rotor rotation.

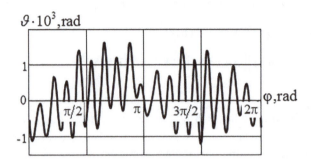

Figure 4.4 Dependences of stator deflection angle on rotor rotation angle without pressure differential in third cylinder ($\eta_3 = 0$).

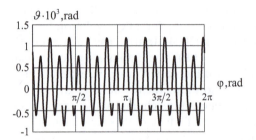

Figure 4.5 Dependence of stator deflection angle on rotor rotation angle without leaks.

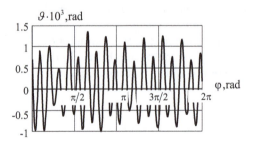

Figure 4.6 Dependence of stator deflection angle on rotor rotation angle with increased deflection in the third cylinder ($\eta_3 = 0.5$).

the same time, significant low-frequency harmonics result from the spectrum of decomposition in the Fourier series.

The graph in Figure 4.6 corresponds to the mathematical simulation of the situation where the third cylinder has completely lost its normal operation ability: $\eta_3 = 0$. It is clearly seen that the deflections noted above become more significant.

In Table 4.1, some verification results of the possibility of practical use of the developed method are shown. Two variants of wear of the third cylinder are considered: partial ($\eta_3 = 0.5$) and full ($\eta_3 = 0$). The motion law of the stator (initial vibration signal)

Table 4.1 Wear parameters ($\eta_i = 1$, $i = 1.9$)

	Cylinder number									Number of harmonics	Notes
	1	2	3	4	5	6	7	8	9		
1	1.000	1.000	0.5	1.000	1.001	0.999	0.999	0.999	0.999	18	
2	1.001	1.001	0	1.001	1.000	0.999	0.999	0.999	0.999	18	
3	0.998	1.002	0.502	0.993	0.996	1.002	0.993	1.002	1.002	18	$\vartheta_{Ci}, \vartheta_{Si}$ − (+10%)
4	1.002	1.004	0.001	0.991	0.997	0.998	1.002	0.992	1.006	18	$\vartheta_{Ci}, \vartheta_{Si}$ − (+10%)
5	0.997	0.994	0.001	0.997	0.995	0.998	0.998	0.995	0.991	9	
6	1.003	0.984	0.002	0.989	0.972	0.978	0.998	0.935	0.933	9	$\vartheta_{Ci}, \vartheta_{Si}$ − (+10%)
7	1.004	0.995	0.004	0.976	0.947	0.955	0.978	0.860	0.919	9	$\vartheta_{Ci}, \vartheta_{Si}$ − (+20%)
8	1.004	0.995	0.004	0.976	0.947	0.955	0.978	0.860	0.919	9	$\vartheta_{Ci}, \vartheta_{Si}$ − (+20%)

was obtained as a result of the calculation of forced oscillations in model (4.1) taking into account the specified detuning. The number of harmonics for diagnostics was taken equal to 18 or 9 (wear parameters are presented by taking to account four significant figures).

4.7 CONCLUSION

1. A method for diagnostics of vibration wear of cylinder-piston couples of an RPH pump with ball pistons is proposed.
2. A model describing vibrations of the pump stator was chosen as a diagnostic model.
3. The parameters characterizing the wear of the cylinder-piston couples are presented.
4. Diagnostic data which are highly informative and easy to acquire and process were chosen.
5. The generalization of the method for vibration signals of a statistical character is given.
6. Computational and experimental research of the dynamic processes with proposed diagnostic method are given.
7. The proposed algorithm has significant value for solving a wide range of diagnostic problems of RPHs with HVTs based on ball pistons.

REFERENCES

Aleksandrov, E.E., Samorodov, V.B. 1984. Mechanical power loss in a radial-piston volumetric hydraulic machine. *Theory of Mechanisms and Machines* 84: 102–106.

Avramov, V.P., Samorodov, V.B. 1986. *Hydrovolumetric Gears in Hydrovolumetric Transmissions of Transport Vehicles*. Kharkov: KhPI.

Avramov, V.P., Samorodov, V.B., Kuzminskiy, V.A. 1981. To the load calculation acting in a hydrostatically unloaded piston of a radial piston hydraulic pump. *Problems of Mechanical Engineering* 11: 51–55.

Avrunin, G.A. et al. 2004. Hydrovolumetric transmission with ball pistons HVT-900: characteristics and technical level. *Mechanics and Mechanical Engineering* 1: 9–16.

Babaev, O.M. et al. 1987. *Volumetric Hydromechanical Transmissions: Calculation and Design.* Moscow: Mashinostroenie.

Borisyuk, M.D. et al. 2008. The new generation of ball-piston transmissions. *Industrial Hydraulics and Kinematics* 1: 66–70.

Cai, H.M., Tian, M.J. 2012. Design of a shaft assignment radial piston pump. *Journal of Mechanical Engineering Science* 510: 9–12.

Jiang, W. 2009. Research of the pressure pulsation within piston chamber in radial piston pump. *Journal of Mechanical Engineering Science* 69–70: 626–630.

Kozlov, L.G., Polishchuk, L.K., Piontkevych, O.V., Korinenko, M.P., Horbatiuk, R.M., Komada, P., Orazalieva, S., Ussatova, O. 2019. Experimental research characteristics of counterbalance valve for hydraulic drive control system of mobile machine. *Przeglad Elektrotechniczny* 95(4): 104–109.

Kukharchuk, V.V., Bogachuk, V.V., Hraniak, V.F., Wójcik, W., Suleimenov, B., Karnakova, G. 2017a. Method of magneto-elastic control of mechanic rigidity in assemblies of hydropower units. *Proc. SPIE 10445, Photonics Applications in Astronomy, Communications, Industry, and High Energy Physics Experiments* 2017 104456A.

Kukharchuk, V.V., Hraniak, V.F., Vedmitskyi, Y.G., Bogachuk, V.V., Zyska, T., Komada, P., Sadikova, G. 2016. Noncontact method of temperature measurement based on the phenomenon of the luminophor temperature decreasing. *Proc. SPIE 10031, Photonics Applications in Astronomy, Communications, Industry, and High-Energy Physics Experiments 2016* 100312F.

Kukharchuk, V.V., Kazyv, S.S., Bykovsky, S.A. 2017b. Discrete wavelet transformation in spectral analysis of vibration processes at hydropower units. *Przegląd Elektrotechniczny* 93(5): 65–68.

Lovtsov, Y.I. 1967. To the calculation of hydraulic losses of a multiple-action radial piston hydraulic motor. *News of Universities. Mechanical Engineering* 5: 39–42.

Loytsyanskiy, L.G., Lure, A.I. 1983. *Theoretical Mechanics Course.* In 2 volumes. Vol. II. Dynamics. Moscow: Nauka.

Ogorodnikov, V.A., Zyska, T., Sundetov, S. 2018. The physical model of motor vehicle destruction under shock loading for analysis of road traffic accident. *Proc. SPIE 10808, Photonics Applications in Astronomy, Communications, Industry, and High-Energy Physics Experiments 2018* 108086C.

Petrov, V.G. 1988. *Hydrovolumetric Transmissions of Self-Propelled Machines.* Moscow: Mashinostroenie.

Polishchuk, L.K., Kozlov, L.G., Piontkevych, O.V., Horbatiuk, R.M., Pinaiev, B., Wójcik, W., Sagymbai, A., Abdihanov, A. 2019. Study of the dynamic stability of the belt conveyor adaptive drive. *Przegląd Elektrotechniczny* 95(4): 98–103.

Shatokhin, V.M. 2008. *Analysis and Parametric Synthesis of Nonlinear Power Transmission of Machines.* Kharkov: NTU KhPI.

Shatokhin, V.M., Shatokhina, N.V., Granko, B.F. 2014. Method of vibration diagnostics of cylinder-piston couples wear of pumps of hydro-volumetric transmissions with ball pistons. *Automation of Production Processes in Mechanical Engineering and Instrument Making: Ukrainian Interdepartmental Scientific and Technical Collection* 48: 65–73.

Tkachuk, N.A., Bibik, D.V. 2018. Development of numerical models for an integrated study of the stress-strain state of hydraulic transmission elements HVT-900. *Bulletin of NTU "KhPI". Series: Machine Science and CAD* 7(1283): 107–120.

Tkachuk, A.V. et al. 2017. Influence of design factors on the stress-strain state of hydrovolumetric transmission bodies. *Mechanics and Mechanical Engineering* 1:78–84.

Vedmitskyi, Y.G., Kukharchuk, V.V., Hraniak, V.F. 2017. New non-system physical quantities for vibration monitoring of transient processes at hydropower facilities, integral vibratory accelerations. *Przegląd Elektrotechniczny* 93(3): 69–72.

Yarmak, N.S. 2004. Study of dynamic processes of volumetric type hydraulic machines, diagnostics and identification of defects. *Mechanics and Mechanical Engineering* 1: 35–45.

Zhao, S., Guo, T., Yu, Y. 2015. Design and experimental studies of a novel double-row radial piston pump. *Journal of Mechanical Engineering* 231(10): 1884–1896.

Chapter 5

Mathematical modeling of dynamic processes in the turning mechanism of the tracked machine with hydrovolume transmission

Vladimir Shatokhin, Boris Granko, Vladimir Sobol, Victor Stasiuk, Maksat Kalimoldayev, Waldemar Wójcik, Aigul Iskakova, and Kuanysh Muslimov

CONTENTS

5.1 Introduction 53
5.2 Analysis of research methods of dynamic processes of power transmissions in vehicles with HVT 54
5.3 Mathematical models of transient regimes in the turning mechanism of a tracked vehicle with the HVT 55
5.4 Experimental and calculation research on choice of parameters of turning mechanism with HVT 59
5.5 Conclusions 63
References 63

5.1 INTRODUCTION

In the domestic and foreign transport engineering industry, there is a stable tendency to develop highly efficient engines and transmissions for wheeled and tracked vehicles by using hydrovolumetric transmissions (HVT). Their advantage is, first of all, the possibility of a smooth change in the speed ratio between the source and the consumer of power and the possibility of zone expanding for the stable operation of turbo-piston diesel engines by introducing hydrovolume machines (HVM) into the compressor drive. The adding of HVT together with differential reduction gears in the turning mechanism of a tracked vehicle (TV) not only improves the technical characteristics of the vehicle (controllability, maneuverability, and mobility) and reduces the driver fatigue but also gives it new quality properties (for example, the turning ability with any fixed radius). Methods for studying dynamic loads in power transmissions of the abovementioned devices are not currently sufficiently developed.

DOI: 10.1201/9781003224136-5

5.2 ANALYSIS OF RESEARCH METHODS OF DYNAMIC PROCESSES OF POWER TRANSMISSIONS IN VEHICLES WITH HVT

Currently, in many papers, dynamic calculations are done separately for the transmission and turning processes of vehicles; if there is a HVT in the power train, the elasticity influence of the connecting shafts and the servo drive on the development of the dynamic processes is not taken into account. In Choi et al. (2013) and Bottiglione et al. (2008), models to evaluate the design and efficiency of a tractor with the hydromechanical transmission are developed, which includes hydraulic machines, a swashplate control system, and a planetary gear; the dynamic model of the tractor, taking into account the force of thrust, resistance to movement, and power takeoff is proposed. Park and Kim (2016) and Wei et al. (2011) are devoted to the analysis of some characteristics of power transfer of the hydromechanical transmission of agricultural tractors. In Shujun (2018), theoretical and experimental studies of the effect of changes in transmission power on torque are presented; a method which provides a rational law of the power change is proposed. Dynamic models of multiengine continuously variable transmissions with HVT are considered in Zhang and Zhou (2014) and Kistochkin (1978).

In Macor and Rossetti (2011), the problem of increasing efficiency for hydromechanical transmission is formulated as a global optimization problem, the variables are the parameters of hydromachines and speed ratios of conventional and planetary gears. The optimization problem is solved by using the particle swarm method, which is effective for eliminating local minima. Rossetti and Macor (2013) and Rossetti et al. (2017) are devoted to the problem of multiobjective optimization, the aim of which is not only the efficiency of hydromechanical transmission but also minimization of the dimensions of the gearbox. Optimization in the Pareto set of compromises is done by using the particle swarm method.

Elements of dynamic research in the turning theory of TVs are considered in Guskov and Oneyko (1984) and Linares et al. (2010). Some models of TV turning, which allow dynamic processes in the engine, transmissions with differential mechanisms, and the HVT and undercarriage as a single system to be studied, are considered in Shatokhin 2008.

When using the HVT as the device that provides the vehicle turning, it is necessary to take into account the fact that the limiting value of the angular velocity of turning is often limited not so much by engine power, as by the power transmitted to the HVT. The turning process on overdrive gears leads to an increase in fluid pressure and triggering of the relief valves. In this case, the drive does not provide the necessary turning radius of the vehicle. The absence of appropriate mathematical models of the turning process makes it difficult to study the choice of the required drive parameters and hydromachines (Vedmitskyi et al. 2017, Kukharchuk et al. 2017b).

The aim of this chapter is to develop mathematical models of the vehicle turning to determine the parameters of the hydromachines and differential reduction gears which provide the necessary angular velocity of the turning process. It is necessary to solve such problems to develop a model of turning processes in the form of a system of differential equations, taking into account the engine, differential mechanisms, adjusted and unregulated hydromachines, and power consumers; to determine the character of the influence of the parameters of the differential reduction gears and various schematics of their layout on the characteristics of the kinematics; to determine the relation of the relative turning radius with the tilt angle of the hydraulic pump swashplate; to determine the dependence of the pressure in the discharge chamber on the characteristic volume of hydromachines; to study the influence of the installation of a double satellite gear on the turning parameters and moments in hydraulic motors; and to perform computational and experimental studies of dynamic processes during the vehicle turning (Kukharchuk et al. 2016, Ogorodnikov et al. 2018).

5.3 MATHEMATICAL MODELS OF TRANSIENT REGIMES IN THE TURNING MECHANISM OF A TRACKED VEHICLE WITH THE HVT

The kinematics diagram of a power transmission with a two-shaft diesel engine is shown in Figure 5.1 without a low-frequency drive circuit to the turbine and compressor. Its distinctive characteristic is the turning mechanism (HVT TM) and two differential reduction gears included in the hydrovolumetric transmission. The diagram shows the engine exhaust shaft, differential mechanisms, adjusted (pump) and unregulated (motor) hydromachines, power consumers, gears, and elastic inertia-free elements with stiffness coefficients c_i ($i = 1.4$). Without loss of information about changes in the average values of the motion characteristics and processes in hydromachines, we can consider the connecting shafts to be absolutely rigid shafts. This fact allows us to further simplify the computational process of the problem. As generalized coordinates, we consider the pressure difference Δp in line of the HVT and the turning angles: ϑ_{cs}, crankshaft and ϑ_m, unregulated hydromachine.

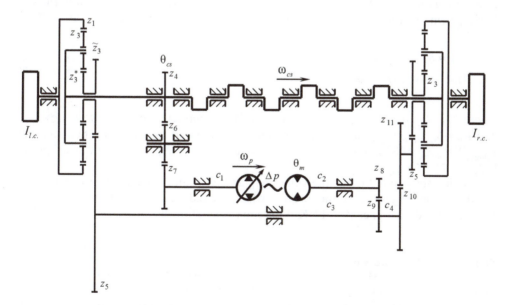

Figure 5.1 Kinematics diagram of turning mechanism with hydrovolumetric transmission (HVT).

After standard procedure of forming the Lagrange equations, the system of differential equations of motion can be written:

$$\left.\begin{array}{l} I_1\ddot{\theta}_{cs} + I_0\ddot{\theta}_m = M_E - (M_L + M_R)\alpha_1 + i_4 M_{FP}; \\ I_0\ddot{\theta}_{cs} + I_2\ddot{\theta}_m = M_{FM} + \alpha_2(i_2 M_L - i_0 M_R); \\ k_{EL}\dfrac{d(\Delta p)}{dt} + k_L \Delta p + q_p e_p i_4 \dot{\theta}_{cs} + q_m e_m \dot{\theta}_m = 0, \end{array}\right\} \quad (5.1)$$

where $I_1 = I_{i.s.} + I_{e.s.} + I_{m.g.} + 2I_{carr} + I_{z4} + i_5^2 I_{z6} + i_4^2 (I_{z7} + I_p) + \alpha_1^2 (2I_{z1} + I_{l.c.} + I_{r.c.})$;

$I_0 = \alpha_1\alpha_2 \left[i_0 (I_{z1} + I_{r.c.}) - i_2 (I_{z1} + I_{l.c.}) \right]$; $I_2 = I_m + I_{z8} + i_3^2(I_{z9} + I_{z5} + I_{z10}) + i_2^2 I_{z_3^*} +$
$i_1^2 I_{z11 z5} + i_0^2 I_{z3} + \alpha_2^2 \left[i_2^2 (I_{z1} + I_{l.c.}) + i_0^2 (I_{z1} + I_{r.c.}) \right]$;

$I_{i.s.}, I_{e.s.}, I_{m.g.}, I_{carr}, I_p, I_m, I_{zk}, I_{l.c.}, I_{r.c.}$ is the inertia moments of inlet and exhaust crankshafts, main gear, carrier of the differential mechanism, hydraulic pump, hydraulic motor, gear with teeth number z_k, left and right power consumers, respectively;

$\alpha_1 = 1 + \alpha_2$; $\alpha_2 = \dfrac{z_3}{z_1}$; $i_0 = i_1\dfrac{z_5}{\tilde{z}_3}$; $i_1 = i_3\dfrac{z_{10}}{z_{11}}$; $i_2 = i_3\dfrac{z_5}{\tilde{z}_3}$; $i_3 = \dfrac{z_8}{z_9}$; $i_4 = \dfrac{z_4}{z_7}$ are the speed ratios; M_E is the engine torque; M_{FP} and M_{FM} are the torques acting on the pump and the motor from the fluid side; and M_L and M_R are the drag torques on the epicyclic gears of the left and right differential mechanisms.

In Table 5.1, the teeth numbers of the gears are given.

The third equation of system (5.1) is an equation of the continuity of fluid flow in the lines of the hydromachines (Pasynkov & Gajcgori 1967). The formulas for the torques acting on the hydromachines' shafts from the fluid side can be written in the following form (Kistochkin 1978, Pasynkov & Gajcgori 1967, Kukharchuk et al. 2017a):

$$\left.\begin{array}{l} M_{FP} = q_p e_p \Delta p - k_{hp}\dot{\theta}_p, \\ M_{FM} = q_m e_m \Delta p - k_{hm}\dot{\theta}_m, \end{array}\right\} \quad (5.2)$$

where $k_{EL}, k_L, q_p, q_m, k_{hp}$, and k_{hm} are the HVT characteristics; $e_p = tg\gamma_p / tg\gamma_{p\max}$ and $e_m = tg\gamma_m / tg\gamma_{m\max}$ are the parameters of pump and motor power control; and $\gamma_{p\max} = \gamma_{m\max} = 18°$ is the maximum tilt angle of the pump and motor swashplates $(-\gamma_{p\max} \leq \gamma_p \leq \gamma_{p\max})$.

Under object conditions, the values M_L and M_R can be determined by the magnitudes of the drag forces of the torques, which are reduced to the left and right drive sprockets (Vedmitskyi et al. 2017, Kukharchuk et al. 2016, 2017a):

Table 5.1 Teeth numbers of gears

Gear designations	Z_1	$Z_1^* = Z_3$	Z_3	Z_4	Z_5	Z_6	Z_7	Z_8	Z_9	Z_{10}	Z_{11}
Number of teeth	89	25	76	40	24	21	36	54	33	19	19

$$M_L = R_{d \cdot s} \cdot \frac{F_L}{i_k}; M_R = R_{d \cdot s} \cdot \frac{F_R}{i_k} \tag{5.3}$$

where $R_{d.s.}$ is the drive sprocket radius; F_L and F_R are the drag forces of the left and right tracks; and $i_k = \omega_e/\omega_{d.s.}$ is the speed ratio from the epicycle of the differential reduction gear to the drive sprocket on k-th gear.

During the problem study, most attention was paid to the analysis of stationary turning – the definition of the turning radius and the pressure in the lines at a given tilt angle of the swashplate and the movement speed of the object. Such method is due to the absence of reliable information of the dependence of the acting forces on the drive sprockets on the movement speed, ground characteristics, turning radius, and function form $M_R(t)$. Due to the abovementioned fact, the first equation of system (5.1) is not used to study the stationary turning, and from the other two equations, assuming the derivatives are equal to zero, we obtain (Ogorodnikov et al. 2018, Polishchuk, et al. 2019):

$$\left. \begin{array}{l} M_{FM} + \alpha_2 \dfrac{R_{d.s.}}{i_k}(i_2 F_L - i_0 F_R) = 0, \\ k_L \Delta p + q_p e_p i_4 \dot{\theta}_{cs} + q_m e_m \dot{\theta}_m = 0. \end{array} \right\} \tag{5.4}$$

From the first equation (5.4), we obtain the well-known fact: the same drag forces differently load the HVT during vehicle motion in different gears. As increasing of the transmission sequence number k will lead to the speed ratio value i_k decreasing, finally, we will have the loading increasing on the hydromachines lines.

Let us study the dependence of the turning radius of an object on the tilt angle of the swashplate of an adjusted hydromachine, neglecting fluid leaks in the lines. Following the kinematics diagram of the drive (Figure 5.1), the angular velocity of the hydraulic pump can be written: $\omega_p = i_4 \omega_{cs}$. Taking into account the assumptions mentioned above, from the second equation of system (5.4) with the value $e_m = 1$, we can determine the angular velocities of the hydraulic motor, the left sun gear, and the right sun gear, respectively:

$$\omega_m = -e_p i_4 \omega_{cs}; \omega_p = i_4 \omega_{cs} \tag{5.5}$$

$$\omega_{z_3^*} = i_2 \omega_m \tag{5.6}$$

$$\omega_{z_3} = -i_0 \omega_m \tag{5.7}$$

The angular velocity of the epicyclic gears depends on the angular velocities of the crankshaft and the sun gear by the following formula:

$$\omega_e = \alpha_1 \omega_{cs} - \alpha_2 \omega_s \tag{5.8}$$

By taking into account the previous formula, we obtain for the left and right epicycles:

$$\omega_{el} = (\alpha_1 + \beta)\omega_{cs}; \omega_{er} = (\alpha_1 - \beta)\omega_{cs} \tag{5.9}$$

where $\beta = \alpha_2 e_p i_4 i_2$.

Dividing the last formulas by the speed ratio of gearbox i_k, we obtain the angle velocities of driving sprockets. Further, we can determine the velocities of the left and right tracks v_l and v_r, and the angle velocity of a vehicle body turning by the following formula:

$$\omega_t = \frac{v_l - v_r}{B} = \frac{2\beta R_{d.s.}}{i_k B} \omega_{cs} \tag{5.10}$$

where B is the track width.

The relative value of the turning radius of a vehicle can be determined from the expression:

$$\frac{R_{tr}}{B} = \frac{1}{B}\left(\frac{v_l}{\omega_t} - \frac{B}{2}\right) = \frac{\alpha_1}{2\beta} \tag{5.11}$$

where R_{tr} is the object turning radius (graph is presented in Figure 5.2).

For confidence estimation of the given model, we determine the pressure in the HVT lines and the torque transmitted by the HVT based on the experimentally known values of the rotational frequencies of the crankshaft $n_{cs} = 1{,}800\,\text{min}^{-1}$ and the sun gear $n_s = 900\,\text{min}^{-1}$. During movement along sandy ground, the angular velocity of the hydraulic pump and hydraulic motor are, respectively, $\omega_p = i_4 \pi n_{cs}/30 = 2{,}094\,\text{sec}^{-1}$ and $\omega_m = -i_2 \pi n_s/30 = -1{,}825\,\text{sec}^{-1}$. The values of the design parameters of hydromachines can be taken as follows: $k_L = 1.6 \cdot 10^{-11}\,\text{m}^5/\text{N/sec}$, $q_p = q_m = 1.61 \cdot 10^{-5}\,\text{m}^3$, and $k_{hp} = k_{hm} = 0.288\,\text{N/m/sec}$. The following speed ratios have been used: $i_0 = i_2 = 0.517$; $i_4 = 1.111$; $\alpha_1 = 1.281$; and $\alpha_2 = 0.281$.

For the control parameters values $e_p = e_m = 1$ from the continuity equation of the system (5.1), we can determine:

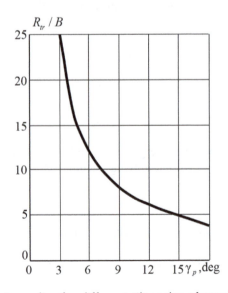

Figure 5.2 Relative turning radius for different tilt angles of swashplate.

$$\Delta p = -\frac{q_p}{k_L}\left(\omega_p + \omega_m\right) = -27.1\,\text{MPa} \tag{5.12}$$

When using the experimental data, the pressure had the value 23.5 MPa. The torque transmitted by the HVT reached the following value $|M_{FM}| = |q_m e_m \Delta p - k_{hm}\omega_m| = 383.86$ Nm.

During movement along turf, $\omega_p = 186.15\,\text{sec}^{-1}$; if $\omega_m = 152.05\,\text{sec}^{-1}$, then $\Delta p = -34.3$ MPa. In this case, the experimental value of the pressure was equal to 39.2 MPa. The HVT transmitted the torque value $|M_{FM}| = 508.6$ Nm. So we can conclude that in both considered cases, the experimental and calculated pressure values were quite close.

Assuming that the drag torque M_d to movement on the sides remains constant during the movement on the k-th and $k+1$-th gears:

$$M_d = \frac{i_k}{\alpha_2 i_2} M_{FM}^{(k)} = \frac{i_{k+1}}{\alpha_2 i_2} M_{FM}^{(k+1)} \tag{5.13}$$

We obtain the dependence between the torques, which characterize the HVT loading:

$$M_{FM}^{(k+1)} = \frac{i_k}{i_{k+1}} M_{FM}^{(k)} \tag{5.14}$$

which was noted above.

5.4 EXPERIMENTAL AND CALCULATION RESEARCH ON CHOICE OF PARAMETERS OF TURNING MECHANISM WITH HVT

During vehicle movement in overdrive gears, increasing loading of the hydromachines occurs according to (5.12). That tends to increase the pressure of the fluid and relief valves. In this case, the drive does not provide the necessary vehicle turning radius. Experiments with turns in overdrive gears have shown pressure swings of the pressure line in the reference design. As a result of the action of the relief valve (design pressure $p_{\max} = 50$ MPa), the turn significantly slowed, and the turn radius increased significantly. In this case, the controller moved from the neutral to an extreme position, which corresponds to the angular variation of the pump swashplate from $\gamma_p = 0$ to $\gamma_{p\max}$. Figure 5.3 shows an oscillogram with the system parameter records during a right turn in fourth gear (Polishchuk et al. 2016, 2016, 2019).

It is possible to eliminate this shortcoming by increasing the characteristic volumes q_p and q_m of hydromachines and changing the layout of the mechanical part of the drive. In the first case, the torque of the drag forces reduced toward the hydraulic motor can be determined by analogy with (5.13) from the first expression of (5.4) (Vedmitskyi et al. 2017, Kukharchuk et al. 2017b):

$$M_d^{(m)} = \frac{\alpha_2 R_{d \cdot s}(i_2 F_L - i_0 F_R)}{i_k} \tag{5.15}$$

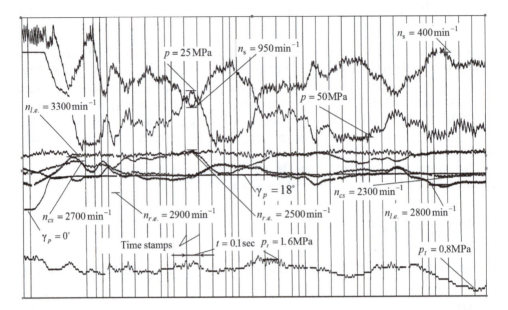

Figure 5.3 Turning process to the right in fourth gear.

We assume this torque and the shaft angular velocity ω_{cs} to be constant. Let us estimate the influence of the parameter $q = q_p = q_m$ on the pressure value Δp. To this end, from the second equation of (5.4), we determine the motor velocity:

$$\omega_m = -\frac{k_L \Delta p}{q} - i_4 \omega_{cs} \quad (5.16)$$

and substitute to the first equation. Then, taking into account the expression for M_{FM} from (5.2), the pressure difference in the lines will be determined from the expression:

$$\Delta p = -\frac{q\left(k_{hm} i_4 \omega_{cs} + M_d^{(m)}\right)}{q^2 + k_{hm} k_L} \quad (5.17)$$

We denote the nominal values of the characteristic volume as q_0 and the pressure difference as Δp^*, and we write the new value q in the form $q = xq_0$. The relative influence q on Δp can be estimated using the expression:

$$\frac{\Delta p}{\Delta p^*} = \frac{q\left(q_0^2 + k_{hm} k_L\right)}{\left(q^2 + k_{hm} k_L\right) q_0} = \frac{x\left(q_0^2 + k_{hm} k_L\right)}{x^2 q_0^2 + k_{hm} k_L} \quad (5.18)$$

The graph of this function for the abovementioned values of the nominal parameters of hydraulic machines is presented in Figure 5.4. It follows that, for example, when increasing the value q by 20% ($x = 1.2$) Δp is decreasing by 14.2%.

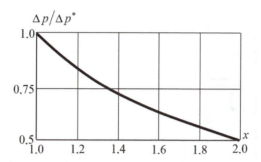

Figure 5.4 Pressure change with increasing of characteristic volume.

Another possibility of improving the turning characteristics is related with a layout change of the drive mechanical part. Figure 5.5 shows a design fragment which has the double satellites gear of differential mechanisms: $z_2 = 21$ and $\tilde{z}_2 = 33$. Furthermore, the number of teeth of the following gear wheels have been changed: $z_1 = 87$, $z_3 = 30$, and $z_8 = 52$.

Let us find the speed ratio $i_{m,t} = \omega_m/\omega_t$ for the basic and new variants.

In the first case, neglecting leaks in the lines of a hydraulic system, the angular velocity of the shaft can be expressed through the angular velocity of the hydraulic motor (see (5.5)), $\omega_{cs} = -\omega_m/i_4$. The value e_p is equal to one.

The angular velocities of epicyclic gears by taking into account (5.9) can be written:

$$\omega_{el} = -\left(\frac{\alpha_1}{i_4} + \frac{\alpha_2}{i_2}\right)\omega_m; \omega_{er} = -\left(\frac{\alpha_1}{i_4} - \frac{\alpha_2}{i_0}\right)\omega_m \tag{5.19}$$

The "minus" sign in these expressions shows the rotation of the crankshaft and hydraulic motor in opposite directions. In the future, we do not take this into account. Following formula (5.10), the angle velocity of the vehicle body can be written:

$$\omega_t = \frac{2\alpha_2 i_2 R_{d \cdot s.}}{i_k B}\omega_m \tag{5.20}$$

Therefore, the speed ratio can be determined:

$$i_{m,t} = \frac{\omega_m}{\omega_t} = \frac{i_k B}{2\alpha_2 i_2 R_{d \cdot s.}} \tag{5.21}$$

Assuming $i_k = 10$, we obtain $i_{m,t} = 241.14$.

In the second case, the formula for the parameter determining $\tilde{\alpha}_2$ becomes different: $\tilde{\alpha}_2 = z_2 z_3^*/z_1 \tilde{z}_2$ (Figure 5.5). With the unchanged i_k based on (5.18), we will have $\tilde{i}_{m,t} = 321.17$. The effect of using a double satellite gearbox is characterized by the ratio:

$$\frac{\tilde{i}_{m,t}}{i_{m,t}} = \frac{i_2 \alpha_2}{\tilde{i}_2 \tilde{\alpha}_2} = 1.332 \tag{5.22}$$

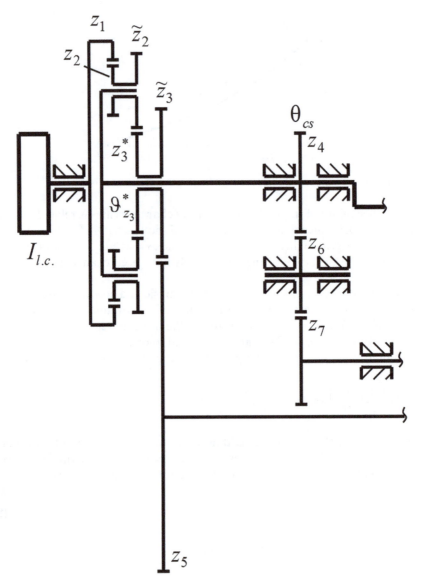

Figure 5.5 Scheme with a double satellite gear.

Thus, using the second scheme leads to decreasing the turning angular velocity of the body and increasing the turning radius. The torque of drag forces reduced toward the hydraulic motor is decreased. So, during the movement on the turf for the basic scheme $\omega_t = 0.6306\,\text{sec}^{-1}$, $R_{tr} = 13.6\,\text{m}$, and $M_{FM} = 508.6\,\text{Nm}$; for the scheme with the double satellite gear $\tilde{\omega}_t = 0.4914\,\text{sec}^{-1}$, $\tilde{R}_{tr} = 16.62\,\text{m}$, and $\tilde{M}_{FM} = 411.4\,\text{Nm}$. In the second case, the turning radius is increased by about 20% and the torque of the hydraulic motor is decreased by the same value.

From the abovementioned studies, it can be concluded that to obtain the necessary turning radius during movement in overdrive gears, it is necessary to increase the characteristic volume of the hydromachines. This cannot be done by changing the mechanical part of the drive.

5.5 CONCLUSIONS

1. Mathematical models of vehicle turning processes to determine parameters of hydromachines and differential reduction gears that provide vehicle turning with the required angular velocity have been developed.
2. A model of turning processes has been created, which takes into account the engine, differential mechanisms, adjusted and unregulated hydromachines, and power consumers.
3. The character of influence of the parameters of differential reduction gears and various schemes of their layout on the kinematic characteristics of turning process is determined.
4. The dependence of the relative turning radius with the tilt angle of the hydraulic pump swashplate is obtained.
5. The dependence of the pressure in the discharge chamber on the characteristic volume of hydromachines is determined.
6. The installation influence of the double satellite gear on the turning parameters and torques in hydraulic motors is studied.
7. The results of computational and experimental studies of dynamic processes during a vehicle turning are presented.
8. The proposed method is valuable for the development and improvement of similar designs.

REFERENCES

Bottiglione, F., De Pinto, S., Mantriota, G. 2014. Infinitely variable transmissions in neutral gear: Torque ratio and power re-circulation. *Mechanism and Machine Theory* 74: 285–298.

Choi, S.H., Kim, H.J, Ahn, S.H. et al. 2013. Modeling and simulation for a tractor equipped with hydro-mechanical transmission. *Journal of Biosystems Engineering* 38(3): 171–179.

Guskov, V.V., Oneyko, A.F. 1984. *Turn Theory of Tracked Vehicles*. Moscow: Mashinostroenie.

Kistochkin, E.S. 1978. Dynamic model of multiengine continuously variable transmissions with hydrovolumetric control loop. *Mashinovedenie* 5: 32–36.

Kukharchuk, V.V., Bogachuk, V.V., Hraniak, V.F., Wójcik, W., Suleimenov, B., Karnakova, G. 2017a. Method of magneto-elastic control of mechanic rigidity in assemblies of hydropower units. *Proc. SPIE 10445, Photonics Applications in Astronomy, Communications, Industry, and High Energy Physics Experiments 2017 104456A*.

Kukharchuk, V.V., Hraniak, V.F., Vedmitskyi, Y.G., Bogachuk, V.V., Zyska, T., Komada, P., Sadikova, G. 2016. Noncontact method of temperature measurement based on the phenomenon of the luminophor temperature decreasing. *Proc. SPIE 10031, Photonics Applications in Astronomy, Communications, Industry, and High-Energy Physics Experiments 2016 100312F*.

Kukharchuk, V.V., Kazyv, S.S., Bykovsky, S.A., Wójcik, W., Kotyra, A., Akhmetova, A., Bazarova, M., Weryńska-Bieniasz, R. 2017b. Discrete wavelet transformation in spectral analysis of vibration processes at hydropower units. *Przegląd Elektrotechniczny* 93(5):65–68.

Linares, P., Méndez, V., Catalán, H. 2010. Design parameters for continuously variable power-split transmissions using planetaries with 3 active shafts. *Journal of Terramechanics* 47(5): 323–335.

Macor, A., Rossetti, A. 2011. Optimization of hydro-mechanical power split transmissions. *Mechanism and Machine Theory* 46(12): 1901–1919.

Ogorodnikov, V.A., Zyska, T., Sundetov, S. 2018. The physical model of motor vehicle destruction under shock loading for analysis of road traffic accident. *Proc. SPIE 10808, Photonics Applications in Astronomy, Communications, Industry, and High-Energy Physics Experiments 2018* 108086C.

Park, Y.J., Kim, S.C. 2016. Analysis and verification of power transmission characteristics of the hydromechanical transmission for agricultural tractors. *Journal of Mechanical Science and Technology* 3: 5063–5072.

Pasynkov, R.M., Gajcgori, M.M. 1967. Calculation of hydrovolumetric transmissions taking into account dynamic loads. *Bulletin of Mechanical Engineering* 10: 48–51.

Polishchuk, L., Bilyy, O., Kharchenko, Y. 2016. Prediction of the propagation of crack-like defects in profile elements of the boom of stack discharge conveyor. *Eastern-European Journal of Enterprise Technologies* 6(1): 44–52.

Polishchuk, L., Kharchenko, Y., Piontkevych, O., Koval, O. 2016. The research of the dynamic processes of control system of hydraulic drive of belt conveyors with variable cargo flows. *Eastern-European Journal of Enterprise Technologies* 2(8): 22–29.

Polishchuk, L.K., Kozlov, L.G., Piontkevych, O.V., Horbatiuk, R.M., Pinaiev, B., Wójcik, W., Sagymbai, A., Abdihanov, A. 2019. Study of the dynamic stability of the belt conveyor adaptive drive. *Przegląd Elektrotechniczny* 95(4): 98–103.

Rossetti, A., Macor, A. 2013. Multi-objective optimization of hydro-mechanical power split transmissions. *Mechanism and Machine Theory* 62(12): 112–128.

Rossetti, A., Macor, A., Scamperle, M. 2017. Optimization of components and layouts of hydro-mechanical transmissions. *International Journal of Fluid Power* 18(6): 1–12.

Shatokhin, V.M. 2008. *Analysis and Parametric Synthesis of Nonlinear Power Transmission of Machines*. Kharkov: NTU KhPI.

Shujun, Y., Yong B., Chengyuan, F. 2018. Full power shift method of hydro-mechanical transmission and power transition characteristics. *Transactions of the Chinese Society of Agricultural Engineering* 34(5): 63–72.

Vedmitskyi, Y.G., Kukharchuk, V.V., Hraniak, V.F. 2017. New non-system physical quantities for vibration monitoring of transient processes at hydropower facilities, integral vibratory accelerations. *Przegląd Elektrotechniczny* 93(3): 69–72.

Wei, C., Yuan, S., Hu, J., Song, W. 2011. Theoretical and experimental investigation of speed ratio follow-up control system on geometric type hydro-mechanical transmission. *Journal of Mechanical Engineering* 47(16): 101–105.

Zhang, M., Zhou, Z. 2014. Modeling and control simulation for the multi-range hydromechanical CVT. Key engineering materials. *Journal of Mechanical Engineering* 621: 462–469.

Chapter 6

Forecasting reliability of use of gas-transmitting units on gas transport systems

Volodumyr Grudz, Yaroslav Grudz, Vasyl Zapuklyak, Lubomyr Poberezhny, Ruslan Tereshchenko, Waldemar Wójcik, Paweł Droździel, Maksat Kalimoldayev, and Aliya Kalizhanova

CONTENTS

6.1 Introduction 65
6.2 Analysis of the problem 65
6.3 Materials and methods of studying 66
6.4 Results and discussion 69
6.5 Conclusions 73
References 73

6.1 INTRODUCTION

Gas pumping units (GPUs) of compressor stations of main gas pipelines are the only energy-intensive elements of the gas transmission system, which are subject to controlling influences to change the mode of operation of the complex. Not only its technical and economic characteristics but also the reliability of the gas supply to consumers depend on the efficiency of the control of the gas supply system. Therefore, the special requirements for improving the efficiency and reliability of gas supply are put at the disposal of the GPUs.

At compressor stations of Ukraine, more than 20 types of GPA of domestic and imported production with different operating periods and technological schemes of work are installed. To increase the reliability of compressor stations, a system of reserving GPU and preventive maintenance is implemented (Grudz et al. 2009).

6.2 ANALYSIS OF THE PROBLEM

As part of the unified gas transportation system, the gas compressor units of compressor stations are integral with their elements, and their operating regimes, efficiency, and reliability are interrelated with the nature of the operation of the linear part, their hydraulic efficiency, and the service system (Grudz et al. 2009; Filipchuk et al. 2018).

At present, the current issue of servicing GPUs of various types at compressor stations is an assessment of the technical state of each unit to establish its term of

DOI: 10.1201/9781003224136-6

failure-free operation, the trend of coefficient of performance, the residual resource, the frequency of preventive maintenance, or the need for complete replacement.

Initial data for calculations toward finding a solution is information about the history of the operation of a specific GPU, the processing of which requires reliable methodological support. Obviously, deterministic mathematical models cannot guarantee a reliable assessment of the technical state of a full array of GPUs of different types and operating periods, since the background of their operation is considered as statistical information. Therefore, to solve the problems, it is expedient to use statistical forecasting (Baykheld and Franken 1988; Wang et al. 2015).

The main task of forecasting is to detect the optimal change in the forecasting of characteristics and parameters to obtain the maximum effect according to a preselected criterion (economic, technical, technological, etc.). At the same time, the forecast serves as the result of forecasting in the form of a set of statements about the future of the investigated process.

The basis of the theory of prediction is prognosis – a scientific discipline that studies the behavior of some systems (predicted), depending on the change in the parameters of others (Brazylovych 1982; Rigdon and Basu 2000; Kobbacy and Murthy 2008; Özekici 2013). These data are needed to predict what will happen to the system function if the behavior of the argument system is known now or in a particular situation.

Retrospection, diagnosis, and prognosis are the three phases of the full cycle of forecasting. At the diagnostic stage, set the initial and allowable characteristics of the parameters and measure their chosen methods of forecasting. The third stage is forecast. In the first stage the tendency of process development is determined, in the second - the state of the process at the moment of forecasting, and in the third - the development of this tendency in the future.

The forecast should be based on the consideration of the actual process of changing the technical condition of machine elements with the detection of the influence of a set of factors, primarily of the managers who are predicting (Doroshenko et al. 2019; Zapukhliak et al. 2019). They are the technical requirements for repairs and maintenance and the periodicity of (diagnosing) the technical condition of the GPU.

The process of state change can be considered to be a change in the state of the parameters without changing the elements, changing their quality or changing the quality of the machine.

6.3 MATERIALS AND METHODS OF STUDYING

The first case describes the normal operation of the elements in the range from the initial to the limiting state; the second describes the failure, the achievement of the limit state, and the loss of efficiency of the elements; and the third describes the loss of efficiency of the unit, its transition to an object of repair (recovery) or write-off (Kryzhanivs'kyi et al. 2004, 2013, 2015). These three processes are dialectically interconnected. Changing the boundaries of the initial and boundary state of the elements affects the frequency of failures, slows or accelerates the transition of the machine to an object of repair or write-off. In turn, the second and third processes can affect the rate of change of the first.

Consideration and forecast of the technical state can be carried out in the following order: processes of transformational changes in the parameters of the state and failure of elements – repair (write-off) of the GPU – defining the cost characteristics of failure and repair – the issuance of projected indicators, including reliability indicators.

Changing the parameters of the technical state is subject to complex dependencies. Therefore, for practical purposes, deviation of the parameters from the nominal values is usually expressed with sufficient precision with simple approximating functions. In developing the methods of forecasting the state of the elements of the unit, it is very important to establish a certain approximating function. The error and complexity of forecasting and ultimately the entire process of fail-safe management and other indicators of reliability depend on its choice (Poberezhnyi et al. 2017; Turner and Simonson 1984; Yavorskyi et al. 2016).

The requirements for mathematical substantiation of the approximating function of the deviation of the parameter are reduced to the following. The function should (1) take into account the physical picture of the deviation of the parameter, including external and internal factors, random velocity and the nature of the parameter change, intercontrol operation, etc. (2) be growing, displaying the integral nature of the deviation of the state parameter of the element, depending on the duration of the service, (3) be simple and universal, characterizing the linear, power, exponential, and other dependencies of the change of the parameter from the time period (service life), and (4) contain a small number of coefficients to facilitate forecasting, compilation of nomograms, tables, and use of simple formulas.

From the analysis of the factors that influence the process of changing parameters and the requirements for the mathematical description of this process, we get some general provisions. The deviation of the state parameter, depending on the work time or the total time, needs to be approximated by a randomized function with increasing implementations. The value of a function at a fixed moment is a positive multivalued value. The implementation of the change in the parameter can be considered as strictly or non-strictly monotone, i.e., as not always increasing function in the range from zero to the marginal deviation of the parameter.

Taking into account the factory and operational factors that influence the change of the parameter, it is possible to investigate its deviation at any time as the sum of two parameters:

$$u_f = u + w \qquad (6.1)$$

where u_f is the actual deviation of the parameter (continuous random variable); u is the theoretical deviation of the parameter under the influence of internal factory factors (essentially a positive continuous random variable); and w is the deviation of magnitude under the influence of external operational factors (continuous random variable).

Random values u and w can take one or another value that is not known to measure. The value of u generates the distribution of the parameter in fixed moments of the development of the averaged results of the work of the element, which characterizes the average operating load; the value of w is the distribution of the deviation of the actual change of the parameter from the averaging curve.

The average values of all items subject to testing obtained on the basis of the results of the first and all subsequent measurements form a series of experimental points on the graph. Built on these points, using the method of least squares, the smooth theoretical curve expresses the nature of a certain process of changing the parameter of aggregation of elements when they work with an averaged operational load. The value of a function at one or another point corresponds to the average value of the random variable $u(t)$. The average deviation of the experimental point from the theoretical curve will be equal to the magnitude that goes to zero with an increasing number of investigated elements or the time of operation of one aggregate of elements.

Instead of equation (6.1), you can write the sum of two random variables:

$$u(t) = V_c f(t) + V_t f_1(t) \tag{6.2}$$

where $f(t)$ and $f_1(t)$ are deterministic nonrandom functions of the parameter characterizing the dependence $u(t)$ from the working time; V_c is the random variable representing the rate of change of a parameter under the influence of internal factors; and V_t is the random value of the deviation w per unit of change in the parameter under the influence of external factors.

The value V_c has the dimension of the unit of measurement of the parameter per unit of output and the value V_t at time t has no dimension. $f(t)$ and $f_1(t)$ are the dimensions, respectively, units of output and parameter.

The first term is an elementary random function. All possible realizations of this function can be obtained from the function of the graph by simply changing the scale along the ordinate axis. An elementary random function is the most simple of the random. In it, V_c is the usual random variable and $f(t)$ is the usual nonrandom function.

The linear random function has the form:

$$u(t) = V_c t + Z(t) \tag{6.3}$$

Functions (6.2) and (6.3) can characterize the change in the parameter of a particular element, that is, one implementation. In the case of V_c, there is a constant and a $Z(t)$ random variable at time t. In the case of smooth or relatively smooth growing implementations of deviation by the parameter from the state of the element, as well as with the approximate consideration of the actual process of changing the parameter of the term, $Z(t)$ can be equated to zero. Then,

$$u(t) = V_c t \tag{6.4}$$

A simple function (6.4) is called the base one. Different variants of random parameter change function are obtained by successively complicating this function.

The coefficient of variation of the random variable obtained with a fixed value t_1 of the elementary random function $V_c f(t_1)$ is the magnitude of the constant and is equal to the coefficient of variation of the random variable V_c.

In formula (6.3), $Z(t) = V_t^* f_1(t)$ is a function of the deviation of the actual values of the parameter from the averaged smooth theoretical curve. At the same V_t^* time, it is possible to consider in time domain as Gaussian-centered stationary or nonstationary process. It is Gaussian because, in any section (at any time), the value of a function is a random variable that is subject to a normal distribution. The mathematical expectation of a random function in any section is zero, so the process is centered. The stationary process is characterized by the same average square deviation of the random variable in any section, as well as the dependence of the correlation function only on the difference of time (time).

The first term of the function (6.2) increases strictly monotonously, depending on the time elapsed. This quality is used for forecasting purposes.

6.4 RESULTS AND DISCUSSION

As already noted, the character of the parameter change of an element is determined by the deterministic function. It can be a different $f(t)$. The criterion for choosing one or another function (linear, power, exponential, fractional-linear, polynomial n-th degree, etc.) is the proximity of the values of the approximating function to the actual implementations of changing the parameter state of the element. There is not enough agreement of the mathematical expectation with the mean experimental curve. It is also necessary to obtain the coordination of the system of theoretical curves with the implementation system. In the absence of close proximity to the system of theoretical curves, there is a sharp increase in the coefficients of variation of parameter change and resource elements, which reduces the efficiency of prediction of machine performance. Thus, the coefficients of variation are used as the criterion of approximation. The coefficient of variation of the resource of the elements is more informative, due to the fact that the result of the calculation over the entire range of parameter changes, taking into account the nature of this change. The coefficient of variation of the parameter change can locally reflect the degree of approximation only at one or several sites.

When approximating the parameter change function, the manufacture of parts of the unit is taken into account, during which there is a short-term sharp increase in the parameter change. However, the greatest interest is not the plot of the piece, but the area of the change in the parameter close to the limit value, as the elements of failure are formed here. Therefore, the greatest degree of approximation is desirable in the range from the end of the product to the achievement of the parameter of the marginal deviation. In most cases, to achieve a sufficient coincidence in the aforementioned range of theoretical and experimental curves, the plot of the product can be neglected, that is, in this section, do not approximate the change of the parameter. Then the character of the change function on the plot of the piece can be conventionally taken as such:

$$u(t) = V_c f(t) + Z(t) + \Delta R \qquad (6.5)$$

It provides a good approximation of the deviation of the parameter from the end of the period of manufacture to the time of reaching the marginal deviation u. In connection with a relatively small change in the parameter during the period of production compared with u, the ΔR variation in the indicator, which by its nature is random, is a magnitude of the second order, which can be ignored. This allows us to consider the indicator ΔR as a deterministic value.

In the case when $Z(t) = 0$ of an elementary random function, the change of parameter $u(t)$ is retained when the term ΔR is transferred to the left part of expression (6.5). For example, linear approximation of the parameter change with the plot area:

$$u_1(t) = V_c t + \Delta R \tag{6.6}$$

When using the power function, change the parameter:

$$u_1(t) = V_c t + Z(t) \approx \Delta R \tag{6.7}$$

At $Z(t) = 0$:

$$u(t) = u_1(t) - \Delta R = V_c t^\alpha \tag{6.8}$$

In formula (6.8), V_c numerically can be considered as the rate of change of the parameter when $t = 1$, reduced α times. Indeed, after differentiating the expression (6.8) by t and with:

$$\frac{\partial [u(t)]}{\partial t} = \alpha V_c \tag{6.9}$$

When $\alpha = 1$ and $Z(t) = 0$, the approximating expression is an elementary random linear function. In this case, the rate of change of the parameter for a particular item during the service life is constant. When $\alpha > 1$ and $0 < \alpha < 1$, the elements have consistently continuous strictly monotonically increasing and decreasing rates of change in the parameter state of the element. The deviation curve of the parameter in the first case will be concave, in the second, convex. It is easy to see that the power-changing feature of the parameter has sufficient versatility. There are few coefficients in this function, and all of them have a clear physical meaning. Therefore, the function is convenient to use for practical forecasting.

The achievement of the parameter of the limiting value causes the element to fail. The density of the distribution of the time to failure is determined on the basis of the theorem of the transformation of random variables. For example, in the basic function:

$$u(t) = u_\Gamma = V_c t \tag{6.10}$$

V_c is a random variable with a distribution density φ_0. Resource of an item having a deviation rate of an element V_c is expressed by a direct function $t = u_\Gamma / V_c$; $u_\Gamma, V_c > 0$. Then the distribution density of the resource at a fixed boundary deviation u_Γ is found as a function of a random argument (Esary et al. 1967):

$$\varphi(t) = \varphi_0 [R(t)][R^*(t)] \tag{6.11}$$

where $R(t)$ is the inverse function and $R^*(t)$ is its derivative.

With normal distribution:

$$\varphi(t) = \frac{u_\Gamma}{\sqrt{2\pi}\sigma_V} \exp\left[-\frac{(u_\Gamma/t - m_V)^2}{2\sigma_V^2}\right] \qquad (6.12)$$

In the Weibull distribution:

$$\varphi(t) = \frac{bK_b^b u_\Gamma}{m_V^b t^2}\left(\frac{u_\Gamma}{t}\right)^{b-1}\exp\left[-\left(\frac{K_b u_\Gamma}{m_V t}\right)^b\right] \qquad (6.13)$$

Taking into account the expression (6.11), the distribution density of an element's resource with a power change function and a normal distribution:

$$\varphi(t) = \frac{u_\Gamma \alpha}{\sqrt{2\pi}\sigma_V t^{\alpha+1}}\exp\left[-\frac{(u_\Gamma/t^\alpha - m_V)^2}{2\sigma_V^2}\right] \qquad (6.14)$$

In the Weibull distribution:

$$\varphi(t) = \frac{bK_b^b u_\Gamma \alpha}{m_V^b t^{\alpha+1}}\left(\frac{u_\Gamma}{t^\alpha}\right)^{b-1}\exp\left[-\left(\frac{K_b u_\Gamma}{m_V t^\alpha}\right)^b\right] \qquad (6.15)$$

The function of the resource distribution of an element in the Weibull distribution as a result of integration (6.15) has the form:

$$F(t) = \exp\left[-\left(\frac{K_b u_\Gamma}{m_V t^\alpha}\right)^b\right] \qquad (6.16)$$

After simple transformations, the average resource of an element:

$$T_{cp} = \left(\frac{K_b u_\Gamma}{m_V}\right)^{1/\alpha}\Gamma\left(1 - \frac{1}{\alpha b}\right) \qquad (6.17)$$

Median deviation and coefficient of variation:

$$\sigma_t = \sqrt{\left(\frac{K_b u_\Gamma}{m_V}\right)^{2/\alpha}\Gamma\left(1 - \frac{2}{\alpha b}\right) - T_{cp}^2} \qquad (6.18)$$

$$v = \sqrt{\frac{\Gamma\left(1 - \frac{2}{\alpha b}\right)}{\left[\Gamma\left(1 - \frac{1}{\alpha b}\right)\right]^2} - 1} \qquad (6.19)$$

The gamma function is valid when the expressions in parentheses are greater than zero. In the simple case, taking into account (6.2) and the term $Z(t)$ from equation (6.7), it can be written as follows:

$$Z(t) = V_t^*(V_c t^\alpha) \tag{6.20}$$

When forecasting on the average statistical change of the parameter of the set of the same name elements, V_c and V_t are random independent values at time t. When predicting the implementation of a change in the parameter of a particular element, V_c is a constant value for this element and V_t and V_t^* are random. In the case of smooth implementations, change the parameter $V_t = 0$. Unlike the constant, for a particular element V_c, it can take different values, V_t changing over time. Therefore, V_t / φ_0, when implementing a change in the parameter, presents the form of non-smooth broken curves.

Given the equation (6.20), function (6.7) has the form:

$$u(t) = V_c t^\alpha + V_t^* V_c t^\alpha = V_c \left(1 + V_t^*\right) t^\alpha \tag{6.21}$$

with the exponential function of changing the parameter:

$$u_1(t) = a e^{V_c t} - \Delta R; \quad t, \alpha, V_c > 0 \tag{6.22}$$

After logarithm of expression (6.22), $\ln[u_1(t) + R] = \ln a + V_c t$.

In such a transformed form, $\ln a$ will characterize the random rate of change of the parameter in the period of manufacture. The density of the resource distribution of an element in the case of a normal distribution of magnitude $\ln a$ will be:

$$\varphi(t) = \frac{\ln u_\Gamma / a}{\sqrt{2\pi}\sigma_V t^2} \exp\left[-\frac{\left(\dfrac{\ln(u_\Gamma / a)}{t} - m_V\right)^2}{2\sigma_V^2}\right] \tag{6.23}$$

By dividing the value V_c by the Weibull law, the average resource of an element can be found by the formula:

$$T_{cp} = \left(\frac{K_b \ln(u_\Gamma / a)}{m_V}\right)^{1/\alpha} \Gamma\left(1 - \frac{1}{b}\right) \tag{6.24}$$

By analogy, other approximating deviations of the parameter of the function are written and output estimates of the resource of the element are derived. However, the application of various approximation functions has a significant disadvantage along with the known advantages (increasing accuracy of approximation and forecast). Each function requires its methods of calculation, prediction of the condition of machines, and the use of appropriate formulas, tables, and nomograms, which greatly complicates the forecasting process. Therefore, after choosing and finding the coefficients of any of the approximating expressions, it transforms into one definite function for which the prediction machine is developed.

Forecasting with the optimization of aggregate indicators is possible with the following logical sequence: processes of changing the parameters of the state, failure of component parts; repair and maintenance of the unit, determining the cost characteristics of repair and maintenance of the machine; and the issuance of predicted indicators. In this case, it is necessary to also have the functions of changing the parameters of the technical condition of the machine.

Optimization of some indicators in forecasting is carried out by managing others – the predicted ones. Technical requirements for repair and maintenance operations, in particular, allowable deviations of parameters and intercontrol working out of component parts, serve as maintenance indicators for maintenance and repair. The analysis of the factors influencing the reliability and durability of repaired machine elements shows that it is possible to directly and purposefully change these two indicators in repair and maintenance. The latter causes the obligatory consideration of the influence of the allowable deviation of parameters and intercontrolling of the elements on the process and the effects of changes in parameters.

Changing the status of machine elements is accidental. The consequences of changing the parameters, observed in the form of failure or preventive recovery (replacement) of the element, also are probabilistic. Therefore, each element of the machine has a probability of failure, preventive replacement (regulation), and average resource. These characteristics depend on the control indicators.

6.5 CONCLUSIONS

The proposed methods based on the use of stochastic mathematical models using the background information of the operation of gas pumping units at compressor stations as a source information allow us to estimate the real technical condition of each GPU and to predict its residual resource and the probability of failure-free operation. The calculations made will allow for the adoption of specific decisions concerning the nature of the further service of the equipment of the compressor stations, the choice of strategies for controlling the parameters of the technical condition, the planning of preventive repairs, or the replacement of GPUs.

REFERENCES

Baykheld, F., Franken, P. 1988. *Nadezhnost y tekhnycheskoe obsluzhyvanye. Matematycheskyy pokhod. Per. s nem.* Moscow. Radyo y sviyaz.

Brazylovych, E.Yu. 1982. *Modely tekhnycheskoho obsluzhyvanyya slozhnykh system*: 1–231. Moscow: Vysshaya shkola.

Doroshenko, Y., Doroshenko, J., Zapukhliak, V., Poberezhny, L., Maruschak, P. 2019. Modeling computational fluid dynamics of multiphase flows inelbowand T-junction of the main gas pipeline. *Transport* 34(1): 19–29.

Esary, J. D., Proschan, F., Walkup, D. W. 1967. Association of random variables, with applications. *The Annals of Mathematical Statistics* 38(5): 1466–1474.

Filipchuk, O., Grudz, V., Marushchenko, V., Myndiuk, V., Savchuk, M. 2018. Development of cleaning methods complex of industrial gas pipelines based on the analysis of their hydraulic efficiency. *Eastern-European Journal of Enterprise Technologies* 2/8: 92–102.

Grudz, V. Y., Tymkiv, D. F., Mykhalkiv, V. B., Kostiv, V. V. 2009. *Obsluhovuvannya i remont hazoprovodiv.* Ivano-Frankivsk: Lileya-NV.

Kobbacy, K. A. H., Murthy, D. P. 2008. *Complex System Maintenance Handbook*: 1–600: London: Springer-Verlag.

Kryzhanivs'kyi, E. I., Hrabovs'kyi, R. S., Fedorovych, I. Y., Barna, R. A. 2015. Evaluation of the kinetics of fracture of elements of a gas pipeline after operation. *Materials Science* 51(1): 7–14.

Kryzhanivs'kyi, E. I., Hrabovs'kyi, R. S., Mandryk, O. M. 2013. Estimation of the serviceability of oil and gas pipelines after long-term operation according to the parameters of their defectiveness. *Materials Science* 49(1): 117–123.

Kryzhanivs'kyi, E. I., Rudko, V. P., Shats'kyi, I. P. 2004. Estimation of admissible loads upon a pipeline in the zone of sliding ground. *Materials Science* 40(4): 547–551.

Özekici, S. 2013. Reliability and maintenance of complex systems. *Springer Science, Business Media* 154: 1–596.

Poberezhnyi, L. Y., Marushchak, P. O., Sorochak, A. P., Draganovska, D., Hrytsanchuk, A. V., Mishchuk, B. V. 2017. Corrosive and mechanical degradation of pipelines in acid soils. *Strength of Materials* 49: 539–549.

Rigdon, S.E., Basu, A.P. 2000. *Statistical Methods for the Reliability of Repairable Systems*: 8–10. New York: Wiley.

Turner, W. J., Simonson, M. J. 1984. A compressor station model for transient gas pipeline simulation. In *PSIG Annual Meeting. Pipeline Simulation Interest Group, PSIG Annual Meeting*, 18–19 October, Chattanooga, TN: 1–13.

Wang, P., Yu, B., Deng, Y., Zhao, Y. 2015. Comparison study on the accuracy and efficiency of the four forms of hydraulic equation of a natural gas pipeline based on linearized solution. *Journal of Natural Gas Science and Engineering* 22: 235–244.

Yavorskyi, A. V., Karpash, M. O., Zhovtulia, L. Y., Poberezhny, L. Y., Maruschak, P. O., Prentkovskis, O. 2016. Risk management of a safe operation of engineering structures in the oil and gas sector. *Proceedings of the 20th International Conference 'Transport Means 2016'*, At Juodkrantė, Lithuania: 370–373.

Zapukhliak, V., Poberezhny, L., Maruschak, P., Grudz, V., Stasiuk, R., Brezinová, J., Guzanová, A. 2019. Mathematical modeling of unsteady gas transmission system operating conditions under in sufficient loading. *Energies* 12(7): 25–31.

Chapter 7

Development of main gas pipeline deepening method for prevention of external effects

Vasyl Zapuklyak, Yura Melnichenko, Lubomyr Poberezhny, Yaroslava Kyzymyshyn, Halyna Grytsuliak, Paweł Komada, Yedilkhan Amirgaliyev, and Ainur Kozbakova

CONTENTS

7.1 Introduction 75
7.2 Analysis of the methods and materials 76
7.3 Results and discussion 83
7.4 Conclusion 84
References 86

7.1 INTRODUCTION

In accordance with the normative documents, there are underground, semi-underground, terrestrial, and aboveground schemes for laying pipelines. The underground layout of laying of oil and gas pipelines is the most widespread (about 98% of the total length of the linear part). The underground layout of the laying is characterized by the fact that the upper production tubes are located below the markings of the daytime surface of the soil. Under such laying plans, after the end of construction, agricultural land is restored, there is no influence of solar radiation and atmospheric precipitation on the pipeline which is in stable temperature conditions, and mechanical damage from the effect of agricultural machinery and the impact of explosive action of ammunition, in the event of military aggression, is made impossible.

However, it is known that main gas pipelines during operation may be partly disclosed as a result of soil erosion (Figure 7.1a), and they also have the ability to lose stability due to various factors (pressure, temperature, water saturation of the soil, etc.) and to climb upward (to come out or float) (Figure 7.1b and c), changing their position in relation to the relevant rules. Also, it is known that sometimes the laying of pipelines in the course of their construction is not done to the correct project marks, due to the "low culture" of construction, that is, when the pipeline at individual sites laid in a trench, the depth of which does not correspond to the project and regulatory documents (Figure 7.1d). According to the results of the in-tube diagnostics, it has been established that almost every section of the main gas pipelines inspected contains places where the depth of the pipeline does not correspond to the project and regulatory documents (Sarvanis et al. 2017, Yavorskyi et al. 2016).

DOI: 10.1201/9781003224136-7

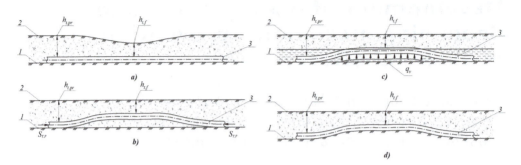

Figure 7.1 Causes of inconsistency of the depth of the pipeline design and normative depth. 1, bottom of the trench; 2, surface of the soil; 3, pipeline; $h_{t \cdot pr}$, design (normative) depth of the pipeline; $h_{t \cdot f}$, actual depth of occurrence; $S_{T, P}$, forces from temperature and pressure; and q_v, pushing force of water-saturated soil.

Further operation of such pipelines comes with the risk of the pipeline migrating to the surface and mechanical damage to the insulation and pipe body (Kryzhanivs'kyi et al. 2004, 2015, Poberezhnyi et al. 2017). Therefore, to ensure the reliable operation of the pipeline which has raised areas, these areas should be lowered to the standard depth. That is, it is necessary to carry out work on subsiding (deepening) the pipeline in the areas which reveal the discrepancy between the actual depth and the relevant norms.

7.2 ANALYSIS OF THE METHODS AND MATERIALS

In the existing technical literature concerning the repair of main pipelines (Maruschak et al. 2015, Kryzhanivs'kyi et al. 2013, Yavorskyi et al. 2017), the problem of pipelines being submerged at transitions through water obstacles, where water erosion of the soil occurs, is mainly considered (Poberezhnyi et al. 2017). In this case, major overhaul by the method of submersion is recommended for erosion and sagging of the coastal part of the underwater transition or on a significant part of the channel, which arose as a result of changes in hydrological conditions, erroneous prediction of re-formation of the channel, or displacement of the pipeline after laying (Zabela et al. 2001, Stanev et al. 2004, Filatov et al. 2015).

However, little attention is paid to the problems of submerging the linear part under normal operating conditions, i.e., not on transitions through water obstacles and bogs (Spiridenok et al. 2016, Yuzevych et al. 2017).

Repair of sections of the linear part of the main pipeline at an abnormal depth, which are not related to transitions through water obstacles and bogs, is carried out in the following ways:

1. repair without plumbing the pipeline – pumping the soil with fixing the backfilled soil;

2. repair with substitution – at the same time replacing the insulation;
3. repairs with subassembly, transfer into a parallel trench;
4. repair by piping with the use of soil jumpers (supports);
5. repairs with deepening, with emptying of the pipeline.

These methods of performing repair work on the linear part of the main pipeline at an abnormal depth are described in detail in Spiridenok et al. (2016).

In the works (Spiridenok et al. 2016, Yuzevych et al. 2017), it is noted that the technological operations during the piping are carried out in the following sequence:

- clarification of the pipeline's position;
- removal of the fertile layer of soil, moving it to a temporary dump, and planning the lane of the route in the zone of motion of repair machines;
- trench design according to the scheme of the markings, which ensure the pipeline deepening according to the working project;
- checking the technical condition of the pipeline, checking the transverse welded joints, and reinforcing them if necessary;
- lifting and maintenance of the pipeline by pipelayers;
- cleaning of the pipeline from the old insulating coating;
- application of a new insulating coating;
- control of the quality of the insulating coating;
- moving and laying the pipeline in the new trench;
- sand backfilling and final filling of the trench;
- reclamation of the fertile soil layer.

However, the mathematical models of the stress-strain state of the pipeline during repair are not given (or shown only for symmetric lifting schemes), which would establish some parameters of repair works: the length of the part being repaired; allowable distances between pipelayers; machines for repairing insulating coatings and coordinates of their placement; and permissible depth of plowing. It is well known that compliance with these parameters is a guarantee of reliability and safety of carrying out repair works.

Consequently, the necessity to develop a unified method for planting pipelines of different diameters under normal conditions and the method for assessing the stress-strain state of the pipeline is obvious.

Taking into account the reasons for the inconsistency of the depth of the pipeline with the design and the regulatory depth and the fact that such areas of the main pipeline do not always require the replacement of insulation, we have proposed a slightly different list of ways to perform repairs that do not provide for the complete emptying of the pipeline:

1. If soil erosion is due to insufficient depth (Figure 7.1a), then it would be advisable to fill the pipeline with imported soil as this will not affect the stressed state of the pipeline.
2. If the cause of the ejection is the forces arising in the pipeline from the temperature and from the internal pressure (which may be related to the previous factor) (Figure 7.1b), then the hollow can be carried out with the lifting under the pipe and

the installation of the pipeline to the design mark under its own weight, and in the technological calculation, it is necessary to take into account the previous tense state.

3. If the cause of the ascent is the effect of water-saturated soil (Figure 7.1c), then the depression should be carried out with lifting under the pipe and the installation of the pipeline to the design mark with the help of a loader, and in the calculation, it is necessary to set their required number. It is also possible to hitch under the pipe with simultaneous pumping of water from the trench, which will slowly lower the pipeline to the design mark.
4. If the cause of insufficient depth is the "low culture" of the construction works (Figure 7.1d), that is, when the pipeline during construction was erected in a trench with different depths in separate sections, then the depression can be carried out by hoisting the pipe and installing the pipeline to the project mark under its own weight.

Consequently, in the last three cases, the deepening can be carried out by hoisting the pipeline. Taking into account the fact that often the reason for the noncompliance of the depth of the pipeline with the design or the normative is not the loss of stability due to the impact of certain force factors but the noncompliance with the regulatory requirements and the project by the construction organizations in the construction process, consider the fourth way of performing the work, which may be basic for two other cases (except for the first one). The scheme of work execution is shown in Figure 7.2. It is obvious that the pipeline, when developing the trench and tucking under it, can subside to the standard depth under its own weight (Figure 7.2a). However, depending on Δh, there may be a case where the strength of the pipeline will not be ensured under such a scheme. Therefore, in this case, it is necessary to maintain the pipeline with one (Figure 7.2b) or several pipelayers.

The calculation scheme of the stressed state of the pipeline is depicted in Figure 7.3. In this scheme, it is first necessary to establish the length of the exposed area l, on which the pipeline under the action of its own weight will begin to descent to the design mark, that is, it will begin to lie at the bottom of the trench (Chapetti et al. 2001, Liu et al. 2009, Zapukhlyak et al. 2016).

First, we will simulate the stress-strain state of the pipeline for the process in Figure 7.2a. In this case, it is necessary to check the strength of the pipeline, taking into account the pressure and temperature of the product.

$$l = \sqrt[4]{\frac{72 E I \Delta h}{q_{tr}}} \tag{7.1}$$

where E is the Young's modulus; I is the moment of inertia of the pipeline; and q_{tr} is the distributed load calculated from the metal weight of the pipe itself, the weight of the transportable product and the weight of the insulation.

Now you can write the equation to determine the transverse forces, moments, and deflections in each section of the pipeline:

$$Q(x) = R_A - q_{tr} x \tag{7.2}$$

Main gas pipeline deepening method 79

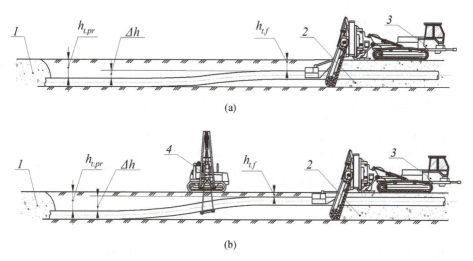

Figure 7.2 Diagrams for carrying out pipeline loading under its own weight (a) and with the support of a pipelayer (b) 1, zone in which the pipeline is in the project position; 2, zone in which the development of soil and excavation under the pipeline is necessary; 3, machine for trench development and stone-picking; 4, pipelayer; $h_{t \cdot pr}$, design (normative) depth of the pipeline; $h_{t.f.}$ actual depth; and Δh, difference between the design and the actual depth of the pipeline.

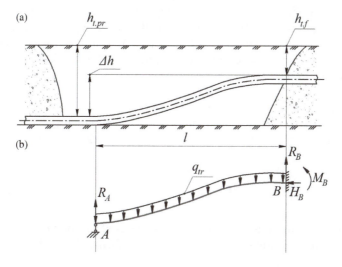

Figure 7.3 Design scheme for laying pipelines to the design mark.

$$M(x) = R_A \cdot x - \frac{q_{tr} x^2}{2} \tag{7.3}$$

$$\theta(x) = \frac{R_A x^2}{2EI} - \frac{q_{tr} x^3}{6EI}, \tag{7.4}$$

$$y(x) = -\Delta h + \frac{R_A x^3}{6EI} - \frac{q_{tr} x^4}{24EI} \tag{7.5}$$

Having determined the reaction $R_A = q_{tr} l/3$ using equations (7.2–7.5), the diagrams, respectively, of transverse forces, moments, and angles of rotation of the planes and displacements are constructed. Further, analyzing the diagrams of moments and transverse forces in the figure, the maximum moment and the maximum value of the reaction of the support are determined, respectively, and the performance of the condition of strength is checked (Liu et al. 2009):

$$\frac{M_{\max}}{W} \leq \psi_4 R_2 - |\sigma_{N(t,P)}| \tag{7.6}$$

Obviously, in some cases, the condition (7.6) is not fulfilled. In this case, consider the scheme where one or two pipelayers are used, as shown in Figure 7.2b. Moreover, the pipelayer raises the pipeline to a height not greater than Δh. The design scheme of the pipeline is shown in Figure 7.4.

To calculate the stress-strain state of the pipeline for the above scheme, it is necessary to specify the following parameters:

- value of the distance of the placement of the pipelayer from point B, which should be $b < l$ (l is length of the pipeline is determined from equation (7.1));
- value of the lifting height of the pipeline above the bottom of the trench $h_p = h(a) + h_{lif}$, and $h_p < \Delta h$ ($h(a)$ is the height of the pipeline placement above the trench bottom at a distance $a = l-b$; $h(a) = \Delta h - y(a)$ ($y(a)$ is the deflection of the pipeline determined by equation (7.5)); and h_{lif} is the height of lifting of the pipeline by the pipe-laying device.

It should be noted that $y_p(a) = y(a) + h_{lif}$.

Next, you need to set the parameters b and hp, at which the bending moments in the pipeline will be tolerable. The permissible value of the moment can be determined from (7.6):

$$M_{per} \leq \left(\psi_4 R_2 - |\sigma_{N(t,P)}|\right) \cdot W \tag{7.7}$$

Following are the equation of deflections and angles of rotation of the elastic line of the beam, taking into account the efforts of lifting P (Pisarenko et al. 2004):

$$\begin{cases} y(x) = y_0 + \theta_0 x + \dfrac{R_A x^3}{6EI} - \dfrac{q_{tr} x^4}{24EI} + \dfrac{P(x-a_p)^3}{6EI}, \\ \theta(x) = \theta_0 + \dfrac{R_A x^2}{2EI} - \dfrac{q_{tr} x^3}{6EI} + \dfrac{P(x-a_p)^2}{2EI}, \end{cases} \tag{7.8}$$

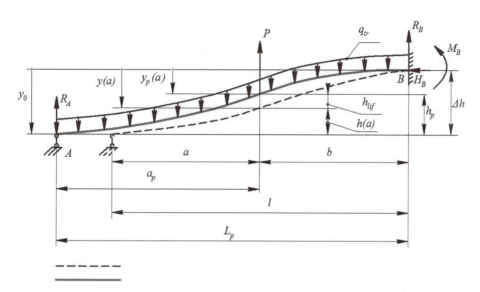

Figure 7.4 Design scheme of the pipeline when deepening it using a pipelayer: bending line of the pipeline without lifting by the pipelayer and bending line of the pipeline with lifting by the pipelayer P is the effort on the pipe of the pipelayer.

where $y_0 = -\Delta h$ and $\theta_0 = 0$ are the geometric initial parameters.

We know that $y(a_p) = y_p(a)$; $y(L_p) = 0$; and $\theta(L_p) = 0$, so, from system (7.8), we have a system of three equations:

$$\begin{cases} y_p(a) = -\Delta h + \dfrac{R_A a_p^3}{6EI} - \dfrac{q_{tr} a_p^4}{24EI}, \\ 0 = -\Delta h + \dfrac{R_A L_p^3}{6EI} - \dfrac{q_{tr} L_p^4}{24EI} + \dfrac{Pb^3}{6EI}, \\ 0 = \dfrac{R_A L_p^2}{2EI} - \dfrac{q_{tr} L_p^3}{6EI} + \dfrac{Pb^2}{2EI}. \end{cases} \quad (7.9)$$

From the third equation of system (7.9), we have:

$$P = -\frac{R_A L_p^2}{b^2} + \frac{q_{tr} L_p^3}{3b^2} \quad (7.10)$$

If we substitute equation (7.10) into the second equation of system (7.9), we obtain:

$$R_A = \frac{6EI\Delta h + \dfrac{q_{tr} L_p^3}{3}\left(\dfrac{3}{4}L_p - b\right)}{L_p^2 (L_p - b)} \quad (7.11)$$

Given that $L_p = a_p + b$, (7.11) will look as follows:

$$R_A = \frac{6EI\Delta h}{a_p(a_p+b)^2} + \frac{q_{tr}\left(\frac{3}{4}a_p^2 + \frac{1}{2}a_p b - \frac{1}{4}b^2\right)}{3a_p} \tag{7.12}$$

Suppose that $y_p(a) = -k\Delta h$ and $0 < k < \frac{y(a)}{y(0)}$, then, substituting equation (7.12) into the first equation of system (7.9), we obtain:

$$72EI\Delta h = \frac{q_{tr}a_p^2 b(2a_p - b)(a_p + b)^2}{(1-k)(a_p+b)^2 - a_p^2} \tag{7.13}$$

The value a_p from equation (7.13) can be determined by the graph-analytic method.

Now you can write the equation for determining transverse forces, bending moments, angles of rotation, and displacements in each section of the pipeline:

$$Q(x) = R_A - q_{tr}x + P \tag{7.14}$$

$$M(x) = R_A x - \frac{q_{tr}x^2}{2} + Px \tag{7.15}$$

$$\theta(x) = \frac{R_A x^2}{2EI} - \frac{q_{tr}x^3}{6EI} + \frac{Px^2}{2EI}, \tag{7.16}$$

$$y(x) = -\Delta h + \frac{R_A x^3}{6EI} - \frac{q_{tr}x^4}{24EI} + \frac{Px^3}{6EI}. \tag{7.17}$$

Using equations (7.14, 7.15, and 7.17), diagrams are constructed, respectively, of the transverse forces, moments, and angles of turns and displacements.

Further, analyzing the diagrams of moments and transverse forces in the figure, the maximum bending moment and the maximum value of the reaction response are determined, respectively: M_{max} and R_{max}. However, it is now enough to fulfill the condition:

$$M_{max} \leq M_{per} \tag{7.18}$$

Approval of the given model can be carried out according to the following algorithm:

1. Calculation of the main parameters of the pipelines (cross-sectional area, moment of inertia, and moment of resistance of the cross-section), load of its weight, and other influences (weight of the product and insulation).
2. Determination of the maximum displacement of the pipeline from the design mark Δh.

3. Determining the length of the open area where the pipeline will lower to the bottom of the trench under its own weight l by the formula (7.1) and the R_A reaction.
4. Calculation of transverse forces, bending moments, angles of turns and displacements, respectively, by the formulas (7.2–7.5), and the construction of a scene is carried out.
5. Verification of the strength of the inequality (7.6) or (7.7). If the condition is fulfilled, the calculation is complete. In the case of non-fulfillment of the condition, a decision is made on the necessity of carrying out the works in the subsoil using a pipelayer.
6. By the method of successive approximations, the measurement of b and h_p is carried out under which condition (7.7) is satisfied. The value is determined from equation (7.13) by the graph-analytic method.
7. The determination of the reaction of R_A and the effort on the pipe of the pipe-laying pipe P, respectively, from equations (7.11) and (7.10).
8. Equations (7.14–7.17) determine the transverse forces, bending moments, angles of rotation, and displacements in each section of the pipeline and construct the corresponding diagrams.
9. Checking the strength of the inequality (7.7).

7.3 RESULTS AND DISCUSSION

We will perform these calculations for an imaginary gas pipeline with the following technical parameters:

- diameter and thickness of the wall of the gas pipeline = 219 × 8 mm; material – steel 20; working pressure – 75.0 kgf/cm² (7,357,500 Pa); and pressure during hollowing – 15 kgf/cm² (1,471,500 Pa);
- design depth of the pipeline $h_{t \cdot pr} = 0.8$ m;
- actual depth of the gas pipeline $h_{t \cdot f} = $ (0.35; 0.4; 0.55; 0.65; 0.25; 0.6) m;
- mechanical properties of the steel pipeline: *stress limit* = 437 MPa and *yield limit* = 295 MPa.

The calculation can be done in the Mathcad environment. Figure 7.5 shows the diagrams for the two variants of submerging in the worst-case scenarios, in terms of strength, where $h_{t \cdot f} = 0.25$ m. Given that the permissible moment $M_{per} = 44,071$ N/m, from the diagram of the moments (Figure 7.5a), it is evident that the condition of strength is not satisfied for a scheme without a pipe-laying machine ($M_{max} = -51,913$ N/m), and for a scheme with a pipe-laying machine $M_{max} = -27,581$ N/m (Figure 7.5b).

The result of the calculation is reduced in Table 7.1 in the form of advisory parameters for the implementation of works of pipeline deepening. In the table № 1 is the scheme of work without pipelayer and № 2 is the scheme of work with the pipelayer.

It should be noted that the result of the calculation for the scheme of work without the pipelayer was confirmed using the finite difference method in the Cpipe program. According to the calculation in Mathcad, the maximum stress in the gas pipeline was $\sigma_{max} = |M_{max}|/W = |-51,913|/2.69 \cdot 10^{-4} = 192 \cdot 10^6$ Pa, which corresponds to a difference of 2.4% of the maximum stresses specified in the program Cpipe (Figure 7.6).

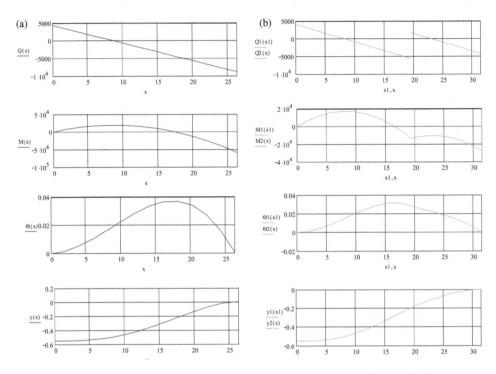

Figure 7.5 Curves of transverse forces, moments, angles of rotation, and displacement for the scheme of work without the pipelayer (a) and with the pipelayer (b).

7.4 CONCLUSION

1. The causes of inconsistencies in the depth of pipelines were analyzed. It was established that prior to the beginning of the work on pipeline loading, it is necessary to determine the reasons for the loss of stability, and only then choose a scheme of works.
2. Mathematical models for calculating the stress-strain state and strength of the pipeline during its natural subsiding and using a pipelayer have been developed.
3. The developed models have been tested on an imaginary DN 200 pipeline and recommendations provided for the pipeline lowering method. The model was tested using the Cipipe program;
4. According to the developed method, a complex of repair works was carried out on the project depth of a gas pipeline with a diameter of 1,420 mm.

Table 7.1 Recommendations for the implementation of works of pipelines deepening

Number of sites	$h_{t.f.}$, m	$h_{t.pr}$, m	Δh, m	Scheme of execution of works	b, m	h_p, m	Length of plot, m	P, kN	$|M_{max}|$, kN/m	M_{per}, kN/m
1	0.35	0.8	0.45	№2	7÷20	0.35	26.9÷41.2	5.5÷12.3	31.7÷28	44.07
2	0.4	0.8	0.4	№2	7÷20	0.3	25.8÷40.4	4.4÷11.8	33.5÷25.8	44.07
3	0.55	0.8	0.25	№1	-	-	21.63	-	39.05	44.07
4	0.65	0.8	0.15	№1	-	-	19.04	-	30.24	44.07
5	0.25	0.8	0.55	№2	7÷20	0.45	28.6÷42.6	7.6÷13	28.6÷31.9	44.07
6	0.6	0.8	0.2	№1	-	-	20.5	-	34.92	44.07

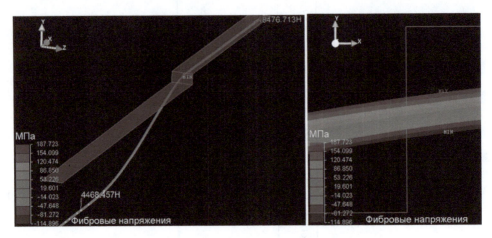

Figure 7.6 Calculation of stresses on the pipeline section which is lowered with one end stuck in the soil.

REFERENCES

Chapetti, M.D., Otegui, J.L., Manfredi, C., & Martins, C.F. 2001. Full scale experimental analysis of stress states in sleeve repairs of gas pipelines. *International Journal of Pressure Vessels and Piping* 78(5): 379–387.

Filatov, A.A., Nikonenko, A.D., Veliyulin, I.I., Polyakov, V.A., Aleksandrov, D.V., & Veliyulin, E.I. 2015. Formirovaniye napryazhenno-deformirovannogo sostoyaniya truboprovoda podvodnogo perekhoda MG na etapakh remonta metodom podsadki (in Russian). *Gazovaya promyshlennost* S2(724): 6–9.

Krizhanivskiy, Y.I., Taraievskyi, O.S., & Makovkin, O.M. 2015. Research on pipeline elements strength with stress raisers in the area of slide. *Metallurgical and Mining Industry* 7(10): 202–204.

Kryzhanivs'kyi, E.I., Hrabovs'kyi, R.S., & Mandryk, O. M. 2013. Estimation of the service ability of oil and gas pipelines after long-term operation according to the parameters of their defectiveness. *Materials Science* 49(1): 117–123.

Kryzhanivs'kyi, E.I., Rudko, V.P., & Shats'kyi, I.P. 2004. Estimation of admissible loads upon a pipeline in the zone of sliding ground. *Materials Science* 40(4): 547–551.

Liu, B., Liu, X.J., & Zhang, H. 2009. Strain-based design criteria of pipelines. *Journal of Loss Prevention in the Process Industries* 22(6): 884–888.

Maruschak, P., Panin, S., Danyliuk, I., Poberezhnyi, L., Pyrig, T., Bishchak, R., & Vlasov, I. 2015. Structural and mechanical defects of materials of offshore and onshore main gas pipelines after long-term operation. *Open Engineering* 5(1): 365–372.

Pisarenko G.S., Kvitka, O.L., & Umansky Y.S. 2004. *Resistance of Materials*. Kiev, Ukraine: Higher School.

Poberezhnyi, L.Y., Marushchak, P.O., Sorochak, A.P., Draganovska, D., Hrytsanchuk, A.V., & Mishchuk, B.V. 2017. Corrosive and mechanical degradation of pipelines in acid soils. *Strength of Materials* 3(2): 1–11.

Pobereznyi, Ya. L., Poberezhna, Ya.L., Maruschak, P.O., & Panin, S. V. 2017. Assessment of potential environmental risks from salines oil subsidence. *IOP Conference Series: Earth and Environmental Science* 50(1): 175–187.

Sarvanis, G.C., Karamanos, S. & Karamanos, S. 2017. Analytical model for the strain analysis of continuous buried pipelines in geohazard areas. *Engineering Structures* 152(1): 57–69.

Spiridenok, L.M. & Kiselev, D.A., & Grinevich, A.A. 2016. Analiz sposobov vosstanovleniya normativnoy glubiny zaleganiya magistral'nykh truboprovodov (in Russian). *Vestnik Polotskogo gosudarstvennogo universiteta. Seriya F, Stroitel'stvo. Prikladnyye nauki* 8(1): 87–114.

Stanev, V.S., Gumerov, A.G., Gumerov, K.M., & Rakhmatullin, S.I. 2004. Strength assessment of the main pipeline section in view of hydroimpact . *Neftyanoe Khozyaistvo - Oil Industry* 4/2004: 112–114.

Yavorskyi, A.V., Karpash, M.O., Zhovtulia, L.Y., Poberezhny, L.Y., & Maruschak, P.O. 2017. Safe operation of engineering structures in the oil and gas industry. *Journal of Natural Gas Science and Engineering* 46(1): 289–295.

Yavorskyi, A.V., Karpash, M.O., Zhovtulia, L.Y., Poberezhny, L.Y., Maruschak, P.O., & Prentkovskis, O. 2016. Risk management of a safe operation of engineering structures in the oil and gas sector. *Transport Means* 16(1): 370–373.

Yuzevych, L., Skrynkovskyy, R., & Koman, B. 2017. Development of information support of quality management of underground pipelines. *EUREKA: Physics and Engineering* 11(4): 49–60.

Zabela, K.A., Kraskov, V.A., & Moskvich, V.M., 2001. *Safety of Crossings of Water Barriers by Pipelines.* Moscow: Nedra.

Zapukhlyak, V.B., Mel'nichenko, Y.G., & Kuz', A.R. 2016. Problems of lowering the existing pipeline to the design mark during repair. *Proc. XI International Educational, Scientific and Practical Conference "Pipeline Transport -2016".* Ufa 6(1): 21–244.

Chapter 8

Increasing surface wear resistance of engines by nanosized carbohydrate clusters when using ethanol motor fuels

Olga O. Haiday, Volodymir S. Pyliavsky, Yevgen V. Polunkin, Yaroslav O. Bereznytsky, Olexandr B. Yanchenko, Andrzej Smolarz, Paweł Droździel, Saltanat Amirgaliyeva, and Saule Rakhmetullina

CONTENTS

8.1 Introduction 89
8.2 Analysis of the literature sources 90
8.3 Materials and methods used in the study 91
8.4 Results and discussion 92
8.5 Conclusion 98
References 98

8.1 INTRODUCTION

The lifespan of automotive vehicles is determined in most cases by the operational durability of the engines to parametric failure. Parametric failure means the reduction of the operating parameters of the unit below the permissible level. The basic unit of internal combustion engines is the cylinder-piston group, the state of which limits the life of motor vehicles. This unit reaching a critical state as a result of wear-out usually happens after operation for 100–200,000 km (1–2,000 service meter hours). With today's intensive work of motor vehicles (on average about 50,000 km/y) after 2–3 years of use, it is necessary to carry out major repairs or replacement of the engine.

The deterioration process of engine performance has a gradual cascading character: at first, precision frictional couplings of the fuel and supply equipment, which are greased with fuel, wear-out. The lifespan of such parts does not exceed hundreds of hours of work. Then the wear of the upper part of the cylinder begins, the lifetime of the friction couplings of the cylinder ring (which is lubricated in the mode of the limit oil film) reaches a thousand hours of operation, and the lifetime of the friction joint of the crankshaft (which is completely immersed in motor oil) can reach several thousand hours.

Due to long-term operation of automobile engines, wear-out, corrosion and hardening of the working surfaces, and deterioration of the fuel equipment efficiency, the reliability of the operation of the ignition system and the thermodynamic conditions of

DOI: 10.1201/9781003224136-8

the combustion process of the fuel mixture in the engines (pressure, temperature, and heat transfer conditions) take place.

As a result of the wear of the fuel supply system units and the wear of a cylinder-piston group with lower compression in the combustion chamber, the conditions of homogenization of the injected fuel mixture are violated. When the fuel in the combustion chamber is distributed non-homogenously, the combustion of the fuel–air mixture occurs in heterophase mode, which, due to the limited duration of the process (especially at high velocities of the crankshaft), leads to incomplete combustion of fuel with increasing fuel consumption, loss of engine power, and increased toxicity of emissions (Pilyavsky & Kovtun 2007).

Continuous increase of environmental requirements for engine emissions during combustion of fuels requires reduction of sulfur, heavy metals, and polycyclic aromatic compounds in fuel. Modern motor fuels after a high degree of purification have poor lubrication and anti-wear properties. As a result, when working on such fuels, the reliability and car accidents increase sharply.

The problem of corrosion and wear of engines was even more acute due to the introduction of environment-friendly motor fuels with a high content of ethanol. The disadvantages of ethanol motor fuels include high corrosive aggressiveness and low anti-wear properties. When working with such fuels, as a result of deterioration and jamming, the fuel pumps for injection engines are rapidly disrupted (Bozhko et al. 2016, Thangavelu 2016, Barakat 2016). The most notable problems mentioned during operation are in the last-generation engines with separate injection of fuel directly into the combustion chamber. In such engines, fuel pumps of plunger type operate at high pressures (5 MPa) and with poor lubricating properties of the fuel, so their lifetime is very short (less than one hundred hours). The cost of such spare parts is too high – from $100 to $1,000.

8.2 ANALYSIS OF THE LITERATURE SOURCES

The most widely studied and used are additives to motor oils. Additives of different types may be added to modern motor oils. To improve the viscosity-temperature properties of the oil, viscosity additives (up to 3% by weight); to reduce the temperature of hardening, depressor additives (up to 1% by weight); to reduce the formation of varnishes and scum on the parts of the engines, detergents (3%–10% by weight); and to reduce wear and corrosion, anti-wear and anticorrosive additives (2%–3% by weight) are used (Kuzmina et al. 2015).

In motor fuels, anti-wear and anticorrosive additives have not been used extensively so far; researchers and petrochemical companies have not paid much attention to them. The necessity of the development and introduction of such fuel additives has emerged in the last 10 years in connection with the fundamental change in the design of engines (the transition from obsolete carburetors to engines with fuel injection), increased environmental requirements for the operation of engines, more intensive operation, and the increase in the number of vehicles.

The anti-wear and anticorrosive additives used for motor oils have proved to be unsuitable for fuel due to the toxicity, low efficiency, and instability of their fuel solutions.

In addition to the requirements for the effectiveness of additives in their main action, it is necessary that they are completely dissolved in the fuel and lubricants and not filtered by the cleaning devices of the engine. Adding additives to the main petroleum product (fuel or oil) should not lead to their settling into sediment with prolonged storage, changes in temperature, or stratification when water gets into the tank. In addition, it is important that, by improving some operating properties, the additives do not degrade other properties.

In world practice, for improving the tribological characteristics of lubricants, in particular engine oils, zinc dialkyldithiophosphate is widely used as a suppository additive (Haidai 2016). But this additive is toxic, as are products of its thermal destruction due to the presence of aggressive elements – sulfur, phosphorus, and heavy metals.

Dispersions of nanosized particles of different origin (disulfide and trisulfide of molybdenum or tungsten, and copper nanoparticles) have been used as anti-wearing additives. But in motor fuels, such compounds cannot be used because they increase the amount of deposits on engine parts and, after thermal decomposition in the combustion chamber, can cause abrasive wear of precision friction surfaces and increase the toxicity of engine emissions.

Finding nontoxic additives for motor fuels is a problem that needs to be solved. Among these additives, attention is drawn to nanosized carbon materials, the use of which in the composition of fuel and lubricants in low concentrations (up to 0.1% by mass) leads to an improvement of their tribological characteristics, in particular, the reduction of frictional force, reduction of wear, and increase of anti-wearing stability of frictional knots (Street et al. 2004, Mang et al. 2010, Selyutin et al. 2015).

In our previous studies, the effect of fullerene C_{60} and products of its modification on the anti-wear properties of engine fuel components was investigated, and it was shown that in the presence of 0.01% wt. fluorinated fullerene $C_{60}F_{48}$, self-ignition resistance of isooctane increases 2.5 times (Polunkin et al. 2012). However, the high cost of fullerene prevents the economical use of these substances as additives. Therefore, further research was aimed at finding cheaper (more economical) analogues of fullerenes. They proved to be carbon nanoparticles. They represent a multilayer structure, the distance between which is 0.335 nm, which is approximately equal to the distance between two graphite planes (0.334 nm). The structure of carbon nanoparticles contains six- and five-degree structures with carbon atoms arranged at the vertices, which have two single and one double bond with adjacent carbon atoms with delocalized electrons along the entire molecule. Graphite layers in the structure of this nanomaterial consist of a large number of defects and cavities. The cavities can be filled with semi-clathrate and pentagonal carbon rings in different ways to form a quasi-spherical bulb of amorphous or crystalline structure (Kukharchuk et al. 2016, Mykhailiv et al. 2017, Sekoai et al. 2019). The undoubted advantage of applications on the basis of the proposed carbon spherical clusters is their environmental friendliness due to the absence of such compounds of aggressive elements (sulfur, phosphorus, and heavy metals), unlike standard anti-drainage applications for lubricants.

8.3 MATERIALS AND METHODS USED IN THE STUDY

Synthesis of carbon spheroidal nanoclusters was carried out using high-frequency discharge pulse synthesis in a propane-butane medium under atmospheric pressure at a pulse voltage at the output of a generator of 6–10 kV and a pulse frequency of the output voltage of 1–100 kHz (Polishchuk et al. 2018, 2019, Kozlov et al. 2019).

For the experimental evaluation of the size of the domains formed by the dispersion medium molecules around the carbon nanoparticles, the method of dynamic coherent laser light scattering was used.

The study of the effect of nanocarboxylic compounds on the dynamic strength of low-viscosity polar and nonpolar liquids was carried out using the ASTM D2783 method on a four-point friction machine in terms of critical load (Ogorodnikov et al. 2004, 2018a, 2018b, Dragobetskii et al. 2015).

The bench tests of the resource of fuel pumps of cars during their work with different fuels were performed in a specially created thermostated stand for 80 hours, which is equivalent to a run of the car of 8,000 km at a speed of 100 km/h (Vasilevskyi 2014b, Kukharchuk et al. 2017, Vasilevskyi et al. 2017).

The state of worn-out friction surfaces of fuel pumps after working in different types of automotive gasoline was monitored by analyzing profilograms and microprofilograms of the working steel shafts and bronze bushings of the pumps. Profilograms were removed using a laser scanning differential-phase profilometer of the measuring system NAU-01 in the research laboratory of nano-tribology technologies of the National Aviation University (Kiev) (Vasilevskyi 2013, 2014a).

8.4 RESULTS AND DISCUSSION

Attempts to introduce carbon nanoclusters into the fuel were carried out for the first time. The problem of using such structures as part of motor fuels is their poor solubility in low-viscosity liquids. To improve the solubility of the resulting nanoclusters in the oxygenated fuel components, a chemical modification of the surface of the carbon clusters by bromination was carried out.

Modified CNOs–Br nanoclusters were dissolved in a solvent, which was used as the absolute ethanol or oxygenated fuel E-85 to obtain an additive to petroleum and mixed fuels.

The effect of liquid lubricants on the wear of frictional knots is determined by two properties:

- Ability to prevent damage to contact surfaces due to the implementation of the hydrodynamic friction mode.
- Ability to minimize damage under extreme friction.

To evaluate these properties of the lubricant, different characteristics are used. The ability of the lubricant to provide a hydrodynamic regime is characterized by its bearing capacity, and reducing the damage to the surface under extreme friction is estimated by the index of scuffing. Figure 8.1 shows the obtained values of the carrier capacity of ethanol (1) and the mixture of ethanol fuel E80 (2) in comparison with the corresponding parameter for motor gasoline (3) (DSTU 7687 2015). The control point shows the value of this parameter for friction conditions in motor oil (4). As can be seen from the above data, when there is friction in an ethanol environment, catastrophic damage to contact surfaces (scuffing) is realized at a load that is 2 times lower than that for a mixed fuel E-80, 6 times lower than for gasoline A-80, and 20 times lower than for motor oil. That is why it is necessary to use anti-wear additives in the ethanol fuel composition to prevent damage to fuel equipment.

Increasing surface wear resistance of engines

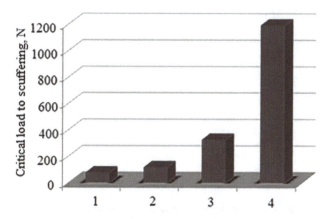

Figure 8.1 Critical load of the liquid using: 1, ethanol; 2, ethanol fuel E-80; 3, hydrocarbon gas A-80; and 4, ELF SAE 5W40 engine oil.

We made an attempt to improve the lubricating properties of ethanol motor fuel by increasing the viscosity of the fuel by introducing high-molecular weight alcohols. Table 8.1 shows the physicochemical characteristics of the studied alcohols, as well as the significance of their critical load to the scuffing we measured.

As can be seen from the comparison of these data, no correlation between the carrying capacity of the investigated alcohols and characteristics such as viscosity and boiling point is observed. At the same time, a correlation between carrying capacity and liquid density was found – the higher the density of the alcohol, the higher its bearing capacity. The highest bearing capacity in the range of the investigated alcohols was found in glycerol. However, it cannot be used as an additive to motor fuels for gasoline engines, since, according to technical requirements (DSTU 7687 2015), the boiling point of the components of these fuels should not exceed 210°C while the boiling point of glycerin is 290°C (Table 8.1).

According to our results, the bearing capacity of fuels almost linearly increases with higher concentrations of higher alcohols and cycloalcohols (Figure 8.2).

Among the studied alcohols, the best results were shown by 2-furylcarbinol: its introduction into ethanol fuel E-85 at a concentration of 5%–20% by weight increases the self-ignition resistance of fuel by two times, which will increase the lifespan of engines when using such fuel.

Table 8.1 Physicochemical characteristics and carrying capacity of alcohols

Alcohol	Dynamic viscosity, Pa/s	Boiling point, °C	Density, g/cm^3	Critical load to scuffing, N
Isopropanol	0.024	82.4	0.78	50
Ethanol	0.012	78.4	0.79	50
Benzyl alcohol	0.050	205.8	1.05	300
2-Furyl carbinol	0.046	171.0	1.13	500
Glycerol	9.45	290.0	1.29	940

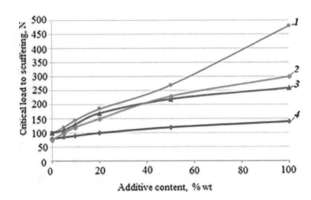

Figure 8.2 Influence of various applications on anti-clamping properties of ethanol fuel E-85: 1, 2-furylcarbinol; 2, n-octanol; 3, benzyl alcohol; and 4, C_3–C_5 alcohols.

A significant step in the development and use of anti-wear materials was the research and introduction into the production of additives on the basis of spherical carbon clusters – fullerenes. The first studies of the influence of these materials on the tribological properties of motor fuels were performed by us at V.P. Kukhar IBOPC NAS of Ukraine in 2008–2011.

Table 8.2 shows the results of the use of fullerene C_{60} and some of its derivatives (azo compounds, hydroxylated, and halogenated fullerenes) (Polunkin et al. 2012) with high solubility in hydrocarbon and ethanol fuels compared to the known industrial additive DF-11 (bis ((O, O-dialkyldithiophosphate) of zinc).

As can be seen from the values obtained for the critical loading parameters of ethanol fuels in comparison with hydrocarbon (Table 8.2), at low concentrations (0.0001%–0.02%, by mass) of fullerene C_{60}, hydroxylated fullerene $C_{60}(OH)n$, azafullerene, as well as industrial additive DF-11, the self-ignition resistance of ethanol fuel practically does not change.

The anti-scuffing properties of ethanol fuels are substantially improved with the use of amidated and halogenated fullerenes as an additive. Fluorinated fullerene $C_{60}F_{48}$ proved to be the most effective self-ignition-preventing additive; in addition, its effect was concentrated in a concentration-dependent manner. Its introduction into ethanol fuel at a concentration of 0.02%–0.05% by weight in 3–5 times improves the antiknock properties of both ethanol and hydrocarbon fuels.

The main drawback of derivatives of fullerene for use in fuel is their high cost. Similar to fullerenes, we synthesized the chemical properties of CNS–Br_n nanocarbon-brominated clusters.

The durability of metal friction units during operation in fuel and lubricants depends on both the load level and the adsorption and electrochemical properties of the liquid medium.

Figures 8.3 and 8.4 show the effect of carbon nanoclusters on the change in the coefficient of friction and on the change in the electrode potential in the friction pair.

The results of studying the influence of samples of carbon nanoclusters on tribocorrosion with friction in benzyl alcohol show (Figures 8.3, 8.4 and Table 8.3) that the incorporation of a small amount of carbon nanoparticles into the composition of the polar lubricating fluid leads to a reduction in the tribocorrosion activity of the medium.

Table 8.2 Comparison of various additives

Additive	Application concentration, % wt.	Critical load, N	
		In isooctane	In ethanol fuel E-85
Fullerene C_{60}	0.001[a]	80	120
OL281B	0.003[a]	150	120
Hydroxylated fullerene C_{60}	0.02[a]	-	100
Amidated fullerene C_{60}	0.01[a]	200	200
Halogenated fullerene $C_{60}F_{48}$	0.01	200	200
	0.02	500	300
	0.05	500	500
	0.01	-	500
DF-11 bis (O, O-dialkyldithiophosphate) of zinc	0.01	150	100
	0.02	300	100
	0.05	400	100
	0.1	500	150
	0.5	600	300
	0.75	600	300
	1.0	550	230

[a] Boundary solubility.

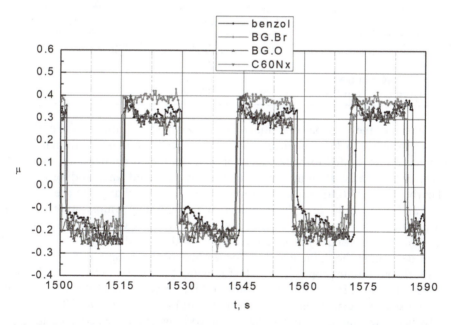

Figure 8.3 Change of the coefficient of friction in the friction pair when nanoclusters of nitrogen-containing fullerene $C_{60}N_x$ and modified multilayer carbon nanoclusters – carbon nano-onions (CNOs) – are added to the model lubricant liquid benzyl alcohol.

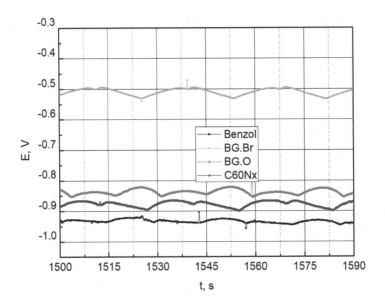

Figure 8.4 Change of the electrode potential in a friction pair when carbon nanoclusters are introduced into the benzyl alcohol liquid substrate.

Table 8.3 Changing the value of the channel of wear of the tribological tests in a pure substrate of benzyl alcohol (BA) and in the substrate with the addition of carbon nanoclusters, CNOs (concentration 0.001% by weight)

Environment	Value of the channel of wear, μm
BA	320
Oxidized CNOs in BA	307.9
Brominated CNOs in BA	303.9

The size of the electrode potential (from −0.95 to −0.5 V) and the size of the channel of wear (from 320 to 303 μ) change significantly when brominated carbon monospacial nanoclusters are added to benzyl alcohol; less significant changes are observed with the use of oxidized carbon monospacial nanoclusters; and when adding single-spherical carbon clusters (nitrogen-containing fullerenes, C_{60}), changes in the electrode potential and the magnitude of the tribocorrosion of the steel samples were not observed (in the latter case, the values of these characteristics were close to the corresponding values for friction conditions in benzyl alcohol without additives).

To determine the influence of the developed additives on fuel for the life of fuel pumps, the anti-wear properties for the following fuels were investigated: I – high octane gasoline A-95, II – ethanol fuel E-85, and III – ethanol fuel E-85 with the addition of a nanosized spherical cluster additive (E85* in Figure 8.5) (Haidai 2016).

By bench testing, it was found that when working with mixed ethanol fuels with the addition of spheroidal carbon nanoparticles (Figure 8.5, curve 1), the wear of a fuel pump is less than its wear when working with high octane gasoline A-95 (Figure 8.5, curve 2), as well as with mixed ethanol fuels without adding the additive (Figure 8.5, curve 3).

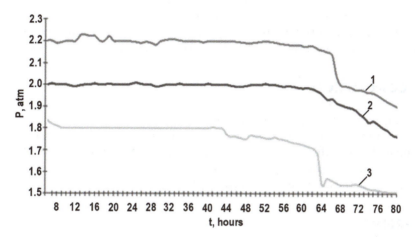

Figure 8.5 Dynamics of change of working pressure of a fuel pump during operation on various fuels: 1, E-85*; 2, A-95; and 3, E-85.

The state of deterioration of the friction surfaces of fuel pumps after working in different components of automotive gasoline was controlled using a laser scanning differential-phase profilograph profilometer (Table 8.4).

Table 8.4 Deterioration of the friction surfaces of fuel pumps

Work environment	General view of the shaft	Radial clearance in roller bearings	Friction surfaces of the plug
Not working		19	
A-95		24.5	
E-85 + CNOs-Br$_n$		26	
E-85		28.5	

It was found that the most significant friction wear is observed in the case of the use of the E-85 blend fuel without the addition of additives. However, when working on oxygenated fuel with nanoparticles, the surface has fewer gaps and no cutaways, indicating reduced wear over extended use (Haidai 2016).

8.5 CONCLUSION

The influence of additives based on nanosized carbon clusters on the increase of wear resistance of the friction surfaces of engines with the use of ethanol motor fuels was investigated. It was discovered that the presence of fuel additive contributes to an increase in its carrying capacity by one and a half times the base fuel with a corresponding decrease in the damage of the metal surface of friction pairs.

REFERENCES

Barakat, Y. 2016. Fuel consumption of gasoline ethanol blends at different engine rotational speeds. *Egyptian Journal of Petroleum* 25: 309–315.

Bozhko, E.A., Yesilevsky, S.A. & Chernyavsky, E.B. 2016. Increasing the carrying capacity of ethanol as a component of an alternative motor fuel: experiment and molecular modeling. In E.A. Bozhko, S.A. Yesilevskiy & E.B. Chernyavskiy (eds), *Additional Information of the National Academy of Sciences of Ukraine*: 79–86. Vidavnichy dim "Akamperíodika" NAS of Ukraine.

Dragobetskii, V., Shapoval, A., Mos'pan, D., Trotsko, O. & Lotous, V. 2015. Excavator bucket teeth strengthening using a plastic explosive deformation. *Metallurgical and Mining Industry* 4: 363–368.

DSTU 7687. 2015. *Gasoline car Euro. Tekhnichni umovi.* Kyiv: DP "UkrNDNTS".

Haidai, O. 2016. Improving of performance properties of ethanol motor fuels through use of additives based on nanoscale carboxylic clusters. *EUREKA: Physical Sciences and Engineering* 6: 3–10.

Kozlov, L.G., Polishchuk, L.K., Piontkevych, O.V., Korinenko, M.P., Horbatiuk, R.M., Komada, P., Orazalieva, S. & Ussatova, O. 2019. Experimental research characteristics of counter balance valve for hydraulic drive control system of mobile machine. *Przeglad Elektrotechniczny* 95(4): 104–109.

Kukharchuk, V.V., Bogachuk, V.V., Hraniak, V.F., Wójcik, W., Suleimenov, B. & Karnakova, G. 2017. Method of magneto-elastic control of mechanic rigidity in assemblies of hydropower units. *Proceedings of SPIE*: 104456A.

Kukharchuk, V.V., Hraniak, V.F., Vedmitskyi, Y.G., et al. 2016. Noncontact method of temperature measurement based on the phenomenon of the luminophor temperature decreasing. *Proceedings of SPIE*: 100312F.

Kuzmina, G.N., Naumov, A.G., Parenago, O.P. & Pautov, A.V. 2015. Lubricating properties of oil SOTS, containing tribologically active additives. *Friction and Wear* 36(4): 409–414.

Mang, Th., Bobzin, K. & Bartels, Th. 2010. *Industrial Tribology: Tribosystems, Friction, Wear and Surface Engineering, Lubrication.* Hoboken, NJ: Wiley.

Mykhailiv, O., Zubyk, H. & Plonska-Brzezinska, M.E. 2017. Carbon nano-onions: Unique carbon nanostructures with fascinating properties and their potential applications. *Inorganica Chimica Acta* 222: 49–66.

Ogorodnikov, V.A., Dereven'ko, I.A. & Sivak, R.I. 2018. On the Influence of Curvature of the Trajectories of Deformation of a Volume of the Material by Pressing on Its Plasticity

Under the Conditions of Complex Loading. *Materials Science* 54(3): 326–332. Doi: 10.1007/s11003-018-0188-x.

Ogorodnikov, V.A., Savchinskij, I.G. & Nakhajchuk, O.V. 2004. Stressed-strained state during forming the internal slot section by mandrel reduction. *Tyazheloe Mashinostroenie* 12: 31–33.

Ogorodnikov, V.A., Zyska, T. & Sundetov, S. 2018. The physical model of motor vehicle destruction under shock loading for analysis of road traffic accident. *Proceedings of SPIE* 108086C. Doi: 10.1117/12.2501621.

Pilyavsky, V.S. & Kovtun, G.A. 2007. Increasing the resource of automobile engines. In V.S. Pilyavskiy & G.A. Kovtun (eds), *Povysheniye resursa avtomobil'nykh dvigateley*: 123. IBOPC NAS of Ukraine.

Polishchuk, L.K., Kozlov, L.G., Piontkevych, O.V., et al. 2018. Study of the dynamic stability of the conveyor belt adaptive drive. *Proceedings of SPIE* 1080862. Doi: 10.1117/12.2501535.

Polishchuk, L.K., Kozlov, L.G., Piontkevych, O.V., et al. 2019. Study of the dynamic stability of the belt conveyor adaptive drive. *Przeglad Elektrotechniczny* 95(4): 98–103.

Polunkin, E.V., Kameneva, T.M., Pilyavsky, V.S., et al. 2012. Antioxidant and anti-seize properties of halogenated fullerenes. *Catalysis and petrochemistry* 20: 70–74.

Sekoai, P.T., Moro Ouma, C.N., du Preez, S.P., et al. 2019. Application of nanoparticles in biofuels: An overview. *Fuel* 237: 380–397.

Selyutin, G.E., Kuznetsov, V.L. & Mishakov, I.V. 2015. Lubricants based on carbon nanomaterials for operation under high loads and low temperatures. *Bulletin of the Siberian Branch of the Academy of Military Sciences* 35: 49–52.

Street, K.W., Marchetti, M., Vander Wal, R.L. & Tomasek, A. 2004. Evaluation of the tribological behavior of nano-onions in Krytox 143AB. *Journal of Tribology* 16: 143–149.

Thangavelu, S.K. 2016. Review on bioethanol as alternative fuel for spark ignition engines. *Renew Sustain Energy Reviews* 56: 820–835.

Vasilevskyi, O.M. 2013. Advanced mathematical model of measuring the starting torque motors. *Technical Electrodynamics* 6: 76–81.

Vasilevskyi, O.M. 2014a. Calibration method to assess the accuracy of measurement devices using the theory of uncertainty. *International Journal of Metrology and Quality Engineering* 5(4). Doi: 10.1051/ijmqe/2014017.

Vasilevskyi, O.M. 2014b. Methods of determining the recalibration interval measurement tools based on the concept of uncertainty. *Technical Electrodynamics* 6: 81–88.

Vasilevskyi, O.M., Kulakov, P.I., Ovchynnykov, K.V. & Didych, V.M. 2017. Evaluation of dynamic measurement uncertainty in the time domain in the application to high speed rotating machinery. *International Journal of Metrology and Quality Engineering* 8(25). Doi: 10.1051/ijmqe/2017019.

Chapter 9

Method of experimental research of steering control unit of hydrostatic steering control systems and stands for their realization

Mykola Ivanov, Oksana Motorna, Oleksiy Pereyaslavskyy, Serhiy Shargorodskyi, Konrad Gromaszek, Mukhtar Junisbekov, Aliya Kalizhanova, and Saule Smailova

CONTENTS

9.1 Introduction 101
9.2 Analysis of literary data and problem statement 101
9.3 Purpose and objectives of the research 104
9.4 Materials and methods of research 104
9.5 Conclusion 110
References 111

9.1 INTRODUCTION

Currently, hydrostatic steering control systems are used in a variety of self-propelled special machines, due to a number of advantages of such systems. Many scientific schools are working to improve existing or create new steering systems (Zardin et al. 2018, Danfoss 2020). Such works on the modernization and research of steering control units of hydrostatic steering systems are being carried out at Vinnitsa National Agrarian University (Ivanov et al. 2014).

One of the important issues that arise when creating new samples of technical objects, and especially such scientifically intensive ones as steering control units, is to conduct experimental research of their prototype samples (Pogorelyi 2004). The experimental studies are aimed at determining the parameters included in the mathematical model of the steering system and checking the adequacy of this model, as well as determining the reliability and convenience (organoleptic quality assessment) of the steering control unit. In addition, there is a problem of conducting comparative tests of steering control units of different manufacturers.

9.2 ANALYSIS OF LITERARY DATA AND PROBLEM STATEMENT

Currently, there are various stands for experimental research of systems of hydrostatic steering control in general. The scheme of the most universal stand is given in Figure 9.1

DOI: 10.1201/9781003224136-9

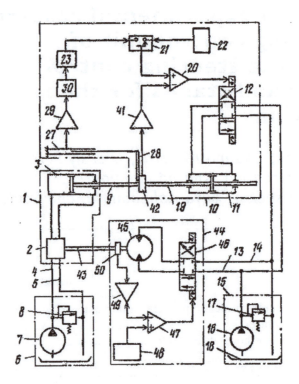

Figure 9.1 Scheme of a steering system test bench utilizing follow-up electrohydraulic distributors.

(Kolosov et al. 1986). The tested hydraulic steering control system 1 contains a steering control unit 2 and an executive hydraulic cylinder 3. The steering control unit is connected by hydrolines 4 and 5 with a power supply 6, consisting of a pump 7 and a safety valve 8. The rod 9 of the executive hydraulic cylinder 3 is connected to the unit 10 which creates a load on the steering system.

The load, in the form of force, is created by an additional hydraulic cylinder 11 whose working cavities are connected to the output channels of the electrohydraulic tracking distributor 12. The electrical system for creating the control signal by the distributor 12 comprises the following basic elements, such as the displacement sensor 27 of the stock of the executive hydraulic cylinder, the force sensor 42, and the block 22, which generates an electrical signal that sets the magnitude and law of the load change on the steering system. This allows you to create a counter load (the load force operates in the opposite direction of movement of the stock of the executive hydraulic cylinder) or associated load (load force operates in one direction with the displacement of the stock of the executive hydraulic cylinder). These forces can be permanent or change according to certain laws (harmonious, triangular, accidental, or any other).

The input shaft 43 of the steering control unit 2 is connected to the control signal generation unit 44. The control signal in the form of a rotation of the input shaft of the pump-dosing unit is formed by the hydromotor 45. The cavities of the hydromotor are connected to the output channels of the electrohydraulic tracking distributor 46, which controls the speed of rotation of this hydromotor. The electrical control signal generating system of the distributor 46 comprises such basic elements as an angular velocity sensor 50 and a block 48 that generates an electrical signal that sets the value and law of the change of control signal to the steering system. This allows you to set the value and direction of the constant speed of rotation of the input shaft of the pump feeder or set the speed of rotation, which varies according to certain laws (harmonic, triangular, random, or any other).

This stand allows you to explore any static and dynamic characteristics of the steering system or to investigate the endurance of the system. The disadvantage of such a booth is that when testing steering control units with different proportioning volumes of the dosing unit, it is necessary to change the hydraulic cylinders 3 and 11. In addition, the limited piston movement of these hydraulic cylinders limits the testing time on the stand in a particular operating mode of the steering control unit. The higher the speed of the input shaft, the smaller this time is.

There is another test bench (Figure 9.2) (Cherbakov et al. 2011), which contains the steering control unit 15 and the executive hydraulic cylinders 16, in which steering systems are tested. The control signal is set from the steering wheel, and the opposite cavities of the executive hydraulic cylinders are used to create the load. Counter loading is formed by opposing and by-passing – by pressure in the direction of displacement given by the distributor 9 according to the direction of rotation of the steering wheel. The load size is given by chokes 5 and 6. This booth has a lower

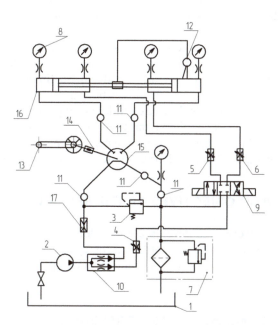

Figure 9.2 Scheme of a test bench for a steering system in which load chokes are used: 1, tank; 2, pump; 3, safety valve; 4, 5, and 6, throttles; 7, filter; 8, pressure gauge; 9, spreader; 10, flow divider; 11, pressure sensor; 12, displacement sensor; 13, speed sensor; 14, torque sensor; 15, steering control unit; 16, hydraulic cylinders; and 17, cost sensor.

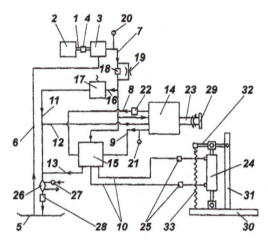

Figure 9.3 Scheme of a test bench for the testing of individual units of hydropower steering systems: 1, drive shaft; 2, electric motor; 3, pump; 5, tank; 7, 8, and 9, pressure lines; 10, outgoing lines; 11, 12, and 13, drain lines; 14, steering control unit; 15, flux amplifier; 17, safety valve; 18, faucet; 19, throttle; 20 and 21, pressure sensors; 22 and 25, cost sensors; 23, drive shaft; 24, hydraulic cylinder; 29, torque sensor; 31, stand; 32, lever; and 33, spring.

level of functionality than the previous one, but it is also intended to test systems of hydropower steering in general. Therefore, it has the same disadvantages as the previous stand.

A well-known stand for the testing of individual units of static steering systems (Chernoivanov et al. 2000), including for the testing of dispensing pumps 14, is shown in Figure 9.3. However, to create a load, a hydraulic cylinder 24 is also used, which together with a lever 32 and a spring 33 allows both a counter and an associated load to be generated. Although this hydraulic cylinder is not part of a static steering system, it limits the amount of fluid supplied from the steering control unit and the duration of the steering control unit operating time in a certain operating mode. Therefore, this stand also has drawbacks that are characteristic of the previous stands.

9.3 PURPOSE AND OBJECTIVES OF THE RESEARCH

The purpose of this work is to develop a methodology for experimental studies of a steering control unit without the use of hydraulic cylinders to create a load. The next task of the work is to develop a booth scheme for the implementation of the proposed methodology of experimental research.

9.4 MATERIALS AND METHODS OF RESEARCH

The load on a hydrostatic steering system is a force on the executive hydraulic cylinder of this system, which is traditionally modeled during experimental tests. This force is significantly different for steering systems of various technological machines – by mass, capacity, purpose, etc. Therefore, hydraulic cylinders with different effective areas of the piston of the executive hydraulic cylinder are used, complete with steering control units with different working volumes. But for loading pumps, there is a pressure drop in its output channels (Ivanov et al. 2016), which allows us to evaluate the performance of these products by modeling the pressure drop regardless of their working volume. Therefore, the differential pressure of Δp in its output channels is considered in the future as a universal load parameter of the steering control unit (Ogorodnikov et al. 2018b, Polishchuk et al. 2019, Kozlov et al. 2019).

The nature of the load for the hydropower steering system is related to the direction of the force on the executive hydraulic cylinder in accordance with the direction of rotation of the steering wheel. In accordance with this load, the steering system is divided into counter and associated. During the experimental research of the steering control units on the stand, it is necessary to generate both types of load.

In the laboratory of the Machinery and Equipment of Agricultural Production department of Vinnitsa National Agrarian University, a special experimental booth was created for the study of pump-dosing devices of hydrovolume steering systems. It allows the operation of the dispensing pump to be analyzed in detail at different operating modes of the system, providing control of the value and direction of the control signal and the load throughout the range of values that arise in practice (Ogorodnikov et al. 2004, 2018a).

The stand has a system for generating a control signal in the form of a rotation of the input shaft of the pump-dosing unit and a system for generating the load in the form of a pressure difference in its output channels. The electrohydraulic circuitry of the stand is shown in Figure 9.4. Since various methods of generating the counter and associated loads are used, a stand with a variable structure was implemented. In accordance with this, various schemes of the stand have been developed for the implementation of loads of different types. Figure 9.4a shows a complete scheme of the stand, in which the system of counter load generation (SFZN) is implemented. Figure 9.4b shows only the part of the booth circuit in which the system of generation of the associated load (SFPN) is implemented. In this case, other systems of the stand are similar to those depicted in Figure 9.4a.

The stand contains the steering control unit (SCU), the pumping station for the power supply of all stand systems, the system for generating the control signal (SPSS), the load generation system, as well as the system for registration and processing of information.

The SCU (Figure 9.4a) by pressure channel P is connected to the pump H1 and the channel T to the drainage channel of the pumping station. This reproduces the work of the SCU with the power system on the self-propelled agricultural machine. The pump H1 is made for regulated performance. This allows you to adjust the flow rate of this pump to a value that corresponds to the pump's power consumption for the steering system with the SCU of the test specimen size.

The system of generating a control signal sets the rotational motion of the input shaft of the pump-dosing unit, which simulates the rotation of the steering wheel on the self-propelled agriculture machine. For this purpose, a hydromotor *HM1* is used, the shaft of which is connected through a rigid coupling to the input shaft of the pump dispenser. The working cavities of the *HM1* hydromotor are connected to the pumping station channels, which includes the output channel of the pump *H2* and the drain channel. The pump *H2* is made for regulated performance. This allows you to adjust

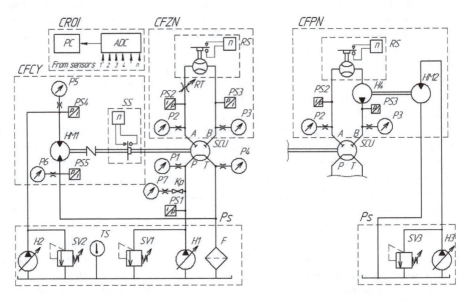

Figure 9.4 Hydraulic diagram of the booth to determine the characteristics of the steering control unit: (a) in the generation of counterloads and (b) in the generation of the associated load.

the speed of rotation of the shaft of the hydraulic motor *HM1*, which sets the desired speed of rotation to the input shaft of the pump-dosing unit.

The output channels *A* and *B* of the SCU are connected to the load generation system. The generation of different types of load is implemented by changing the structure of the stand. The system for generating the counter (passive) load (Figure 9.4a) is implemented as a regulated throttle *RT*, which is connected to the output channels *A* and *B* of the dispensing pump. Adjusting the area of this throttle allows you to change the resistance of the flow of the working fluid through the outlets of the SCU. Due to this, at the output of the SCU, a load is generated in the form of a differential pressure Δp in its output channels, which reproduces overcoming a certain counter load.

The system generating the associated (active) load (Figure 9.4b) is implemented as a pump H4, the suction and output channels of which are connected to the output channels A and B of the dispensing pump. The pump shaft *H4* is driven by the *HM2* hydromotor, and the rotational speed is regulated by setting the flow rate of the pump *H3* of the pumping station. Adjusting the pump speed *H4* allows you to adjust a certain amount of negative pressure difference in the output channels of the dispensing pump, which reproduces a certain associated load.

This method of generating the counter and associated load for the SCU prevents you from having to use load hydraulic cylinders that interact with the executive hydraulic cylinders of the steering system (Chernoivanov et al. 2000, Motorna 2015). This significantly simplifies the construction of the stand and does not require the use of hydraulic cylinders for the implementation of both the system of hydroturbine steering and the hydraulic cylinder load creation system.

In addition, such a stand allows you to reproduce a certain mode of operation of the SCU over a long period of time, in contrast to the traditional stand, when this process lasts for several seconds at high speeds of rotation of the input shaft. This allows, at the stage of the experimental verification of the work of the prototype, the values of the parameters controlled on the stand and the quality of the work of the prototype to be analyzed, and the conformity of the state of the hydrosystem to expectations to be assessed.

The bench is equipped with control and measuring equipment which registers the parameters of the system in steady and dynamic states. The parameters measured on the stand in a steady state are given in Table 9.1. For visual measurements of the pressure $P1-P7$ in the steady state, bench pressure gauges of type DM, accuracy class 0.6, are installed. Different sections of the hydrosystem use pressure gauges with different measuring ranges.

Two pressure gauges are installed in the pressure duct of the SCU: one is P1 with a measurement range of up to 25 MPa, designed to measure the pressure on the operating modes of the steering system, and another is P7 that is connected through the faucet Kp, which has a measurement range of up to 1.6 MPa and is designed to measure the pressure in the absence of a control signal at the average position of the spool valve of the SCU.

For measurement of pressure during the analysis of dynamic processes on the stand, strain gauge pressure sensors with built-in amplifiers are installed, which record changes in pressure over time in the corresponding sections of the hydrosystem. Pressure sensors $PS1-PS5$ on different sections of the hydraulic system are used with

Table 9.1 List of parameters measured on an experimental bench

Name	Unit of measurement	Value Min	Value Max	Measurement error, %
Flush of the input shaft of the steering control unit	deg	0	30	1.5
Rotation speed of the input shaft of the steering control unit	rpm	5	100	1
Torque transmitted to the input shaft of the steering control unit	N/m	0	120	4
Adjusting torque limitation on the input shaft of the steering control unit	N/m	5	120	4
Loading pressure difference in the output channels of the steering control unit	MPa	0	±25	1
Excess pressure applied to the outlet duct of the steering control unit	MPa	0	25	0.6
Pressure at the pressure line:	MPa			
Pump H1		0	25	
Pump H2		0	12	0.6
Pump H3		0	25	
Consumption at the pressure line:	l/min			
Pump H1		8	126	
Pump H2		5	31.8	1.5
Pump H3		6	63	
Working temperature	°C	10	100	2

different nominal pressure. The PS1, PS2, and PS3 Danfoss (Denmark) model MBS 3050 sensors (Figure 9.5a) have a nominal pressure of $p_{nom} = 16$ MPa, and pressure sensors the PS4 and PS5 ADZ-NAGANO (Germany and Japan) model ADZ-SML-20 sensors (Figure 9.5b) have a nominal pressure of $p_{nom} = 10$ MPa.

A photo of the stand for testing a SCU with a working volume of 160 cm³/rotation during the counterload is shown in Figure 9.6.

The control and measuring system of the stand contains a sensor of angular velocity of the input shaft of the SCU. The sensor is of frequency pulse type and is for measuring the frequency of electric impulses. A low-frequency frequency meter of grade Ch3-49 is used.

In the stand, the flow sensor *DF* allows you to control the flow of working fluid in different sections of the hydraulic system. In the scheme, this sensor is shown when measuring the flow of working fluid in the output channels of the SCU, which, in the steering system, is fed to the executive hydraulic cylinder under the action of the control signal.

Figure 9.5 Photos of pressure sensors: (a) model MBS 3050 and (b) model ADZ-SML-20.

Figure 9.6 Photo of the booth for testing the dosing pump during countercurrent action: 1, steering control unit; 2, flow sensor for working fluid; 3, 4, 5, and 7, pressure gauges; 6, 9, and 10, pressure sensors; 8, throttle creating counter strike; 11, hydromotor, which sets the control signal; and 12, sensor of the speed of rotation of the input shaft

For the registration of dynamic signals, a computerized system for recording and processing the measured results of the studied parameters is used (Figure 9.7) using a personal computer, an analog-to-digital converter, and a program for measuring the results of the measurement. The block diagram of the system for recording the measurement results is shown in Figure 9.7a, and its photograph is shown in Figure 9.7b.

Analog electrical signals from the pressure sensors are fed to the input channels of the analog-to-digital converter (ADC). An ADA-1406 ADC by LLC HOLIT Delta Systems, Ltd. (Ukraine, Kyiv) is used. Each input has individual settings for switching the mode and transmission ratio. The converter of continuous electric signals in the digital format itself is based on a 14-bit integrated circuit. In this case, sufficient performance (measurement frequency up to 350 MHz) and accuracy (error up to 3%) of dynamic signal transformation are provided. To export data to a personal computer, a parallel data channel is used to achieve high performance.

On the computer, the PowerGraph software is used to receive, record, store, and process the data, which provides individual setup and calibration of channels and allows any signal measurement units to be used. The program allows you to register a series of measurements in the form of independent data blocks, each of which has a continuous array of random channel data.

Under constant parameters of the system, the values of the pressure on the corresponding sections of the hydraulic system, the speed of rotation, and the flow of the working fluid, which describe the static characteristics of the steering system, are recorded on the readings of the devices.

During the study of the dynamic characteristics of the SCU, transient processes were recorded at different modes of its operation. Figure 9.8 shows an oscillogram that is registered at a rotation speed of the input shaft of 80 rpm and a counter load in the outlet ducts of the pump-dosing unit $\Delta n n = 10$ MPa.

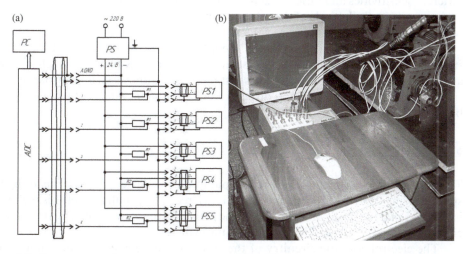

Figure 9.7 Computerized metering-recording system: (a) block diagram and (b) photograph.

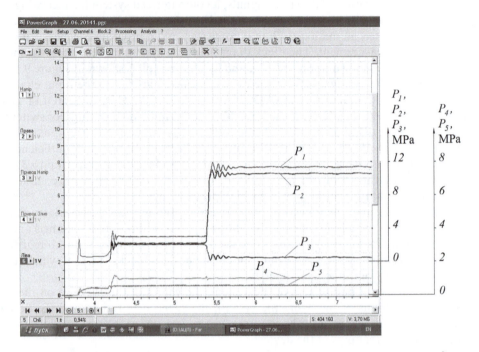

Figure 9.8 Oscillogram of the steering control unit with a working volume of 160 cm³ at a speed of rotation of the input shaft of n = 80 rpm and countercurrent actions Δp = 10 MPa.

The oscillogram shows the changes in the pressure P1, P2, P3, P4, and P5 in the corresponding sections of the hydraulic system of the steering control. The vertical axis on the oscillogram shows voltage, but pressure sensors of different types have different sensitivities and different zero points in accordance with the tare graphs. For convenience, on the left side of the oscillogram, the scales of the correspondence of the recorded signals with the pressure values are plotted.

For the organoleptic assessment of the quality of the pump-dosing device on the stand, the hydraulic motor GM1 is disconnected and in its place, the self-propelled agricultural machine's steering wheel is installed, which allows an appropriate estimation of the operation of the pump dispenser under different modes of its operation to be obtained.

9.5 CONCLUSION

The method of experimental research of pump-dosing devices for systems of hydrostatic steering control of special machines is offered. The peculiarity of this technique is that the load for the SCU is created in the form of a pressure difference in its output channels without the use of loading hydraulic cylinders.

The electrohydraulic circuitry of the stand was developed for the implementation of the proposed test method for SCUs. On the stand, a system for generating a control signal in the form of a rotation of the input shaft of the pump-dosing unit and a system for generating the load in the form of a pressure difference in its output channels are implemented. To generate a counter or associated load, the stand is made with a variable structure. The characteristic of control and measuring equipment is given.

Such a performance of the stand allows the quality of the operation of the pump dosing to be evaluated, regardless of their working volume, and the parameters of the SCU to be determined in conditions that correspond to different modes of operation of the steering system. In addition, it simplifies the analysis and research of the behavior of the SCU over a long period of time, which is not limited to the progress of the executive hydraulic cylinder of the steering system or the hydraulic cylinder loading system.

REFERENCES

Cherbakov, S.H., Mukushev, V.S. & Zhdanov, S. K. 2011. Sovershenstvovanie obieemnykh gidroprivodov rulevogo upravleniia dorozhno-stroitelnykh mashin: monografiia. In *Improving the Volumetric Hydraulic Actuators Steering Road-Building Machines*. Omsk: SibADI.

Chernoivanov, V.I., Northern, A.E., Kolchin, A.V., Kargiev, B.Sh. & Dankov, A.A. 2000. Stand for testing the elements of the hydraulic unit steering. Patent 2173414 of the Russian Federation, MKI F 15 B 19/00, No 2000 116467/06, stated. 06.27.2000.

Danfoss – Engineering Tomorrorsdzw, today. Products. Steering. Accessed 17 March 2021. https://www.danfoss.com/en/products/steering/dps/steering-components-and-systems/.

Ivanov, M.I., Motorna, O.O., Kozak, Y.M. & Pereiaslavskyi, O.M. 2016. The testbench and the experimental research methodology of the steering unit parameters of the self-propelled machinery hydrostatic steering system under the counter load. *Promyslova hidravlika i pnevmatika – Industrial Hydraulics and Pneumatics* 3(53): 66–74.

Ivanov, N., Motorna, O.O., Sereda, L. & Pereyaslavsky, A. 2014. Improvement of the design of the pump-dosing system of the hydraulic power steering of self-propelled machines. *MOTROL: Commission of Motorization and Energetics in Agriculture: Polish Academy of Sciences* 16(5): 103–114.

Kolosov, L.P., Potapov, D.S., Pereyaslavskii, A.N. & Yareshko, V.M. 1986. Stand for testing the steering mechanisms of vehicles. A. s. 1280376 SSSR, MKI G 01 M 17/06, No 3937388/31-11, zayavl. 06.08.85, publ. 30.12.86, Biul. No 48, 1986.

Kozlov, L.G., Polishchuk, L.K., Piontkevych, O.V., Korinenko, M.P., Horbatiuk, R.M., Komada, P., Orazalieva, S. & Ussatova, O. 2019. Experimental research characteristics of counter balance valve for hydraulic drive control system of mobile machine. *Przeglad Elektrotechniczny* 95(4): 104–109.

Motorna, O.O. 2015. Choice of quality indicators for a comprehensive assessment of the functioning of the hydro-volume steering system of self-propelled agricultural machines. *Promyslova hidravlika i pnevmatika – Industrial Hydraulics and Pneumatics* 2(48): 71–75.

Ogorodnikov, V.A., Dereven'ko, I.A. & Sivak, R.I. 2018. On the influence of curvature of the trajectories of deformation of a volume of the material by pressing on its plasticity under the conditions of complex loading. *Materials Science* 54(3): 326–332. Doi: 10.1007/s11003-018-0188-x.

Ogorodnikov, V.A., Savchinskij, I.G. & Nakhajchuk, O.V. 2004. Stressed-strained state during forming the internal slot section by mandrel reduction. *Tyazheloe Mashinostroenie* 12: 31–33.

Ogorodnikov, V.A., Zyska, T. & Sundetov, S. 2018. The physical model of motor vehicle destruction under shock loading for analysis of road traffic accident. *Proceedings of SPIE* 108086C. Doi: 10.1117/12.2501621.

Pogorelyi, L.V. 2004. *Use of Agricultural Equipment: Scientific and Methodological Basis for Assessing and Predicting the Reliability of Agricultural Machinery*. Kyiv: Feniks.

Polishchuk, L.K., Kozlov, L.G. & Piontkevych, O.V., et al. 2019. Study of the dynamic stability of the belt conveyor adaptive drive. *Przeglad Elektrotechniczny* 95(4): 98–103.

Zardin, B., Borghi, M. & Zanasi, F.G.N. 2018. Modelling and simulation of a hydrostatic steering system for agricultural tractors. *Energies* 11(230): 20.

Chapter 10

Possibility of improving the dynamic characteristics of an adaptive mechatronic hydraulic drive

Leonid Kozlov, Yurii Buriennikov, Volodymyr Pyliavets, Vadym Kovalchuk, Leonid Polonskyi, Andrzej Smolarz, Paweł Droździel, Yedilkhan Amirgaliyev, Ainur Kozbakova, and Kanat Mussabekov

CONTENTS

10.1 Introduction 113
10.2 Analysis of the results of investigations of mechatronic hydraulic drives 113
10.3 Object and method of research 114
10.4 Research results of mechatronic hydraulic drive 120
10.5 Conclusions 123
References 124

10.1 INTRODUCTION

In recent times, mobile and technological machines for various applications contain a wide range of mechatronic hydraulic drives based on adjustable pumps, proportional hydraulic equipment, and controllers (Finzel & Helduser 2008, Kozlov 2011, Stamm von Baumgarten et al. 2008). In such hydraulic systems, simultaneous operation of several consumers from one pump is provided with the possibility of proportional control over the movement parameters of each of them. The supply of an adjustable pump is adapted with the total consumption by a hydraulic motor. It provides minimization of power losses in the work of mechatronic drives and high hydraulic efficiency. Application of controllers provides parameter adaptation of hydraulic motor motion to the change of external conditions which considerably increases the operational efficiency of mechatronic hydraulic drives.

10.2 ANALYSIS OF THE RESULTS OF INVESTIGATIONS OF MECHATRONIC HYDRAULIC DRIVES

Such mechatronic hydraulic drives are multi-coordinate, have one pump and several hydro-motors, the work of which must be coordinated by additional cross couplings, which are implemented by controllers and additional hydraulic connections. In distinguished works by several authors (Dyakonitsa & Sugachevsky 2014, German-Galkin

2001, Shumigay et al. 2013), it is noted that the work of multi-coordinate systems with cross couplings deteriorates significantly due to the interaction of various branches of the system. This is manifested in the decrease of stability margins and deterioration of static, dynamic, and energy characteristics of the systems (Kozlov et al. 2019, Mashkov et al. 2014, Ogorodnikov et al. 2018, Titov et al. 2017, Tymchyk et al. 2018).

In a number of papers devoted to the study of automatic regulation systems with transient couplings (Kostarev & Sereda 2018, Kwasco et al. 2014, Pavlov 2018), measures are proposed to ensure sustainable operation and to improve characteristics through the development of special measures. This improvement of work processes is provided through the development of regulators, including digital ones, and the development of methods for their adjustment (Nizhnik & Neluba 2015, Povarchuk 2017, Repnikova et al. 2010).

It is of scientific and practical interest to solve the problem of developing and adjusting a digital adaptive regulator for a mechatronic hydraulic drive of a mobile machine. Such a task should be solved taking into account the peculiarities of working processes in hydraulic drives operating in wide ranges of changes in speed modes of hydraulic motors and loads on them. In addition, the solution of such a task should take into account the simultaneous operation of two hydraulic motors in coordinated speed modes with several simultaneously operating regulators available, with one of them being digital (Dragobetskii et al. 2015, Ogorodnikov et al. 2004, 2018).

10.3 OBJECT AND METHOD OF RESEARCH

Figure 10.1 shows a diagram of mechatronic hydraulic drive developed by the authors. The hydraulic drive comprises an adjustable pump 1 with a regulator 2 and throttles 3 and 4. Two branches of the hydraulic drive are connected in parallel to the pump 1. The first branch feeds the hydraulic actuator 9. The working fluid enters the hydraulic actuator 9 through the adjustable throttle 5 and the hydraulic distributor 7. The second branch feeds the hydraulic cylinder 10. The working fluid enters the hydraulic cylinder 10 through the valve 14, the adjustable throttle 6, and the hydraulic distributor 8. The hydraulic drive also includes a controller 11 which receives signals from sensors 12 and 13 installed at the inlet to the hydraulic actuator 9 and at the inlet to the hydraulic cylinder 10, respectively. Based on signals from sensors 12 and 13, signals are generated that enter the electromagnets of adjustable throttles 5 and 6 through amplifiers 15 and 16. The logic valve 17 determines the largest pressure on the inlet to the hydraulic actuator pc_1 and to the hydraulic cylinder pc_2 and passes it to control the regulator 2 of the pump 1.

The mechatronic hydraulic drive works in the following way. Working fluid moves from pump 1 to hydraulic actuator 9, which provides the main motion of a work tool (for example, a drill head when drilling wells of a shallow depth). The presence of an adjustable throttle 5, a regulator 2, and a logic valve 17 enables the pump 1 to supply a stable value of flow Q_{n1} to the hydraulic actuator 9. The flow Q_{n1}, which determines the rotational speed ω_1 of the hydraulic actuator 9, will be maintained constant, regardless of the magnitude of the load moment m on the shaft of the hydraulic actuator 9. By changing the magnitude of the area f_{x1} of the window of the adjustable throttle 5 by the signal U_1 from the controller, it is possible to implement a program-driven control over the rotational speed of the hydraulic actuator 9, which drives the work tool.

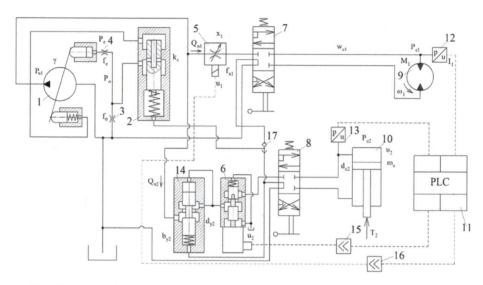

Figure 10.1 Diagram of mechatronic hydraulic drive.

The working fluid from the pump 1 is also fed to the hydraulic cylinder 10, which provides the tool feed. The presence of the valve 14, the adjustable throttle 6, and the hydraulic distributor 8 enables the value of flow Q_{n2} to be maintained and supply to the cylinder 10 to remain constant, regardless of the load size T_2 on the hydraulic cylinder 10 at the pressure ratio $pc_1 > pc_2$. This will ensure a constant feed rate of the work tool $s = v_2$. The ratio of the feed rate s and the rotational speed of the work tool ω_1 will be determined by adjusting the areas of the working windows f_{x1} and f_{x2} of the adjustable throttles 5 and 6. This ratio may change in the process of operation of the hydraulic drive when changing the external conditions of the hydraulic drive operation (changing the resistance during drilling or changing the depth of the work tool immersion). The change in the ratio of the feed rate s and the rotational speed of the work tool ω_1 is provided by the controller 11 on the basis of an algorithm determined by the developer and the pressure signals pc_1 and pc_2, which arrive at the controller during the process of performing a working operation. In this way, the controller 11 will act as an adaptive regulator, which brings the operation of the mechatronic hydraulic drive into accordance with the change in the external conditions of its operation.

The mechatronic hydraulic drive ensures the movement of an actuation mechanism. Dynamic processes will be determined by the interaction of the hydraulic drive, the actuation mechanism, and external stimuli. In Figure 10.2, the structural diagram of the adaptive mechatronic hydraulic drive is presented.

One of the branches of the hydraulic drive provides the basic movement of the drill tool with the help of a hydraulic actuator, and the transmission function of the basic movement mechanism is described by the function:

$$\frac{1}{m_c s + b_c}.$$

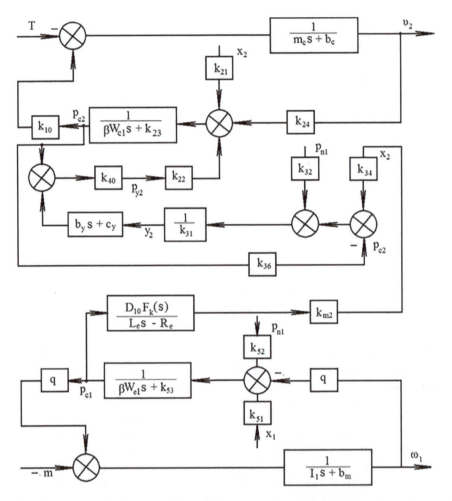

Figure 10.2 Structural diagram of adaptive mechatronic hydraulic drive.

The other branch of the hydraulic drive provides the feed of a drill head with the help of the hydraulic cylinder, and the transmission function of the mechanism is described by the transfer function:

$$\frac{1}{I_1 s + b_m}.$$

The branch of the hydraulic drive, which drives the hydraulic actuator, acts as a back coupling and forms the driving force $pc_2 \cdot k_{10}$ on the hydraulic cylinder. In the stable mode of operation, the driving force $pc_2 \cdot k_{10}$ is equal to the force applied to the work tool brought to the stock of the hydraulic cylinder. At the input to the branches of the hydraulic drive, there is a pressure p_{n1}, and this signal is a common input for both circuits of the structural scheme describing the operation of the mechanisms of the basic movement and feed. The signal pc_1 is formed by a pump with a regulator, and the pattern of its change is determined by their dynamic properties and excitations acting on the adjustable pump. Between the branches of the adaptive mechatronic hydraulic drive, there is cross coupling, which is implemented by the controller. The transmission function of this back coupling is determined as:

$$\frac{k_{m2} \cdot D_{10} \cdot F_k(s)}{L_e \cdot s + R_e}$$

and depends on the dynamic properties of the pressure sensor, the controller, the amplifier, and the electromagnet of the adjustable throttle. The cross coupling forms the x_2 signal that enters the adjustable throttle of the hydraulic drive branch, which makes the hydraulic cylinder move and feeds the work tool. The use of a coupling of this type allows for the adjusting of the feed rate depending on the pressure at the input to the hydraulic actuator.

A nonlinear mathematical model is developed for the proposed mechatronic hydraulic drive, being compiled taking into account such assumptions and simplifications. The concentrated parameters of the hydraulic drive elements are considered, the temperature of the working fluid during the transition process is taken to be constant, the wave processes are not taken into account, the flow factors through the throttling and spool elements are constant, operating modes are without cavitation, the volume of hydraulic lines during the transition process is not changed, the pressure losses in hydraulic lines are not taken into account, the forces of dry friction do not depend on the velocity of the elements, the amplifier blocks were modeled by proportional branches, and the signal at the output of the controller is considered to be analogue.

The mathematical model of the mechatronic hydraulic drive includes the equations of the moments acting on the pump faceplate 1 (equation 10.1) acting on the shaft of the hydraulic actuator 9 (equation 10.6); the equation of continuity of flow between pump 1, regulator 2, adjustable throttle 5, and valve 14 (equation 10.3); between the regulator 2 and the throttle 3 (equation 10.4); between the throttle 4 and the servo cylinder of pump 1 (equation 10.2); between the adjustable throttle 5 and the hydraulic actuator 9 (equation 10.5); between the valve 14 and the adjustable throttle 6 (equation 10.7); between the adjustable throttle 6 and hydraulic cylinder 10 (equation 10.8); the equation for forces acting on the hydraulic cylinder 10 (equation 10.9); for the forces acting on the valve of the regulator 2 (equation 10.10); for the forces acting on the valve 14 (equation 10.11); the voltage drop equation on the electromagnet of the adjustable throttle 6 (equation 10.12); and the dependence of the adjustable throttle displacement on the magnitude of the voltage on the electromagnet (equation 10.13). The dependence of the resistance moment on the faceplate of pump 1 on the feed rate and pressure is expressed by equation (10.14). The dependence of the reduced coefficient of compliance on the parameters of the pipeline is given by equation (10.15), and the dependence of the elasticity modulus of the working fluid in the presence and volume of insoluble air by equation (10.16).

$$I\frac{d^2\gamma}{dt^2} = p_{n1}f_5 l - p_e f_4 l - b_\gamma \frac{d\gamma}{dt} + M_c; \tag{10.1}$$

$$\mu f_0 \sqrt{\frac{2|p_0 - p_e|}{\rho}} \sin(p_0 - p_e) = \beta_p W_e \frac{dp_e}{dt} - f_4 \frac{d\gamma}{dt} l \cdot \cos\gamma \tag{10.2}$$

$$F_7 d_8 k_1 n_n tg\gamma = \mu\pi d_{x1} x_1 \sqrt{\frac{2|p_{n1}-p_{c1}|}{\rho}} \sin(p_{n1}-p_{c1}) + \mu\pi d_{y2} y_2 \sin(\alpha) \sqrt{\frac{2|p_{n1}-p_{y2}|}{\rho}} \times$$
$$\times \sin(p_{n1}-p_{y2}) + \beta_n W_{n1} \frac{dp_{n1}}{dt} + \mu k_z z \sqrt{\frac{2|p_{n1}-p_0|}{\rho}} \sin(p_{n1}-p_0) \tag{10.3}$$

$$\mu k_z z \sqrt{\frac{2|p_{n1}-p_0|}{\rho}} \sin(p_{n1}-p_0) = \mu f_e \sqrt{\frac{2|p_0-p_e|}{\rho}} \sin(p_0-p_e) + \mu f_0 \sqrt{\frac{2p_0}{\rho}} + \beta_p W_0 \frac{dp_0}{dt} \tag{10.4}$$

$$\mu\pi d_{x1} x_1 \sqrt{\frac{2|p_{n1}-p_{c1}|}{\rho}} = q\omega_1 + \beta_n W_{c1} \frac{dp_{c1}}{dt} \tag{10.5}$$

$$I_1 \frac{d\omega_1}{dt} = p_{c1} q - m - b_M \omega_1 - M_T \sin g(\omega_1) \tag{10.6}$$

$$\mu\pi d_{y2} y_2 \sin(\alpha) \sqrt{\frac{2|p_{n1}-p_{y2}|}{\rho}} \sin(p_{n1}-p_{y2})$$
$$= \mu\pi d_{x2} x_2 \sqrt{\frac{2|p_{y2}-p_{c2}|}{\rho}} \sin(p_{y2}-p_{c2}) + + \beta_p W_{y2} \frac{dp_{y2}}{dt} \tag{10.7}$$

$$\mu\pi d_{x2} x_2 \sqrt{\frac{2|p_{y2}-p_{c2}|}{\rho}} \sin(p_{y2}-p_{c2}) = \vartheta_2 \frac{\pi d_{c2}^2}{4} + \beta_n W_{c2} \frac{dp_{c2}}{dt} \tag{10.8}$$

$$m_c \frac{d\vartheta_2}{dt} = p_{c2} \frac{\pi d_{c2}^2}{4} - T_2 - b_c \vartheta_2 - T_T \sin g(\vartheta_2) \tag{10.9}$$

$$p_{n1} \frac{\pi d_z^2}{4} = p_{c1} \frac{\pi d_z^2}{4} + c_z(z+H_z) + b_z \frac{dz}{dt} \tag{10.10}$$

$$p_{y2} \frac{\pi d_{y2}^2}{4} = p_{c2} \frac{\pi d_{y2}^2}{4} + c_y(H_y - y_2) - b_y \frac{dy_2}{dt} \tag{10.11}$$

$$p_{c1} k_4 k_c F_k(i_{p1}) = L_e \frac{di_{m2}}{dt} + i_{m2} R_e \tag{10.12}$$

$$\left(L_e \frac{di_{m2}}{dt} + i_{m2} R_e\right) k_m = x_2 \tag{10.13}$$

$$M_c = m_0 + m_1 Q_{n1} + m_2 p_{n1} + m_3 Q_{n1}^2 + m_4 p_{n1}^2 + m_5 p_{n1} Q_{n1} \tag{10.14}$$

$$\beta_n = \frac{1}{E_p} + \frac{d_{mp}}{\delta_{mp} E_{mp}(p)} \tag{10.15}$$

$$E_p = \frac{1}{\beta_p} = \frac{W_f/W_a + 1}{W_f/W_a + E_{p0}p_0/p^2} \tag{10.16}$$

where $p_{n1}, p_{c1}, p_{c2}, p_{y2}, p_0, p_e$ are the pressures at the outlet of pump 1, inlets of the hydraulic actuator 9 and the hydraulic cylinder 10, outlet of valve 14, in the control system of pump 1, in the servo cylinder of pump 1; z, x_2, y_2 are the position coordinates of regulator 2, adjustable throttle 6, and the valve 14; ω_1, ϑ_2 are the rotational speed of the shaft of hydraulic motor 9 and the speed of the piston of the hydraulic cylinder 10; γ is the rotation angle of the faceplate of pump 1; f_0, f_e, f_{x2}, F_7 are the area of throttles 3 and 4, the area of the spools of the adjustable throttle 6, and the area of the pump pistons; $d_{c2}, d_{x2}, d_{y2}, d_z, d_8$ are the diameters of the hydraulic cylinder 10, the spool of the adjustable throttle 6, valve 14, regulator 2, and the diameter of the contact circle of the pump pistons with the pump faceplate; $i_{m1}, i_{m2}, i_{p1}, i_{p2}$ are the magnitudes of currents in the windings of the electromagnets and the output of the pressure sensors; k_m, k_n, k_c, k_1 are the proportionality coefficients of the forces of the electromagnets of the adjustable throttles, amplifiers, and pressure sensors, the number of pistons in the pump 1; L_e, R_e are the inductance and active resistance of the electromagnet windings; c_z, c_y are the stiffness of springs of the regulator 2 and the valve 14; T_2, m are the forces of reduced loads on the rod of the hydraulic cylinder 10 and the shaft of the hydraulic actuator 9; μ is the flow rate through throttling and spool elements; ρ is the working fluid density; l is the action arm of the servo cylinders of the pump 1; I, I_1 are the moments of inertia of the pump faceplate and moving parts of the shaft of hydraulic actuator 9; m_c is the mass of moving elements brought to the rod of the hydraulic cylinder 10; $W_{n1}, W_0, W_{c1}, W_{y2}, W_{c2}, W_e$ are the volumes of hydraulic lines at the outlet of pump 1, between the regulator 2 and the throttle 5, at the inlet of the hydraulic actuator 9, at the outlet of valve 14, at the inlet to the hydraulic cylinder 10, between the throttle and the servo cylinder of the pump 1; n_n is the shaft rotary speed of the pump 1; $F_k(i_{p1})$ is the transition function of the controller for the signal supplied to the amplifier 15; H_z, H_y are the previous compression of the springs of the regulator 2 and the valve 14; $m_0, m_1, m_2, m_3, m_4, m_5$ are the coefficients of dependence of the resistance moment on the faceplate of pump 1 on the supply and pressure values; M_c is the moment of resistance on the faceplate of pump 1; β_p is the reduced coefficient of compliance of the gas–liquid mixture; β_n is the reduced coefficient of the rubber-metal pipelines and gas–liquid mixture; q is the working volume of the hydraulic actuator 9; b_c, b_z, b_y are the coefficients of viscous friction in the hydraulic cylinder 10, damping of the spool of the regulator 2, and valve 14; $E_{p0}, E_p, E_{mp}(p)$ are the elasticity modulus of the working liquid, reduced elasticity modules of the gas–liquid mixture, and rubber-metal pipelines; δ is the pipeline wall thickness; W_f is the volume of liquid in the gas–liquid mixture at pressure value p; W_a is the volume of gas in the gas–liquid mixture at atmospheric pressure.

Equations of the mathematical model were solved by utilizing the Rosenbrock numerical method in the MATLAB-Simulink environment with an absolute accuracy of $\varepsilon_a = 10^{-6}$ and a relative accuracy of $\varepsilon_a = 10^{-3}$. The transient process was simulated in a mechatronic hydraulic drive with a step change in the load moment on the shaft of the hydraulic actuator 9 from 200 to 800 Nm.

Calculation of the transient processes in the mechatronic hydraulic drive was carried out at such values of variables: $p_{n1} = 10 \cdot 10^6 \, \text{N/m}^2$; $p_{c1} = 8.2 \cdot 10^6 \, \text{N/m}^2$; $p_{c2} = 4.0 \cdot 10^6 \, \text{N/m}^2$; $p_{y2} = 4.4 \cdot 10^6 \, \text{N/m}^2$; $p_0 = 5.2 \cdot 10^6 \, \text{N/m}^2$; $p_e = 5.2 \cdot 10^6 \, \text{N/m}^2$; $\gamma = 0.07 \, \text{radians}$; $z = 1.6 \cdot 10^{-3} \, \text{m}$; $\omega_1 = 7.0 \, \text{s}^{-1}$; $y_2 = 5.3 \cdot 10^{-5} \, \text{m}$; $v_2 = 0.05 \, \text{m/s}$; and $U_2 = 1.4 \, \text{V}$.

The variable moment of resistance to the hydraulic actuator shaft in the transient process was $\Delta m = 600 \, \text{Nm}$.

10.4 RESEARCH RESULTS OF MECHATRONIC HYDRAULIC DRIVE

The mechatronic hydraulic drive (see Figure 10.1) can be used in two modes. The first mode is characterized by the independent work of the two branches of the hydraulic drive. In this mode, the speed modes of the hydraulic actuator 9 and the hydraulic cylinder 10 are independent and determined by the setting of the controller 11, which determines the opening values of the working windows of the adjustable throttles 5 and 6; the pressure signals from sensors 12 and 13 are not used in this case. The pressure value of p_{n1} at the pump outlet 1 will be determined by a greater actuating load from the hydraulic actuator 9 and the hydraulic cylinder 10. The supply Q_{n1} from the pump 1 will be equal to the sum of flows consumed by the hydraulic actuator 9, the hydraulic cylinder 10, and the regulator 2. The process of parameter adjustment of the flow coming to the more loaded hydraulic actuator 9 is provided by an adjustable pump with a regulator 2 and a valve 14. The hydraulic drive branch actuating the hydraulic actuator 9 interacts in this mode with a branch actuating the hydraulic cylinder 10 only through the adjustable pump 1. The working processes occurring in the branch actuating the hydraulic actuator 9 affect the nature of the working processes in the branch actuating the hydraulic cylinder 10. In Figure 10.3, the transition process in a mechatronic hydraulic drive is represented during a stepwise change in the torque load moment on the shaft of the hydraulic actuator 9 with a constant value of the load T_2 at the rod of the hydraulic cylinder 10.

Changing the torque m at 600 Nm leads to a sharp change in the pressure value of p_{c1} and the shaft rotation speed ω_1 of the hydraulic actuator 9. The change in the values of p_{c1} and ω_1 leads to the fact that in the transitional process, the pressure p_{c2} at the outlet to the hydraulic cylinder 10 also varies as well as the velocity v_2, with which the piston of the hydraulic cylinder 10 is moving. Upon completion of the transient process, the values of p_{c2} and v_2 are set as they were before the moment M_1 was changed on the shaft of hydraulic actuator 9. That is, the perturbation of the values describing the motion of the hydraulic cylinder 10 occurs only in the transient process and is due to the interaction of the branches of the mechatronic hydraulic drive through a common power source, which is the adjustable pump 1.

In the second mode of operation of the hydraulic drive, a change in the velocity of the hydraulic cylinder 10 is provided depending on the pressure value of p_{c1} at the input of the hydraulic actuator 9. With an increase in the pressure of p_{c1}, which is determined by the torque value of m on the hydraulic actuator, the velocity v_2 of the piston movement of the hydraulic cylinder 10 decreases proportionally. When the pressure value of p_{c1} decreases, the velocity v_2 increases. This process is provided by the controller 11, which implements a cross coupling in the hydraulic drive with a transmission function:

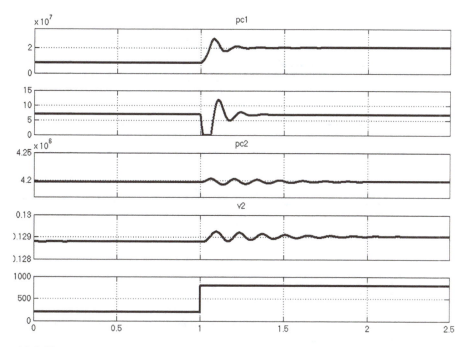

Figure 10.3 Transition process without a cross coupling between the branches of mechatronic hydraulic drive.

$$\frac{D_{10}F_k(s)}{L_e s + R_e}$$

With a step change in the torque value of m from 200 to 800 N/m (Figure 10.4), the transition process in the branch that feeds the hydraulic cylinder 10 passes more dynamically with greater overshoot according to the pressure value of p_{c2}. The regulation time t_p also increases in comparison with the case where there is no cross coupling between the branches of the hydraulic drive (Figure 10.3). The pressure overshoot value of p_{c2} increases from 12% to 90% and the adjusting time of t_p from 0.5 to 1.05 s.

The negative effect of cross coupling on the dynamic characteristics of a mechatronic hydraulic drive can be reduced by introducing an adjustment of the cross coupling signal. This control is provided by a controller that changes the cross coupling signal according to the required law. The cross coupling is proposed to be formed as $U_2 = k_4 k_c \left(p_{c1} - k_p \frac{dp_{c1}}{dt} \right)$. The controller filed a cross coupling signal with a delay ΔT_1 to the adjustable throttle 6. The delay value in the research process varied in the range of $\Delta T_1 = (0.005...0.06)$ s. The value of the coefficient k_p of the correctional component of the cross coupling signal varied in the range of $k_p = (40...320) \cdot 10^{-4}$ seconds.

In Figure 10.5, the dependence of the adjusting time on k_p and ΔT_1 is presented in the branch that feeds the hydraulic cylinder if there is an adjustment of cross coupling signal. The value of adjustment coefficient k_p of the correction component has

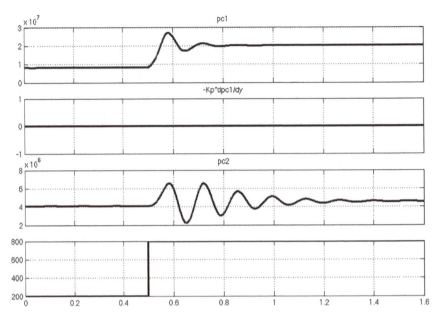

Figure 10.4 Transition process in the interaction of mechatronic hydraulic drive branches through adjustable pump and controller.

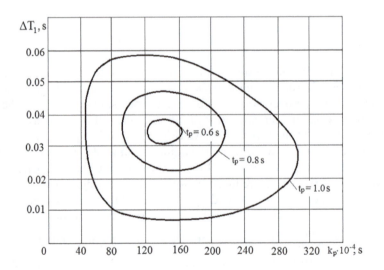

Figure 10.5 Influence of ΔT_1 and k_p on regulation time t_p.

a mixed effect on the regulation time. At values $k_p < 120 \cdot 10^{-4}$ seconds and for values $k_p > 160 \cdot 10^{-4}$ seconds, the value of the regulation time increases. The regulation time t_p in the mechatronic hydraulic drive is minimized at values $k_p = (120...160) \cdot 10^{-4}$ seconds. At the value of the delay ΔT_1 of the cross coupling signal $\Delta T_1 > 0.03$ seconds and

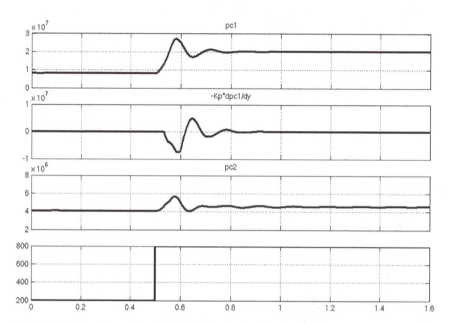

Figure 10.6 Transient process in a mechatronic hydraulic drive with a cross coupling, which is implemented by an adaptive regulator with the use of signal adjustment.

$\Delta T_1 < 0.04$ seconds, the regulation time value of t_p is minimized. Coupling from the range of $\Delta T_1 = (0.03...0.04)$ seconds and $k_p = (120...160) \cdot 10^{-4}$ seconds will provide the time of regulation t_p in the mechatronic hydraulic drive that is close to 0.6 seconds.

The research also found that when changing the value of the delay ΔT_1 of the cross coupling signal by more than 0.04 seconds and less than 0.02 seconds, the overshoot in the mechatronic hydraulic drive increases. With values $\Delta T_1 = (0.025...0.04)$ seconds, an overshoot of $\sigma = 20\%$ can be provided. For this purpose, the value k_p from the range $k_p = (200...280) \cdot 10^{-4}$ seconds should also be chosen. The change in the coefficient $k_p < 200 \cdot 10^{-4}$ seconds and $k_p > 280 \cdot 10^{-4}$ seconds leads to an increase in the overshoot value of σ.

In Figure 10.6, the transition process in a mechatronic hydraulic drive is presented, and it is calculated according to the mathematical model when a cross coupling signal is adjusted. At values of the coefficient of the correction component $k_p = 250 \cdot 10^{-4}$ seconds and of the delay of the correction component $\Delta T_1 = 0.02$ seconds, the time of adjustment is $t_p = 0.7$ seconds, and the value of overshoot does not exceed 40%.

To simultaneously reduce the regulation time t_p and the overshoot value σ, when the coefficient k_p and the delay value ΔT_1 change, an optimization problem should be solved.

10.5 CONCLUSIONS

1. A cross coupling occurs in a mechatronic hydraulic drive as a change in the value of movement speed of the hydraulic cylinder rod when the load on the hydraulic actuator changes, and it degrades the dynamic characteristics.

2. The negative effect of cross coupling on the dynamic characteristics can be reduced by adjusting the cross coupling signal.
3. Adjustment of the cross coupling signal is provided by introducing a correction component $k_p - \dfrac{dp_{c1}}{dt}$ and the delay value ΔT_1 of the correction component.
4. To reduce the regulation time in a mechatronic hydraulic drive, it is recommended to choose a value of k_p in the range of $k_p = (120...160) \cdot 10^{-4}$ seconds and a delay value of ΔT_1 in the range of $\Delta T_1 = (0.03...0.04)$ seconds. To reduce the overshoot value, it is necessary to use $k_p = (200...280) \cdot 10^{-4}$ seconds, and the delay value of $\Delta T_1 = (0.025...0.04)$ seconds.

REFERENCES

Dragobetskii, V., Shapoval, A., Mos'pan, D., Trotsko, O. & Lotous, V. 2015. Excavator bucket teeth strengthening using a plastic explosive deformation. *Metallurgical and Mining Industry* 4: 363–368.

Dyakonitsa, S.A. & Sugachevsky, I.R. 2014. The use of compensating regulation for the multiply connected control of a multiparameter system. *System Methods Technology* 1: 86–90.

Finzel, R. & Helduser, S. 2008. New electro-hydraulic control Systems for Mobile Machinery. *Proceedings of the Fluid Power and Motion Control FPMC 2008* 1: 311–321.

German-Galkin, S.G. 2001. *Computer Simulation of Semiconductor Systems in Matlab 6.0.* St. Petersburg: Corona Print.

Kostarev, S.N. & Sereda, T.G. 2018. Microcoordinate control system. *IOP Conference Series: Mamaterial Science fnd Engeneering* 450: 1–5.

Kozlov, L. 2011. Energy-saving mechatronic drive of the manipulator. *Buletinul institutului politehnic Din Iasi Fasc.* 3(LXI): 231–239.

Kozlov, L.G., Polishchuk, L.K., Piontkevych, O.V., Korinenko, M.P., Horbatiuk, R.M., Komada, P., Orazalieva, S. & Ussatova, O. 2019. Experimental research characteristics of counter balance valve for hydraulic drive control system of mobile machine. *Przeglad Elektrotechniczny* 95(4): 104–109.

Kwasco, M.Z., Zhurakovskii, Ya.Yu., Milenkiy, V.V. & Nosov, A.V. 2014. Control of evaporation in the apparatus of immersion. *The Bulletin of the National Technical University of Ukraine "Kyiv Polytechnic Institute" Series "Chemical Engineering, Ecology and Resource Savin"* 2: 91–97.

Mashkov, V., Smolarz, A., Lytvynenko, V. & Gromaszek, K. 2014. The problem of system fault-tolerance. *Informatyka, Automatyka, Pomiary w Gospodarce i Ochronie Środowiska* 4: 41–44.

Nizhnik, O.V. & Neluba, D.M. 2015. Smoothing control of multidimensional systems. *Communication* 2: 41–43.

Ogorodnikov, V.A., Dereven'ko, I.A. & Sivak, R.I. 2018. On the influence of curvature of the trajectories of deformation of a volume of the material by pressing on its plasticity under the conditions of complex loading. *Materials Science* 54(3): 326–332.

Ogorodnikov, V.A., Savchinskij, I.G. & Nakhajchuk, O.V. 2004. Stressed-strained state during forming the internal slot section by mandrel reduction. *Tyazheloe Mashinostroenie* 12: 31–33.

Ogorodnikov, V.A., Zyska, T. & Sundetov, S. 2018. The physical model of motor vehicle destruction under shock loading for analysis of road traffic accident. *Proceedings of SPIE* 108086C: 1–5.

Pavlov, A.I. 2018. Automatic control of parameters of a non-stationary object with crosslink. *Automation, Technological and Business Processes* 10(1): 50–54.

Povarchuk, D.D. 2017. Synthesis of the automatic control system of the process of two-stage separation of oil. *Naftogazova energetika* 2: 77–82.

Repnikova, N.B., Pisarenko, A.V. & Furshal, Yu.O. 2010. Synthesis of multidimensional multi-stage control system with unknown vector of states. *Bulletin of the National University "Lviv Polytechnic".– Automation, Measurement and Control* 665: 146–150.

Shumigay, D.A., Ladajuk, A.P. & Smythyukha, Ya.V. 2013. Fast algorithm for the adapters of the PI regulator. *Automation of Technological Business Processes* 8: 97–104.

Stamm von Baumgarten, T., Grösbrink, B., Lang, T. & Harms, H. 2008. A novel system layout for extended functionality of mobile machines. *Proceedings of the Fluid Power and Motion Control FPMC 2008* 1: 13–25.

Titov, A.V., Mykhalevych, V.M., Popiel, P. & Mussabekov, K. 2017. Statement and solution of new problems of deformability theory. *Proceedings of SPIE* 108085E: 1611–1617.

Tymchyk, S.V., Skytsiouk, V.I., Klotchko, T.R., Ławicki, T. & Demsova, N. 2018. Distortion of geometric elements in the transition from the imaginary to the real coordinate system of technological equipment. *Proceedings of SPIE* 108085C: 1595–1604.

Chapter 11

Application of feedback elements in proportional electrohydraulic directional control valve with independent flows control

Dmytro O. Lozinskyi, Oleksandr V. Petrov, Natalia S. Semichasnova, Konrad Gromaszek, Maksat Kalimoldayev, and Gauhar Borankulova

CONTENTS

11.1 Introduction 127
11.2 Literature review and proposed solutions 128
11.3 Materials and methods of studying processes in directional control valve
      with independent flows control 130
11.4 Results and discussion 133
11.5 Conclusion 134
References 135

11.1 INTRODUCTION

Mobile machines are an integral and important part of different types of production. In particular, such machines are used to perform loading and unloading operations during construction, repair, agricultural, and other types of jobs.

The work of such machines in many cases is associated with large loads and requires accurate movement of the executive device and its fixation in some position over a long period of time (Kailei et al. 2016). Often, the working modes of mobile machines are accompanied by significant changes in the magnitude and direction of the load, which is considered to be an adverse condition, as it can cause unmanaged movement of the executive devices (Lübbert et al. 2016, Titov et al. 2017, Tymchyk et al. 2018).

Using hydraulic drives with electrohydraulic directional control elements in mobile machines to perform work operations provides significant power characteristics and small dimensions of the drive and also retains the mobility of the machine. Directional controlling equipment and regulating hydraulic equipment are integral part of the hydraulic drives of mobile machines and perform one of the main tasks – flows control and therefore the movement control of executive devices in all necessary operation modes and also sets the energy expenditures during work operations (Eriksson & Palmberg 2011, Jong-Chan et al. 2016, Ruqi et al. 2018).

DOI: 10.1201/9781003224136-11

11.2 LITERATURE REVIEW AND PROPOSED SOLUTIONS

Generally, the hydro distribution equipment has one spool element for controlling fluid flows to the hydraulic motor, which under normal circumstances is an acceptable solution both in terms of ensuring sufficient controllability and from an economic point of view. However, for hydraulic drives of machines performing loading and unloading operations, this approach is not optimal, which is confirmed by the results of a lot of research (Eriksson & Palmberg 2011, Jong-Chan et al. 2016, Ruqi et al. 2018). Kailei et al. (2016) and Ogorodnikov et al. (2018) suggested a system known as the independent metering valve control be used. Such systems have several control elements and have the ability to independently control flows at the inlet and outlet of the hydraulic motor. The separation of flows control makes it possible to significantly increase the controllability of executive devices and to reduce energy expenditures. The authors of the work have performed comparison research of the power expenditure characteristics of the hydraulic distributors with one control element, manipulator conventional load-sensing system, and with several elements – a multiple-pressure compensatory independent metering system (PCIMS), which confirmed the better efficiency of the PCIMS.

Using several control elements allows the throttle and volume control of the hydraulic motor to be combined, which increases its controllability and reduces energy expenditures. In addition, the hydraulic drives with several control elements have considerably wider variations in both the layout diagrams and functional features, which allows a qualitative "setup" of the system for specific types of work to be performed. This pays particular attention to research on the development of specialized algorithms for controlling the operation of the hydraulic distributor and hydraulic drive in general (Bilynsky et al. 2018, Ruqi et al. 2018, Jong-Chan et al. 2016, Wydra et al. 2017).

Ruqi et al. (2018) have compared the energy consumption of the LS system. In the article, using a three-stage control system (flow / pressure hybrid control) with a hydraulic drive of an excavator is proposed to ensure optimal performance. In that work, comparative studies of energy expenditures of the proposed solution and the usual LS system are carried out. As a result of the research, it was found that the use of independent flows control and the developed control system in aggregate provides a reduction of overall energy losses up to 28% compared to the LS system, which is a significant indicator, since such mobile machines operate with significant energy expenditures.

Energy savings in hydraulic drive with independent flows control can also be achieved using energy recovery (Jong-Chan et al. 2016) and battery systems (Ogorodnikov et al. 2004, Wydra et al. 2017).

In Jong-Chan et al. (2016), studies of work of a hydraulic drive with several control elements of the valve type were carried out. A specialized combined mode of electronic control by using distribution elements and a hydraulic pump was proposed, which ensured high efficiency of the hydraulic drive for all types of loads. In combination with the regenerative mode of work, the system of electronic controls and the control system of the hydraulic drive were managed to achieve a reduction in energy losses of up to 10%.

In the works by Dragobetskii et al. (2015) and Wydra et al. (2017), it is proposed to use both regenerative and recuperative operating modes, in accordance with the magnitude and direction of loading and other parameters of work. A hydraulic battery was proposed as a power storage device, which can also be used to reduce pressure drops in the hydraulic drive, which is an additional means of reducing the energy losses in the hydraulic drive. Accumulated battery power can be used to perform low-power working operations or to reduce the load when performing basic work.

Jianpeng et al. (2018) carried out research on quality problems of the process of controlling the speed and precision of the positioning of the executive devices in addition to energy problems. In particular, the dynamic characteristics and the performance of the control process were investigated.

By using a co-simulation model, the authors managed to obtain imitative research results of sufficiently high accuracy. On the basis of the results of simulation studies, it was possible to achieve high smoothness of the motion of the executive devices and a relatively high accuracy of positioning (with an error of up to 3 mm).

The main disadvantages of the solutions proposed in the analyzed papers are the increased complexity of the systems and, consequently, increased cost, which is due to the increased number of control elements in the design, as well as the complexity of the control system. In addition, mainly high-cost regulated hydraulic pumps are used in the proposed solutions (Lübbert et al. 2016).

In this chapter, a scheme of a proportional electrohydraulic directional control valve with independent flows control (Figure 11.1) is proposed, which has a smaller number of control elements but retains the modular structure and positive features of systems with independent flows control at the input and output of the hydraulic motor (Kozlov et al. 2009).

The scheme includes the main line 1, the line of the first stage 2, the working section 3, the overflow section 4, and the exit lines 5 and 6.

In the working section 3, there are spools of the second stage 7 and 8 with valves of the first stage 9 and 10, controlled check valves 11 and 12 with a servo valve 13 and logic valves 14 and 15.

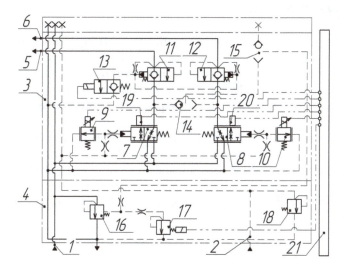

Figure 11.1 Hydraulic scheme of proportional electrohydraulic directional control valve with independent flows control.

The developed proportional electrohydraulic directional control valve can control the movement of the executive device in the forward and reverse directions and provides a float mode (in which the chambers of the hydraulic motor are connected to the drain) and a "neutral" operation mode, in which both chambers of the hydraulic motor are locked due to the controlled check valves 11 and 12.

In addition, the use of the two spools 7 and 8 allows for independent, proportional flows control both at the input and output of the hydraulic motor, as well as carrying out the work operations due to the associated technological load (in this mode, the hydraulic motor is controlled by one of the spools and the other passes the liquid from the pump under the low pressure in the tank, which reduces energy losses) (Kozlov et al. 2009).

In the overflow section, there is a safety overflow valve 16 with a control valve 17 and a safety valve 18 that regulates the flow of the hydraulic motor in accordance with the load on the executive device, transferring the excess fluid in the tank under low pressure, while in the distributors operating in constant flow mode, the transfer of excess fluid is carried out under pressure which was set by the safety valve (16–25 MPa).

Better improvement of performance can be achieved by the use of controllers and an appropriate algorithm for the work of the distributor (Qi et al. 2019).

To increase the proportionality of movements and improve controllability, applying feedback elements (tracking system) is proposed in the form of sensors of position 19 and 20 and a control system, controller 21. The system works as follows: to provide the desired law of motion $y_0(t)$ of the spool of the second stage to the first stage, a specific control signal $x(t)$ is provided while $y_0(t) = k_1 \cdot x(t)$, where k_1 is the conversion coefficient. The error of the real motion law $y(t)$ as to the desired $\Delta = y(t) - y_0(t)$ is fixed by the sensor and with a certain gain k, goes to the tracking system, which adjusts the value of the input signal of $x = x(t) - k\Delta$.

11.3 MATERIALS AND METHODS OF STUDYING PROCESSES IN DIRECTIONAL CONTROL VALVE WITH INDEPENDENT FLOWS CONTROL

For research on the work of the proportional electrohydraulic directional control valve, a design model was created (Figure 11.2). The design model simulates the work of the proportional electrohydraulic directional control valve with the oncoming load (the flow at the output of the hydraulic motor is not regulated).

On the basis of the design model, equations of the mathematical model were developed.

The mathematical model includes flows continuity equations (11.1–11.7) as well as the equations of forces acting on the spools of the second stage 7 and 8, controlled check valves 9 and 10, overflow valve 14, and valve 16 (11.8–11.12) (Kozlov et al. 2009).

$$\mu \cdot f_1 \cdot \sqrt{\frac{2 \cdot |p_{N1} - p_2|}{\rho}} \cdot \sin(p_{N1} - p_2) = \mu \cdot f_2 \cdot \sqrt{\frac{2 \cdot |p_2 - p_1|}{\rho}} \cdot \sin(p_2 - p_1) + \beta \cdot W_A \cdot \frac{dp_2}{dt} +$$
$$+ \mu \cdot \left[\frac{\pi}{2} \cdot ((x_0 - x) \cdot \sin\alpha + 2 \cdot d_{X1}) \cdot (x_0 - x) \cdot \sin\frac{\alpha 1}{2} \right] \cdot \sqrt{\frac{2 \cdot |p_2 - p_3|}{\rho}} \cdot \sin(p_2 - p_3)$$

(11.1)

Application of feedback elements 131

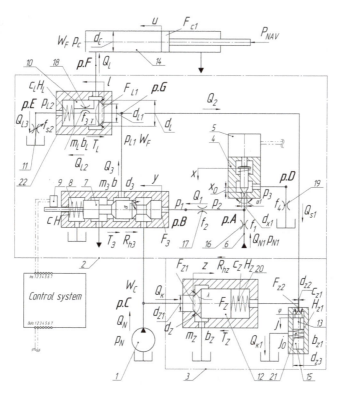

Figure 11.2 Design model of the proportional electrohydraulic directional control valve with independent flows control.

$$\mu \cdot \left[\frac{\pi}{2} \cdot ((x_0 - x) \cdot \sin\alpha + 2 \cdot d_{X1}) \cdot (x_0 - x) \cdot \sin\frac{\alpha 1}{2}\right] \cdot \sqrt{\frac{2 \cdot |p_2 - p_3|}{\rho}} \cdot \sin(p_2 - p_3) =$$
$$= \mu \cdot f_4 \cdot \sqrt{\frac{2 \cdot p_3}{\rho}} + \beta \cdot W_D \cdot \frac{dp_3}{dt}; \quad (11.2)$$

$$\mu \cdot f_2 \cdot \sqrt{\frac{2 \cdot |p_2 - p_1|}{\rho}} \cdot \sin(p_2 - p_1) = F_3 \cdot \frac{dy}{dt} + \beta \cdot W_B \cdot \frac{dp_1}{dt} \quad (11.3)$$

$$\mu \cdot \pi \cdot d_3 \cdot y \cdot \sin\alpha \cdot \sqrt{\frac{2 \cdot |p_N - p_{L1}|}{\rho}} \cdot \sin(p_N - p_{L1}) = \mu \cdot \left[\frac{\pi}{2} \cdot (j \cdot \sin\phi + 2 \cdot d_{Z2}) \cdot j \cdot \sin\frac{\phi}{2}\right] \cdot$$
$$\sqrt{\frac{2 \cdot p_{L1}}{\rho}} + \mu \cdot \left[\frac{\pi}{2} \cdot (l \cdot \sin\gamma + 2 \cdot d_{L1}) \cdot l \cdot \sin\frac{\gamma}{2}\right] \cdot \sqrt{\frac{2 \cdot |p_{L1} - p_C|}{\rho}} \cdot sign(p_{L1} - p_C)$$
$$+ \beta \cdot W_G \cdot \frac{dp_{L1}}{dt} + F_Z \cdot \frac{dz}{dt} \quad (11.4)$$

$$Q_N = \mu \cdot \left[\frac{\pi}{2} \cdot (z \cdot \sin\lambda + 2 \cdot d_{Z1}) \cdot z \cdot \sin\frac{\lambda}{2}\right] \cdot \sqrt{\frac{2 \cdot p_N}{\rho}} + \mu \cdot \pi \cdot d_3 \cdot y \cdot \sin\alpha \cdot \sqrt{\frac{2 \cdot |p_N - p_{L1}|}{\rho}} \times$$
$$\sin(p_N - p_{L1}) + \beta \cdot W_C \cdot \frac{dp_N}{dt} \tag{11.5}$$

$$\mu \cdot f_3 \cdot \sqrt{\frac{2 \cdot |p_c - p_{L2}|}{\rho}} \cdot \sin(p_c - p_{L2}) = \mu \cdot \left[\frac{\pi}{2} \cdot (s \cdot \sin\delta + 2 \cdot d_{L2}) \cdot s \cdot \sin\frac{\delta}{2}\right] \cdot \sqrt{\frac{2 \cdot p_{L2}}{\rho}} \tag{11.6}$$

$$\mu \cdot \left[\frac{\pi}{2} \cdot (l \cdot \sin\gamma + 2 \cdot d_{L1}) \cdot l \cdot \sin\frac{\gamma}{2}\right] \cdot \sqrt{\frac{2 \cdot |p_{L1} - p_C|}{\rho}} \cdot \sin(p_{L1} - p_C) = \frac{du}{dt} \cdot$$
$$F_C + \beta \cdot W_F \cdot \frac{dp_C}{dt} + \mu \cdot f_3 \cdot \sqrt{\frac{2 \cdot |p_c - p_{L2}|}{\rho}} \cdot \sin(p_c - p_{L2}) \tag{11.7}$$

$$m_3 \frac{dV_y}{dt} = p_1 \cdot F_3 - c \cdot (H + y) - b \frac{dy}{dt} - T \cdot \sin\frac{dy}{dt} - R_{h3} \tag{11.8}$$

$$m_Z \frac{dV_Z}{dt} = p_N \cdot F_{Z1} - p_{L1} \cdot F_Z - c_Z \cdot (H_Z + z) - b_Z \frac{dz}{dt} - T_Z \cdot \sin\frac{dz}{dt} - R_{hZ} \tag{11.9}$$

$$p_{L1} \cdot F_{Z2} = c_{Z1} \cdot (H_{Z1} + j + j_0) - b_{Z1} \frac{dj}{dt} \tag{11.10}$$

$$m_L \frac{dV_l}{dt} = p_{L1} \cdot F_{L1} - p_L \cdot F_L + p_C(F_L - F_{L1}) - c_L \cdot (H_L + l) - b_L \frac{dl}{dt} - T_L \cdot \sin\frac{dl}{dt} \tag{11.11}$$

$$m_C \frac{dV_u}{dt} = p_C \cdot F_C - P_{NAV} - bc \frac{du}{dt} - T_C \cdot \sin\frac{du}{dt} \tag{11.12}$$

$$x = x_0(t) - k\Delta \tag{11.13}$$

$$\Delta = y(t) - y_0(t) \tag{11.14}$$

The mathematical model of the proportional electrohydraulic directional control valve with independent flows control is composed of the following basic assumptions and simplifications: some parameters were considered; the value of the flow and pressure at the input in the first stage was constant (pulsation and losses in the hydraulic lines were not taken into account); the lengths of the channels of drain and pressure trunks are relatively small, therefore, the wave processes were not taken into account; the coefficients of the flow through the throttle and the spool elements were considered to be permanent; volumes of the hydraulic lines did not change; pressure losses in the hydraulic lines were not taken into account, because they are insignificant compared to the losses at the local supports (Petrov et al. 2019); and the coefficient of fluidity of the working fluid was taken into account as a value dependent on the pressure (Kozlov et al. 2019, Polishchuk et al. 2018).

The mathematical model indicates the following: x_0 is the initial coordinate of the spool of the servo valve 4; x is the displacement of the spool of the servo valve 4; y is the position of the spool 7; z is the displacement of the valve spool 12; l is the displacement of the spool of the controlled checked valve 10; u is the displacement of cylinder rod 14; s is the displacement of the edge of servo valve 11; j is the displacement of the servo valve 13; j_0 is the displacement of the pusher 15; p_N is the pressure from pump 1; p_{N1} is the pressure in line 6; p_1 and p_2 are pressures operating in valve; p_{L1} and p_{L2} are pressures acting on the controlled checked valve 10; p_C is the pressure in the chamber of the hydraulic cylinder 14; P_{NAV} is the force acting on the rod of the hydraulic cylinder 14; d_{x1}, d_3, d_L, d_{L1}, d_{L2}, and d_{Z1} are the diameters of the channel of the servo valve 4, spool 7, spool of the controlled checked valve 10, input channel of the controlled checked valve 10, input channel of the servo valve 11, and input channel of the valve 12, respectively; α, α_1, γ, λ, and δ are the angles of inclination of the working edges of spool 7, spool of servo valve 4, spool of controlled checked valve 10, valve spool 12, and servo valve cone 11, respectively; f_1, f_2, f_3, f_4, and f_{s2}, are the areas of throttling orifices 16, 17, 18, 19, and 11, respectively; $F_3, F_Z, F_{Z1}, F_{Z2}, F_{L1}$, and F_C are the working areas of the spool 7, end of the valve 12, input channel of the valve 12, input channel of the servo valve 13, input channel of the checked valve 10, and cylinder piston 14, respectively; c, c_Z, c_{Z1}, and c_L are the stiffness of the springs; H, H_Z, H_{Z1}, and H_L are the initial compression of the springs; $Q_N, Q_{N1}, Q_K, Q_{K1}, Q_2, Q_3, Q_L, Q_{L2}, Q_{L3}$, and Q_{s1} are the fluid flows; W_A, W_B, W_C, W_D, W_E, and W_F are the volumes of the fluid in the hydraulic line at points A–F; m_3, m_Z, m_L, and m_C are the masses of the moving elements; b, b_Z, b_{Z1}, b_L, and b_C are the coefficients of viscous friction; R_{h3} and R_{hZ} are the hydrodynamic forces acting on the elements; T, T_Z, T_L, and T_C are the dry friction forces; Q_N is the flow pumped by pump 1; ρ is the working fluid density; μ is the flow rate coefficient; and β is the coefficient of the working fluid compression (Kozlov et al. 2019, Ogorodnikov et al. 2018, Polishchuk et al. 2018).

The mathematical model equation is solved for the following initial conditions: $P_{NAV} = 8{,}000\,\text{N}$, $p_N(0) = 30\cdot10^5\,\text{Pa}$, $p_1(0) = 1.2\,\text{MPa}$, $p_2(0) = 1.4\,\text{MPa}$, $p_3(0) = 1.0\,\text{MPa}$, $p_{L1}(0) = 2.8\,\text{MPa}$, $p_{L2}(0) = 0.05\cdot\text{MPa}$, $p_C(0) = 2.6\,\text{MPa}$, $x(0) = 0\,\text{m}$, $y(0) = 0\,\text{m}$, $z(0) = 0\,\text{m}$, $l(0) = 0\,\text{m}$, and $u(0) = 0\,\text{m}$. When solving of the mathematical model, the following values of the parameters were used: $d_3 = 10\ldots30\cdot10^{-3}\,\text{m}$, $d_x = 8\cdot10^{-3}\,\text{m}$; $d_{x1} = 4\cdot10^{-3}\,\text{m}$; $d_L = 28\cdot10^{-3}\,\text{m}$, $d_{L1} = 20\cdot10^{-3}\,\text{m}$, $d_{L2} = 3\cdot10^{-3}\,\text{m}$, $d_Z = d_{Z1} = 10\ldots30\cdot10^{-3}\,\text{m}$, $f_1 = 0.5\ldots3\cdot10^{-6}\,\text{m}^2$, $f_2 = 0.8\cdot10^{-6}\,\text{m}^2$, $f_3 = 0.5\ldots3\cdot10^{-6}\,\text{m}^2$, $f_4 = 0.5\ldots3\cdot10^{-6}\,\text{m}^2$, $\beta = 1\cdot10^{-9}\,\text{m}\cdot\text{s}^2/\text{kg}$, $\rho = 900\,\text{kg/m}^3$, $\mu = 0.7$; $c = 78.31\cdot10^3\,\text{N/m}$, $H = 2\cdot10^{-3}\,\text{m}$, $c_Z = 7.8\cdot10^2\,\text{N/m}$, $H_Z = 13.6\cdot10^{-3}\,\text{m}$, $c_L = 5\ldots20\cdot10^3\,\text{N/m}$, $H_L = 9.63\cdot10^{-3}\,\text{m}$, $b = 43\ldots121\,\text{kg/s}$, and $Q_N = 1.67\cdot10^{-3}\,\text{m}^3/\text{s}$, etc.

The mathematical model was processed using the MATLAB Simulink software package. Simulink is an interactive tool for modeling, simulation, and analysis of dynamic systems. It is an application of the MATLAB package and completely integrated into it.

11.4 RESULTS AND DISCUSSION

Based on the solution of the mathematical model, transient processes in the proportional electrohydraulic directional control valve were obtained, on the basis of which the working processes in the system were obtained. The influence that the change of the design parameters of the directional control valve had on the characteristics of its

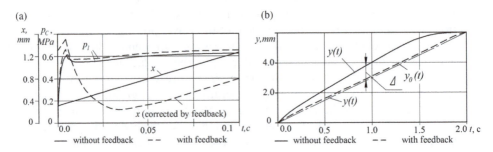

Figure 11.3 Influence of application of the feedback elements in a proportional electrohydraulic directional control valve with independent flows control: (a) correction of the input signal x(t) according to the changes in the system and (b) change of proportionality of moving of the spool of the second stage with and without using feedback elements.

work were investigated. Using feedback elements provides the correction of the input signal $x(t)$ (Figure 11.3a) according to the changes in the system. This improves the proportionality of moving of the spool of the second stage and provides a minimal error Δ from the preset law. Without the using feedback, the minimal error is much higher (Figure 11.3b).

Based on the results of the research, it was found that values of the areas of throttling orifices f_1, f_3, and f_4 and the values of the diameters of the spool 7 d_3 and the overflow valve d_Z as well as the viscous friction coefficient b and the stiffness of the spring c_L have a significant effect on the characteristics of the proportional electrohydraulic directional control valve (Figure 11.4a and b). The analysis of the influence of these design parameters on the level of the overshoot of pressure σ in the system did not reveal any significant deterioration during application of feedback elements, but it can lead to increasing the level of influence of the design parameters on the dynamic characteristics while applying feedback elements by 15%–20% compared to the proportional electrohydraulic directional control valve without feedback elements (Figure 11.4c and d).

Reducing overshooting of the pressure can be achieved by using the following values of the design parameters: $d_3 = 13...20 \cdot 10^{-3}$ m, $d_Z = 13...20 \cdot 10^{-3}$ m, $f_1 = 2...3 \cdot 10^{-6}$ m², $f_3 = 0.8...1.2 \cdot 10^{-6}$ m², $f_4 = 2...3 \cdot 10^{-6}$ m², $c_L = 5...8 \cdot 10^3$ N/m, and $b = 90...121$ kg/s.

11.5 CONCLUSION

1. The scheme of the proportional electrohydraulic directional control valve with independent flows control is proposed, which contains controlled check valves and two spools.
2. It is suggested to apply feedback elements to improve the controllability of the proportional electrohydraulic directional control valve.
3. On the basis of the mathematical model of the proportional electrohydraulic directional control valve with independent flows control, research on the influence of the design parameters of the directional control valve on the dynamic characteristics was carried out.

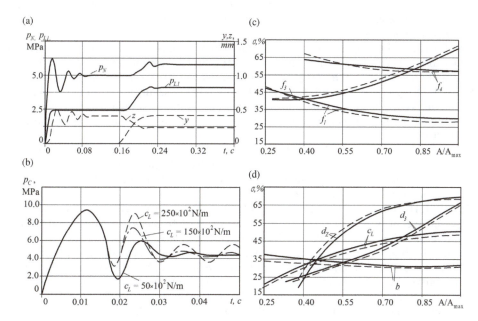

Figure 11.4 Investigation of processes in the proportional electrohydraulic directional control valve: (a) dynamic processes in the valve; (b) example of the influence of stiffness of the spring – c_L on the dynamic processes in the valve; (c and d) relationship between design parameters and overshooting of pressure (solid line, without feedback elements and dashed line, with feedback elements).

4. Using feedback elements in the proportional electrohydraulic directional control valve with independent flows control improves the controllability and does not lead to a significant deterioration of the dynamic characteristics of the hydraulic directional control valve.
5. It was established that the using feedback elements can lead to increasing the level of influence of the design parameters by 15%–20%.
6. Reducing overshooting of the pressure can be provided by using the following values of design parameters: $d_3 = 13...20 \cdot 10^{-3}$ m, $d_Z = 13...20 \cdot 10^{-3}$ m, $f_1 = 2...3 \cdot 10^{-6}$ m^2, $f_3 = 0.8...1.2 \cdot 10^{-6}$ m^2, $f_4 = 2...3 \cdot 10^{-6}$ m^2, $c_L = 5...8 \cdot 10^3$ N/m, and $b = 90...121$ kg/s.

REFERENCES

Bilynsky, Y.Y., Horodetska, O.S., Ogorodnik, K.V., Smolarz, A. & Muslimov, K. 2018. The ultrasonic converter mathematical model of flow rate of flowing environment. *Proceedings of SPIE* 108085T: 1715–1722.

Dragobetskii, V., Shapoval, A., Mos'pan, D., Trotsko, O. & Lotous, V. 2015. Excavator bucket teeth strengthening using a plastic explosive deformation. *Metallurgical and Mining Industry* 4: 363–368.

Eriksson, B. & Palmberg, J.O. 2011. Individual metering fluid power systems: challenges and opportunities. *Proceedings of the Institution of Mechanical Engineers, Part I Journal of Systems and Control Engineering* 225: 196–211.

Jianpeng, S., Long, Q., Xiaogang, Z. & Xiaoyan, X. 2018. Electro-hydraulic velocity and position control based on independent metering valve control in mobile construction equipment. *Automation in Construction* 94: 73–84.

Jong-Chan, L., Ki-Chang, J., Young-Min, K., Lim-Gook, C., Jae-Yoon, Ch. & Byung-Kyu, L. 2016. Development of the Independent Metering Valve Control System and Analysis of its Performance for an Excavator. *ASME/BATH 2016 Symposium on Fluid Power and Motion Control* 6: 1–6.

Kailei, L., Yingjie, G., Zhaohui, T. & Peng, L. 2016. Energy-saving analysis of the independent metering system with pressure compensation for excavator's manipulator. *Proceedings of the Institution of Mechanical Engineers Part I Journal of Systems and Control Engineering* 230(9): 905–920.

Kozlov, L.G., Polishchuk, L.K., Piontkevych, O.V., Korinenko, M.P., Horbatiuk, R.M., Komada, P., Orazalieva, S. & Ussatova, O. 2019. Experimental research characteristics of counter balance valve for hydraulic drive control system of mobile machine. *Przeglad Elektrotechniczny* 95(4): 104–109.

Kozlov, L.H., Dmytro, O. & Lozinskyi, O. 2009. Hydro-drive with proportional electro-hydraulic control. *Patent 41887 Ukraine, MIIK8 F15B 11, patent holder Vinnytsia National Technical University – Ngu200900907*, published 10.06.2009.

Lübbert, J., Sitte, A. & Weber, J. 2016. Pressure compensator control-a novel independent metering architecture. *10th International Fluid Power Conference*, Dresden, Germany 10: 231–246.

Ogorodnikov, V.A., Dereven'ko, I.A. & Sivak, R.I. 2018. On the influence of curvature of the trajectories of deformation of a volume of the material by pressing on its plasticity under the conditions of complex loading. *Materials Science* 54(3): 326–332.

Ogorodnikov, V.A., Savchinskij, I.G. & Nakhajchuk, O.V. 2004. Stressed-strained state during forming the internal slot section by mandrel reduction. *Tyazheloe Mashinostroenie* 12: 31–33.

Ogorodnikov, V.A., Zyska, T. & Sundetov, S. 2018. The physical model of motor vehicle destruction under shock loading for analysis of road traffic accident. *Proceedings of SPIE* 108086C: 1–5.

Petrov, O., Kozlov, L., Lozinskiy, D. & Piontkevych, O. 2019. Improvement of the hydraulic units design based on CFD modeling. *Lecture Notes in Mechanical Engineering* 86: 653–660.

Polishchuk, L., Kozlov, L.G., Piontkevych, O.V., Gromaszek, K. & Mussabekova, A. 2018. Study of the dynamic stability of the conveyor belt adaptive drive. *Proceedings of SPIE* 1080862: 1791–1800.

Qi, Z., Bin, Z., Hui-Ming, B., Hao-Cen, H., Ji-en, M., Yan, R., Hua-Yong, Y. & Rong-Fong, F. 2019. Analysis of pressure and flow compound control characteristics of an independent metering hydraulic system based on a two-level fuzzy controller. *Journal of Zhejiang University-SCIENCE A (Applied Physics & Engineering)* 20: 184–200.

Ruqi, D., Junhui, Z. & Bing, X. 2018. Advanced energy management of a novel independent metering meter-out control system: A case study of an excavator. *IEEE Access* 6: 45782–45795.

Titov, A.V., Mykhalevych, V.M., Popiel, P. & Mussabekov, K. 2017. Statement and solution of new problems of deformability theory. *Proceedings of SPIE* 108085E: 1611–1617.

Tymchyk, S.V., Skytsiouk, V.I., Klotchko, T.R., Ławicki, T. & Demsova, N. 2018. Distortion of geometric elements in the transition from the imaginary to the real coordinate system of technological equipment. *Proceedings of SPIE* 108085C: 1595–1604.

Wydra, M., Geimer, M. & Weiß, B. 2017. An approach to combine an independent metering system with an electro-hydraulic flow-on-demand hybrid-system. *Proceedings of 15th Scandinavian International Conference on Fluid Power* 1: 1–10.

Chapter 12

Optimization of design parameters of a counterbalance valve for a hydraulic drive invariant to reversal loads

Leonid Kozlov, Leonid Polishchuk, Oleh Piontkevych, Viktor Purdyk, Oleksandr V. Petrov, Volodymyr M. Tverdomed, Piotr Kisala, Saltanat Amirgaliyeva, Bakhyt Yeraliyeva, and Aigul Tungatarova

CONTENTS

12.1 Introduction 137
12.2 Analysis of the literary sources of hydraulic drives with
       counterbalance valves 138
12.3 Materials and methods of studying the hydraulic drives invariant to
       alternating loads 139
12.4 Results and discussion 145
12.5 Conclusion 147
References 147

12.1 INTRODUCTION

Hydraulic drives developed at the end of the twentieth century on the basis of nonadjustable pumps, discrete hydraulic distributors, and throttles with check valves are morally obsolete today (Burennikov et al. 2017, Conrad & Jensen 1987, Gubarev & Ganpantsurova 2010). Energy-efficient work is not provided by obsolete hydraulic drives during their multimode operation. The speed of an actuating device of such a hydraulic drive essentially depends on the load and leads to an additional number of movements to bring the bucket to the required position. The effect of the overshoot value on pressure in a morally obsolete hydraulic drive can lead to pipeline failures and emergency situations.

Today, in Ukraine and abroad, special attention is being paid to the development of multimode hydraulic drives that are invariant to reversal loads. They are usually equipped with modern adjustable pumps, proportional hydraulic distributors, and counterbalance valves (CBV) (Gubarev et al. 2010, Purdyk & Brytskiy 2018, Wenfeng et al. 2016). During multimode operation of a hydraulic drive, the adjustable pumps are energy saving in comparison with nonadjustable ones. Unlike discrete hydraulic distributors, the proportional hydraulic distributors operate in a wide range of changes in the supply of working fluid. CBVs provide stabilization of the actuating device movement speed, and energy saving under reverse and concomitant loads compared to throttles with check valves.

DOI: 10.1201/9781003224136-12

The authors simulated and optimized the design of a CBV for a hydraulic drive that is invariant to reversal loads. During the scientific and engineering research, the following tasks are solved:

- static, dynamic, and power characteristics of a hydraulic drive are provided;
- design parameters of a CBV are optimized according to the complex criterion, taking into account the operation of the hydraulic drive under reverse and concomitant loads.

12.2 ANALYSIS OF THE LITERARY SOURCES OF HYDRAULIC DRIVES WITH COUNTERBALANCE VALVES

The designs of CBVs are distinguished according to the type of control: internal, external, and mixed. CBVs with internal control are usually used in stationary machines when working with loads of one mass. More commonly used CBVs have external control (produced by Bosch Rexroth, Hidromek, Eaton, etc.) and mixed control (manufactured by Oleostar, Ponar Wadowice, etc.). The authors designed a CBV that works as an external control CBV with reverse load of a hydraulic drive, while with a concomitant load, it operates as a mixed control CBV.

In Ritelli and Vacca (2013), CBVs with a mixed type of control were considered. The numerical model is validated on the basis of experimental results and permits to derive not only considerations about the energy consumption of counterbalance valves but also their effect in terms of the dynamic behavior of the system. However, the numerical model in question does not take into account the influence of a CBV when applied to the static characteristics of the hydraulic drive.

In Xu and Xinhui (2012) and Zhao et al. (2010), the structure and working principle of a counterbalance valve used in some hydraulic cranes is presented. Based on the AMESim software, a model of the counterbalance valve is built, and a simulation analysis of its performance is completed. The considered hydraulic drive with CBV shows the stability of the system at different frequencies and its good dynamic characteristics. However, the considered CBV design with an external control type is complex compared to the leading firms of Bosch Rexroth and Hidromek.

CBVs with external and mixed control types were considered by Zaehe and Herbert (2015). Justifiable energy efficiency of a CBV was obtained with a mixed type of control. However, these valves have problems with tightness when the executive body fluctuates due to the carriage of goods.

The improved hydraulic drives of industrial tote dumpers and lifters were considered by Jalayeri et al. (2015). They replaced the throttling valves with counterbalance valves. The improved scheme showed good energy efficiency and performance indexes. However, the paper does not include studies of the dynamic and static characteristics of the hydraulic drive.

The design and experimental installation of a hydraulic drive with a CBV were described by Imam et al. (2018). Several schemes of hydraulic drives with CBVs and their effect on performance, dynamic, and energy characteristics of the hydraulic drive are considered. The proposed circuit displays improved performance, besides being capable of energy regeneration. However, the effect of the boom on the hydraulic cylinder is significantly simplified.

In Petrov et al. (2019) and Rahman et al. (1997), simulation modeling of the working fluid flow in the hydraulic equipment of the hydraulic drive was demonstrated. Simulation modeling makes it possible to calculate the dependences of the characteristics of hydraulic equipment to be taken into account in mathematical models of the entire hydraulic drive (Bilynsky et al. 2018, Titov et al. 2017, Tymchyk et al. 2018).

In Sorensen et al. (2016), the hydraulic drive with CBV was studied for an experimental boom model. A nonlinear model of the system was developed and an experimental variation showed good correspondence between the model and the real system. Stability of transient processes was only achieved in a hydraulic drive with a CBV.

In the reviewed papers, there is no clear mechanism for optimizing the CBV parameters for the static, dynamic, and energy characteristics of the hydraulic drive. Therefore, the proposed comprehensive criterion for optimizing the parameters of a CBV is relevant.

12.3 MATERIALS AND METHODS OF STUDYING THE HYDRAULIC DRIVES INVARIANT TO ALTERNATING LOADS

Diagrams of a hydraulic drive that is invariant to reversal loads are shown in Figure 12.1. The hydraulic drive consists of an adjustable pump 1, a feed regulator 2, a nonadjustable throttle 3, an adjustable throttle 4, a hydraulic cylinder 5, a CBV 6, and a hydraulic tank 12. The CBV 6 comprises nonadjustable throttles 7 and 8, the main spool valve 9, a piston 10, and a spring 11. The hydraulic drive has an actuating device, which includes the base 13, a rack 14, a boom 15, a load 16, and hinges 17–19.

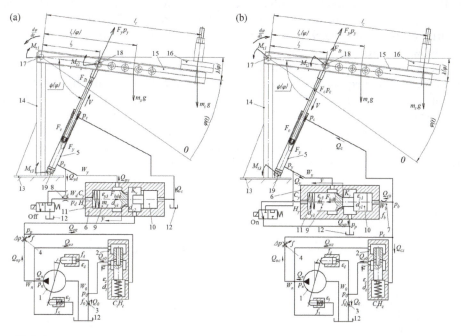

Figure 12.1 Design models of the hydraulic drive being invariant to reversal loads with (a) reversal and (b) concomitant loads.

The hydraulic drive works as follows. At the reverse load, the servo spool is disabled in the "Off" position. The working fluid from the adjustable pump 1 is fed through the working window f of the adjustable throttle 4 and the main spool 9 which is open to the piston chamber of the hydraulic cylinder 5. The rod of the hydraulic cylinder 5 moves and the load 16 is lifted up. The working fluid is drained from the rod chamber of the hydraulic cylinder 5 into the hydraulic tank 12.

The pump controller 2 determines the differential pressure Δp on the adjustable throttle 4 and maintains a constant flow through it. The piston 10 of the CBV 6 is in the right position. The main spool 9 of the CBV 6 provides proportional control of the speed for lifting the load up and secure locking of the load when stopped.

In the case of a concomitant load, the channels of the CBV 6 are recombined due to activation of the servo spool in the "On" position. The left side of the main spool 9 is unloaded, and it reduces the energy cost on its control. The working liquid from the adjustable pump 1 is fed through the nonadjustable throttle 4 to the rod chamber of the hydraulic cylinder 5 and to the piston 10 through the nonadjustable throttle 7. The movement of the piston 10 from the right to the left provides the opening of the main spool 9 and the drain of working fluid from the piston chamber of the hydraulic cylinder 5 into the hydraulic tank 12.

Mathematical models of the hydraulic drive (see Figure 12.1a and b) for reverse and concomitant loads are different and compiled with typical assumptions for hydraulic drives (Polishchuk et al. 2016, Purdyk et al. 2018). In addition, the hydraulic drive with a CBV will work with different output characteristics of the adjustable pump 1. When lifting the load 16 up, the adjustable pump 1 supplies the working fluid in the hydraulic drive under pressure p_n and with Q_n feed, and during lowering – under pressure p_n^* and with Q_n^* feed. The ratio of these characteristics is as follows: $p_n \gg p_n^*$, $Q_n \leq Q_n^*$, which is caused by high-speed reverse movements and pressure up to 3 MPa under the opening control of the main spool 9 of the CBV 6 during lowering of the boom 15 with the load 16 and without it.

The mathematical model of the hydraulic drive (see Figure 12.1a) under the reverse load includes the equation of continuity of flows for the hydraulic lines between the pump 1 and the adjustable throttle 4 (equation 12.1), between the adjustable throttle 4 and the CBV 6 (equation 12.2), between the CBV 6 and the piston chamber of the hydraulic cylinder 5 (equation 12.3), between the nonadjustable throttle 8 and the chamber of the right end of the main spool 9 (equation 12.4) as well as between the flow regulator 2 and the nonadjustable throttle 3 (equation 12.5):

$$0.25 \cdot \pi \cdot d_7^2 \cdot d_8 \cdot k \cdot n \cdot tg(\gamma) = \mu \cdot f \cdot \sqrt{\frac{2|p_n - p_p|}{\rho}} \cdot sign(p_n - p_p) + \\ + \mu \cdot \pi \cdot d_z \cdot z \cdot \sqrt{\frac{2|p_n - p_0|}{\rho}} \cdot sign(p_n - p_0) - \frac{\pi \cdot d_z^2}{4} \cdot \frac{dz}{dt} + \beta_1 \cdot W_n \cdot \frac{dp_n}{dt}; \quad (12.1)$$

$$\mu \cdot f \cdot \sqrt{\frac{2|p_n - p_p|}{\rho}} \cdot sign(p_n - p_p) = Q_y - \frac{\pi \cdot d_z^2}{4} \cdot \frac{dz}{dt} + \beta_1 \cdot W_p \cdot \frac{dp_p}{dt}; \quad (12.2)$$

$$Q_y = F_y \cdot V + \mu \cdot f_d \cdot \sqrt{\frac{2|p_y - p_d|}{\rho}} \cdot sign(p_y - p_d) + \beta_1 \cdot W_y \cdot \frac{dp_y}{dt}; \qquad (12.3)$$

$$\mu \cdot f_d \cdot \sqrt{\frac{2|p_y - p_d|}{\rho}} \cdot sign(p_y - p_d) = \beta_2 \cdot W_d \cdot \frac{dp_d}{dt} - \frac{\pi \cdot d_{y3}^2}{4} \cdot \frac{dy}{dt}; \qquad (12.4)$$

$$\mu \cdot \pi \cdot d_z \cdot z \cdot \sqrt{\frac{2|p_n - p_0|}{\rho}} \cdot sign(p_n - p_0) = \mu \cdot f_0 \cdot \sqrt{\frac{2p_0}{\rho}} + \beta_2 \cdot W_0 \cdot \frac{dp_0}{dt}; \qquad (12.5)$$

and equations of force equilibrium on the spools of the CBV 6 (equation 12.6) and the feed regulator 2 (equation 12.7):

$$m_y \cdot \frac{d^2 y}{dt^2} = p_y \cdot \frac{\pi \cdot d_{y2}^2}{4} - p_d \cdot \frac{\pi \cdot d_{y3}^2}{4} - \frac{\pi \cdot \rho \cdot v \cdot d_{y3} \cdot l_{y3}}{\varepsilon_{y3}} \cdot \frac{dy}{dt} - C_y(H_y + y) - F_{hda};$$

$$(12.6)$$

$$\frac{\pi \cdot \rho \cdot v \cdot d_z \cdot l_z}{\varepsilon_z} \cdot \frac{dz}{dt} = p_n \frac{\pi \cdot d_z^2}{4} - p_p \frac{\pi \cdot d_z^2}{4} - C_z(H_z + z). \qquad (12.7)$$

The equilibrium of moments on the faceplate of the adjustable pump 1 (equation 12.8) and on the boom 15 (equation 12.9) is described by the formulas:

$$J \cdot \frac{d^2 \gamma}{dt^2} = p_n \cdot f_5 \cdot l - p_0 \cdot f_4 \cdot l - \frac{\pi \cdot \rho \cdot v \cdot d_4 \cdot l_4^2}{\varepsilon_4} \cdot \frac{d\gamma}{dt} \cdot \cos(\gamma) - \frac{\pi \cdot \rho \cdot v \cdot d_5 \cdot l_5^2}{\varepsilon_5} \cdot \frac{d\gamma}{dt} \cdot \cos(\gamma) +$$
$$+ \left(m_0 + m_1 \cdot Q_n + m_2 \cdot p_n + m_3 \cdot Q_n^2 + m_4 \cdot p_n^2 + m_5 \cdot Q_n \cdot p_n \right); \qquad (12.8)$$

$$I \cdot \frac{d^2 \phi}{dt^2} = p \cdot F \cdot l \cdot \sin[\psi(\phi)] - m \cdot g \cdot l(\phi) \cdot \cos[\lambda(\phi)] - m \cdot g \cdot l \cdot \cos[\lambda(\phi)] -$$
$$F \cdot sign\left(\frac{d\phi}{dt}\right) \cdot l \cdot \sin[\psi(\phi)] - M \cdot sign\left(\frac{d\phi}{dt}\right) - M \cdot sign\left(\frac{d\phi}{dt}\right) - M \cdot sign\left(\frac{d\phi}{dt}\right).$$

$$(12.9)$$

The mathematical model of the hydraulic drive (see Figure 12.1b) with concomitant loads is essentially different in the equations (12.2–12.4, 12.6) which are, respectively, given below:

$$\mu \cdot f \cdot \sqrt{\frac{2|p_n - p_c|}{\rho}} \cdot sign(p_n - p_c) = -F_c \cdot V + \mu \cdot f_b \cdot \sqrt{\frac{2|p_c - p_b|}{\rho}} sign(p_c - p_b)$$
$$- \frac{\pi \cdot d_z^2}{4} \cdot \frac{dz}{dt} + \beta_1 \cdot W_c \cdot \frac{dp_c}{dt}; \qquad (12.10)$$

$$\mu \cdot K_y \cdot y \cdot \sqrt{\frac{2p_y}{\rho}} + \beta_1 \cdot W_y \cdot \frac{dp_y}{dt} = -F_y \cdot V; \qquad (12.11)$$

$$\mu \cdot f_b \cdot \sqrt{\frac{2|p_c - p_b|}{\rho}} \cdot \text{sign}(p_c - p_b) = \frac{\pi \cdot d_{y1}^2}{4} \cdot \frac{dy}{dt} + \beta_2 \cdot W_b \cdot \frac{dp_b}{dt}; \qquad (12.12)$$

$$m_y \cdot \frac{d^2 y}{dt^2} = p_b \cdot \frac{\pi \cdot d_{y1}^2}{4} - \left[\frac{\pi \cdot \rho \cdot v \cdot d_{y1} \cdot l_{y1}}{\varepsilon_{y1}} + \frac{\pi \cdot \rho \cdot v \cdot d_{y3} \cdot l_{y3}}{\varepsilon_{y3}} \right] \cdot \frac{dy}{dt} - C_y (H_y + y) + F_{hdb}.$$

$$(12.13)$$

In the nonlinear differential equations (12.1–12.13) of the mathematical model, the following abbreviations are adopted: f is the area of the working window of adjustable throttle 4; Q_y is the feed through CBV (Kozlov et al. 2019); F_{hda} and F_{hdb} are the hydrodynamic forces on the main spool 9 under reverse and concomitant loads; μ_f is the friction coefficient of steel against steel; $F\Sigma_t$ is the total frictional force in the hydraulic cylinder 5; μ is the flow rate through the throttling and spool elements; ρ is the working fluid density; β_1 and β_2 are the compression coefficients of the working fluid with flexible pipelines and without them; d_{y2} and d_{y3} are the diameters of the left and right ends of the main spool 9; l_{y3} is the contact length of the main spool 9 with the housing of the CBV 6; f_d and f_o are the areas of the nonadjustable throttles; d_c, d_s, b_c, and b_s are the diameters and widths of the piston and rod seals of the hydraulic cylinder 5; W_n, W_p, W_y, W_0, and W_d are the 1volumes of the hydraulic lines; m_c, m_s, and m_y are the reduced weights of the load 16, boom 15, and the main spool 9 with the spring 11, respectively; g is the free-fall acceleration; C_y and C_z are the stiffness of the springs of the CBV 6 and the feed regulator 2, respectively; H_y and H_z are the previous compression of the springs of the CBV 6 and the feed regulator 2, respectively; ε_4 and ε_5 are the gaps formed between the working surfaces of the plungers and housing of the pump 1; ε_z and ε_{y3} are the gaps formed between the working surfaces: the spool of the pump regulator 2 and its body, and the main spool 9 of the CBV 6 and its body; F_c and F_y are the area of the rod and piston chambers of the hydraulic cylinders 5; J and I_{mc} are the moment of inertia of the pump faceplate 1 and the experimental boom sample; v is the coefficient of viscosity of industrial oil; d_4, d_5, l_4, and l_5 are the diameters of plungers and plunger contact lengths with the sleeves of the adjusting pump 1, respectively; d_z and l_z are the diameter of the spool and spool contact length with the regulator housing of the pump 2; d_7 and d_8 are the diameters of piston of the pump 1 and the contact range of the pump pistons with faceplate; k is the number of pistons in the pump 1; n is the number of shaft revolutions of the pump 1; f_4 and f_5 are the areas of the plungers; z and y are the spool position coordinates of the regulator of the pump 2 and the CBV 6; p_n, p_c, p_p, p_y, p_0, and p_d are the pressure values for the corresponding hydraulic lines; γ, φ, λ, and ψ are the angles of faceplate rotation of the pump 1, boom 15, and hydraulic cylinder 5; m_0, m_1, m_2, m_3, m_4, and m_5 are the dependence coefficients of resistance moment on the magnitude of Q_n and p_n; Q_n is the feed rate of the adjusting pump 1; l_b is the length of the boom 15 from the axis of rotation to the axis of the hydraulic cylinder 5; and M_{t1}, M_{t2}, and M_{t3} are the friction moments occurring in the hinges of the hydraulic cylinder 5 and boom 15.

The solution of nonlinear differential equations (12.1–12.13) is carried out under these initial conditions: $z(0) = 0$; $y(0) = 0$; $V(0) = 0$; $p_n(0) = 4.6 \cdot 10^6$ Pa; $p_p(0) = 4.2 \cdot 10^6$ Pa;

$p_y(0) = 4\cdot 10^6$ Pa; $p_0(0) = 0.1\cdot 10^6$ Pa; $p_d(0) = 3.7\cdot 10^6$ Pa; and $\gamma(0) = 0.05$ radians using the MATLAB Simulink software package. To calculate transient processes, the ode23s function was used, which is based on a one-dimensional modified Rosenbrock method of the second order (Polishchuk et al. 2018). The percentage error made in the process of simulation is 0.1%. The adequacy of the mathematical models was checked and found to be 95% according to Fisher's ratio test (Kozlov et al. 2019).

The variation ranges of design parameters of the CBV during the research of the static, dynamic, and power characteristics of the hydraulic drive are as follows: $d_{y1} = (15...20)\cdot 10^{-3}$ m; $d_{y2} = (15...20)\cdot 10^{-3}$ m; $K_y = (0.5...1.5)\cdot 10^{-3}$ m; $f_b = (0.5...1.1)\cdot 10^{-6}$ m^2; $f_d = (0.5...1.1)\cdot 10^{-6}$ m^2; $m_y = 0.05...0.2$ kg; $C_y = (1...2.5)\cdot 10^4$ N/m; $H_y = (2...7)\cdot 10^{-3}$ m; $l_{y3} = (10...30)\cdot 10^{-6}$ m; $\varepsilon_{y3} = (6...41)\cdot 10^{-6}$ m; $W_b = (0.04...0.1)\cdot 10^{-3}$ m^3; and $W_d = (0.08...0.12)\cdot 10^{-3}$ m^3.

On the x axis (see Figures 12.3–12.6) of the researched characteristics of the hydraulic drive, we mark the values of dimensionless parameters of the CBV:

$$P = P_{real}/P_{max}, \qquad (12.14)$$

where P_{real} is the current parameter value and P_{max} is the maximum parameter value.

The character of influence of the CBV design parameters on the error δ of stabilization of velocity V of the hydraulic cylinder for the hydraulic drive is calculated (see Figure 12.1a and b) under reverse and concomitant loads:

$$\delta = \frac{|V_{max} - V_{min}|}{V_{min}} \cdot 100\%; \qquad (12.15)$$

where V_{max} and V_{min} are the maximum and minimum values of velocity V of the hydraulic cylinder 5 for constant feed on the adjustable throttle 4 (see Figure 12.1a and b), respectively, but when changing the load m_c from 50 to 100 kg. The error δ^* of stabilization for the speed V at the concomitant load of the hydraulic drive is calculated with the negative value of the velocity V of the hydraulic cylinder 5 due to the use of the same coordinate system for the calculation of the hydraulic drive configurations.

The character of the influence of the design parameters of the CBV on the overshoot value σ and the time of the transient process t_p based on the pressure p_n in the

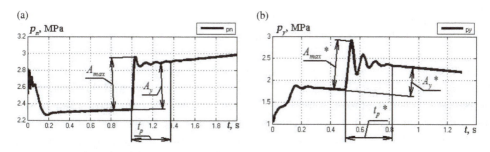

Figure 12.2 Dependence of pressure p_n in the injection hydraulic line and the pressure p_y of the working hydraulic line on time t: (a) under reverse and (b) concomitant loads.

hydraulic line of injection and the pressure p_y in the working hydraulic line under the reverse and concomitant loads, respectively, were researched. Diagrams for calculating the dynamic characteristics of the hydraulic drive are shown in Figure 12.2a and b for reverse and concomitant loads, respectively.

To calculate the value of overshoot σ according to the pressure p_n under reverse load, we use the formula:

$$\sigma = \frac{|A_{max} - A_y|}{A_y} \cdot 100\% \tag{12.16}$$

where A_{max} and A_y are the maximum and fixed pressure value p_n at the pulse change of mass m_c of the load 16 from 50 to 100 kg for the reverse load of the hydraulic drive, respectively.

The energy characteristics of the hydraulic drive were studied by calculating the expended power $N(p_n)$ and $N(p_n)^*$ from the adjustable pump 1 (see Figure 12.1a and b) under the constant mass of the load 16 $m_c = 100$ kg and the velocity $V = 0.25$ m/s of the hydraulic cylinder 5 under reverse and concomitant loads, respectively. The supplied power $N(p_n)$ and $N(p_n)^*$ from the adjustable pump 1 for operations under the same operating conditions of the hydraulic drive were calculated according to the formula:

$$N(p_n) = N(p_n)^* = p_n \cdot Q_n. \tag{12.17}$$

We optimized the static, dynamic, and power characteristics of the hydraulic drive to calculate the optimal combinations of design parameters of the CBV. We optimized these by using the LP search method for the reverse and concomitant loads separately. The comprehensive W_{opt} optimization criterion is the result of total optimization criteria for the reverse W_{opt1} and the concomitant W_{opt2} loads, and it is calculated by the formulas:

$$W_{opt} = W_{opt1} + W_{opt2}; \tag{12.18}$$

$$W_{opt1} = 0.2\left(\frac{\sigma_i}{\sigma_{max}} + \frac{\delta_i}{\delta_{max}}\right) + 0.3\left(\frac{t_{pi}}{t_{pmax}} + \frac{N_i}{N_{max}}\right); \tag{12.19}$$

$$W_{opt2} = 0.2\left(\frac{\sigma_i^*}{\sigma_{max}^*} + \frac{\delta_i^*}{\delta_{max}^*}\right) + 0.3\left(\frac{t_{pi}^*}{t_{pmax}^*} + \frac{N_i^*}{N_{max}^*}\right), \tag{12.20}$$

where i is the number of the experiment; * is the index of parameters at concomitant load; σ_i, σ_{max}, σ_i^*, and σ_{max}^* are the values of pressure overshoot and their maximum overshoot in a series of experiments; δ_i, δ_{max}, δ_i^*, and δ_{max}^* are the error value of stabilization of the velocity V when changing the operating modes and the maximum stabilization error of the velocity V in a series of experiments; t_{pi}, t_{pmax}, t_{pi}^*, and t_{pmax}^* are the time of the transition process and the maximum time of the transition process in a series of experiments; and N_i, N_{max}, N_i^*, and N_{max}^* are the values of the supplied power of hydraulic drive and its maximum supplied power in a series of experiments.

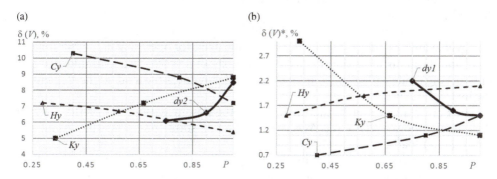

Figure 12.3 Influence of parameters of the CBV on errors δ and δ^* of stabilization of velocity V of the hydraulic cylinder of the hydraulic drive in the case of reverse (a) and concomitant (b) loads.

12.4 RESULTS AND DISCUSSION

The influence of the parameters of the CBV on the errors δ and δ^* of stabilization of velocity V of the hydraulic cylinder of the hydraulic drive in the case of reverse (a) and concomitant (b) loads are shown in Figure 12.3a and b. Under reverse load, according to Figure 12.3a, it was calculated that, with increasing stiffness of the spring H_y and the preliminary compression of the spring C_y, the stabilization error δ of the velocity V of the hydraulic actuator of the hydraulic drive decreases, and with increasing the diameter of the main spool d_{y2} and the gain factor of the working window K_y, it increases.

Under concomitant load, according to Figure 12.3b, we have a reduction in the error δ^* of stabilization of velocity V of the hydraulic cylinder of the hydraulic drive with an increase in the gain factor of the working window K_y and the diameter of the piston d_{y1}. An increase of the stiffness parameters of the spring C_y and the previous compression of the spring H_y leads to an increase in the error δ^* of stabilization of velocity V of the hydraulic actuator of the hydraulic drive.

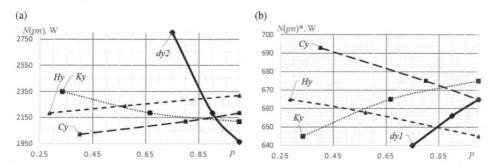

Figure 12.4 Influence of the design parameters of the CBV on the supplied power $N(p_n)$ and $N(p_n)^*$ from the adjustable pump for the reverse (a) and concomitant (b) loads.

The ranges of parameters for the CBV were calculated: $d_{y1} = (18...20) \cdot 10^{-3}$ m; $H_y = (2...5) \cdot 10^{-3}$ m; $K_y = (0.5...1.2) \cdot 10^{-3}$ m; $d_{y2} = (16...19) \cdot 10^{-3}$ m; $C_y = (1...2.5) \cdot 10^4$ N/m; $f_b = (0.5...0.7) \cdot 10^{-6}$ m^2; and $f_d = (0.5...1.0) \cdot 10^{-6}$ m^2, at which the regulation time $t_p < 0.5$ s is provided and overshoot $\sigma < 35\%$ in a hydraulic drive for operating mode with reverse load, and the adjustment time $t_p^* < 0.4$ s and overshoot $\sigma^* < 50\%$ in a hydraulic drive for the working mode with a concomitant load, which reduces the load on the elements of the hydraulic drive.

In Figure 12.4a and b, the influence of the design parameters of the CBV on the supplied power $N(p_n)$ and $N(p_n)^*$ from the adjustable pump for the reverse and concomitant loads, respectively, are shown.

For the reverse load (see Figure 12.4a), the value of the supplied power $N(p_n)$ from the adjustable pump for operations under identical operating conditions decreases, when the stiffness of the spring C_y decreases, the previous compression of the spring H_y decreases, and the diameter of the main spool d_{y2} and the gain factor of the working window K_y increase. For the concomitant load (see Figure 12.4b), the value of the supplied power $N(p_n)^*$ from the adjustable pump for operations under identical operating conditions decreases, when the gain factor of the working window K_y increases, the piston diameter d_{y1} decreases, and when both the stiffness of the spring C_y and the previous compression of the spring H_y increase.

The ratio of the calculated supplied power $N(p_n) > N(p_n)^*$ from the adjustable pump is due to the fact that the lowering of the load takes place under the forces of its own weight, and the CBV only provides control of the load lowering velocity. It is confirmed that during the lowering of the load, the pressure p_n of the adjustable pump does not exceed 3 MPa for different loading ranges.

Insignificant or nonexistent influences of the design parameters of the CBV on the characteristics of the hydraulic drive in Figures 12.3 and 12.4 are not depicted. For optimization, we chose the following design parameters of the CBV: $f_d = (0.5...1.1) \cdot 10^{-6}$ m^2; $f_d = (0.5...1.1) \cdot 10^{-6}$ m^2; $d_{y1} = (15...20) \cdot 10^{-3}$ m; $d_{y2} = (15...20) \cdot 10^{-3}$ m; $K_y = (0.5...1.5) \cdot 10^{-3}$ m; and $H_y = (2...7) \cdot 10^{-3}$ m, which are most influential on the static, dynamic, and energy characteristics of a hydraulic drive that is invariant to reversal loads. To optimize the mode of operation with the mass of the load, $m_c = 100$ kg and the velocity $V = 0.25$ m/s of the hydraulic cylinder were selected.

In the process of optimization, the transient processes for 160 experiments were calculated and 720 values of the complex optimization criterion W_{opt} were obtained. The smallest criterion of optimization for a reverse load $W_{opt1} = 0.384278$ was calculated in 33 experiments. For a concomitant load, the smallest optimization criterion $W_{opt2} = 0.513705$ was calculated in 61 experiments. After summarizing the results of the calculation, it was established that the complex optimization criterion would be the smallest $W_{opt} = 0.91328$ for experiment 32 under reverse load and for experiment 59 under concomitant load. The optimal combination of parameters was $d_{y1} = 20 \cdot 10^{-3}$ m, $d_{y2} = 18 \cdot 10^{-3}$ m, $H_y = 4 \cdot 10^{-3}$ m, $K_y = 1 \cdot 10^{-3}$ m, f_d and $f_b = 0.5 \cdot 10^{-6}$ m^2 provides: the value of overshoot $\sigma(p_n) = 11.1\%$, $\sigma(p_y)^* = 18.5\%$, the transition process time $t_p(p_n) = 0.6$ s, $t_p (p_y)^* = 0.22$ s, the stabilization error of velocity $\delta(V) = 7.8\%$, $\delta(V)^* = 3\%$, power consumption of the hydraulic drive $N(p_n) = 1340$ W, and $N(p_n)^* = 827$ W under reverse and concomitant loads. Transient processes in the hydraulic drive under reverse (a) and the concomitant (b) loads after optimization are shown in Figure 12.5.

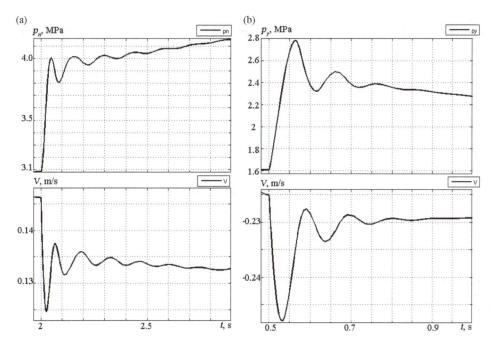

Figure 12.5 Transient processes in the hydraulic drive under reverse (a) and the concomitant (b) loads after optimization.

12.5 CONCLUSION

A complex criterion for parameter optimization of a counterbalance valve in order to optimize the static, dynamic, and power characteristics of a hydraulic drive that is invariant to reversal loads is proposed. The optimal combination of the CBV parameters for the hydraulic drive of the boom – $d_{y1} = 20 \cdot 10^{-3}$ m, $d_{y2} = 18 \cdot 10^{-3}$ m, $H_y = 4 \cdot 10^{-3}$ m, $K_y = 1 \cdot 10^{-3}$ m, and $f_d = f_b = 0.5 \cdot 10^{-6}$ m^2 – provides the value of overshoot $\sigma = 11.1\%$, $\sigma^* = 18.5\%$, the transition process time $t_p = 0.6$ s, $t_p^* = 0.22$ s, the stabilization error of the hydraulic actuator velocity $\delta = 7.8\%$, $\delta^* = 3\%$, the power consumption of the hydraulic drive $N = 1340$ W and $N^* = 827$ W and increases the efficiency of the hydraulic drive by reducing the supplied power and the velocity stabilization error of the hydraulic cylinder in operating modes.

REFERENCES

Bilynsky, Y.Y., Horodetska, O.S., Ogorodnik, K.V., Smolarz, A. & Muslimov, K. 2018. The ultrasonic converter mathematical model of flow rate of flowing environment. *Proceedings of SPIE* 108085T: 1715–1722.

Burennikov, Y., Kozlov, L., Pyliavets, V. & Piontkevych, O. 2017. Mechatronic hydraulic drive with regulator, based on artificial neural network. *IOP Conference Series: Materials Science and Engineering* 209(1): 1–8.

Conrad, F. & Jensen, C.J.D. 1987. Design of hydraulic force control systems with state estimate feedback. *IFAC Proceedings Volumes* 20(5): 307–312.

Gubarev, A. & Ganpantsurova, O. 2010. Coordination of rated power of wind turbine with a bar graph of wind speed by means of flexible control algorithm of the hydraulic drive. *Solid State Phenomena* 164: 111–115.

Gubarev, A., Yakhno, O. & Ganpantsurova, O. 2010. Control Algorithms in Mechatronic Systems with Parallel Processes, *Solid State Phenomena* 164: 105–110.

Imam, A., Rafiq, M., Jalayeri, E. & Sepehri, N. 2018. A pump-controlled circuit for single-rod cylinders that incorporates limited throttling compensating valves. *Actuators* 7: 1–21.

Jalayeri, E., Imam, A., Zeljko, T. & Sepehri, N. 2015. A throttle-less single-rod hydraulic cylinder positioning system: Design and experimental evaluation. *Advances in Mechanical Engineering* 7(5), 1–14.

Kozlov, L.G., Polishchuk, L.K., Piontkevych, O.V., Korinenko, M.P., Horbatiuk, R.M., Komada, P., Orazalieva, S. & Ussatova, O. 2019. Experimental research characteristics of counter balance valve for hydraulic drive control system of mobile machine. *Przeglad Elektrotechniczny* 95(4): 104–109.

Petrov, O., Kozlov, L., Lozinskiy, D. & Piontkevych, O. 2019. Improvement of the hydraulic units design based on CFD modeling. *Lecture Notes in Mechanical Engineering* 86: 653–660.

Polishchuk, L., Kharchenko, Ye., Piontkevych, O. & Koval, O. 2016. The research of the dynamic processes of control system of hydraulic drive of belt conveyors with variable cargo flows. *Eastern-European Journal of Enterprise Technologies* 8(80): 22–29.

Polishchuk, L., Kozlov, L.G., Piontkevych, O.V., Gromaszek, K. & Mussabekova, A. 2018. Study of the dynamic stability of the conveyor belt adaptive drive. *Proceedings of SPIE* 1080862: 1791–1800.

Purdyk, V.P. & Brytskiy, O.L. 2018. Flow control valve with the polymer envelope as control organ. *Mechanics and Advanced Technologies* 1(82): 100–106.

Purdyk, V.P., Brytskiy, O. & Sapognik, V. 2018. Dynamics of adaptive drive equipment for forming billets bricks. *International Journal of Engineering & Technology* 7(4.3): 87–91.

Rahman, M.M., Porteiro, J.L.F. & Weber S.T. 1997. Numerical simulation and Animation of oscillating Turbulent flow in a counterbalance valve. *Energy Conversion Engineering Conference, Honolulu, HI, USA* 1: 1525–1530.

Ritelli, G.F. & Vacca, A. 2013. Energetic and dynamic impact of counterbalance valves in fluid power machines. *Energy Conversion and Management* 76: 701–711.

Sorensen, J.K., Hansen, M.R. & Ebbesen, M.K. 2016. Numerical and experimental study of a novel concept for hydraulically controlled negative loads, *Modeling, Identification and Control* 37(4): 195–211.

Titov, A.V., Mykhalevych, V.M., Popiel, P. & Mussabekov, K. 2017. Statement and solution of new problems of deformability theory. *Proceedings of SPIE* 108085E: 1611–1617.

Tymchyk, S.V., Skytsiouk, V.I., Klotchko, T.R., Ławicki, T. & Demsova, N. 2018. Distortion of geometric elements in the transition from the imaginary to the real coordinate system of technological equipment. *Proceedings of SPIE* 108085C: 1595–1604.

Wenfeng, Z., Shengjie, J., Yaoxiang, Y. & Feng, D. 2016. Flow saturation characteristic of counterbalance valve and application. *Journal of Mechanical Engineering* 52: 205–212.

Xu, J. & Xinhui, L. 2012. Simulation analysis of the counterbalance valve used in cranes based on AMESim software. *Applied Mechanics and Materials* 233: 55–61.

Zaehe, B. & Herbert, D. 2015. New energy saving counterbalance valve. *SAE International Journal of Commercial Vehicles* 8: 583–589.

Zhao, L., Liu, X. & Wang, T. 2010. Influence of counterbalance valve parameters on stability of the crane lifting system. *2010 IEEE International Conference on Mechatronics and Automation, Xi'an* 1: 1010–1014.

Chapter 13

Calculations of unsteady processes in channels of a hydraulic drive

Alexander Gubarev, Alona Murashchenko, Oleg Yakhno, Alexander Tyzhnov, Konrad Gromaszek, Aliya Kalizhanova, and Orken Mamyrbaev

CONTENTS

13.1 Introduction 149
13.2 Analysis of features and varieties of channels of hydraulic drive types 151
13.3 Materials and methods of studying the nonstationary processes in
 channels of the hydraulic drive 152
13.4 Conclusion 159
References 160

13.1 INTRODUCTION

The operating time of a hydraulic drive is a very important indicator in the control systems of complex objects (Hanzha et al. 2017, Sokol et al. 2018). Calculation of this time traditionally takes into account the temperature of the environment, especially when designing objects with means of hydroautomatics (Skorek 2014, Kozlov et al. 2018, Mykheev 1977, Petukhov et al. 1983). Changing the temperature of the fluid affects the performance of the hydraulic drive in several directions, for example, a significant reduction in lubricating properties, loss of power due to pressure loss in the hydraulic drive channels, decrease of useful output due to unexpected operation of the pressure valves, and cessation of movement due to paste-like transformation of the fluid (Malrai et al. 2017, Peña et al. 2015, Guana et al. 2015). Known effects of changes in the fluid are deviations from the calculated values of power, speed, effort, and speed of the initial stage of movement (Kozlov et al. 2018, Hanzha et al. 2017, Balova et al. 2012, Kotyra et al. 2013). The established temperature conditions and stabilized operational modes of operation for hydraulic control systems in industrial applications are typical. These circumstances allow us to calculate and maintain the most favorable temperature parameters of the fluid during operation of the systems. The process of stabilizing heat flows in the system covers the initial or preparatory stage of the system. Cyclic power fluctuations associated with fluctuations in power consumption in a production cycle are a consequence of the production technology. They lead to cyclical fluctuations in temperature fluid. Such changes are predictable and stable for a certain technology; they are taken into account when designing a hydraulic drive. The influence of cyclic changes in heat fluxes is reduced by means of heat exchangers,

DOI: 10.1201/9781003224136-13

accumulators, an increase in the amount of fluid in the system, and the intensification of heat exchange with the surrounding environment (Kozlov et al. 2018, Malrai et al. 2017). The speed and timing of the hydraulic actuators of the control systems are derived from several unstated processes (Hraniak et al. 2018). The first tireless process is the transient process of acceleration and stopping of the drive under the action of loading and inertial efforts. The second tireless process is the replacement of the fluid contained in the hydraulic drive channels with the fluid contained in the hydraulic tank. This process characterizes the change through the length of the channels of fluid with one parameter for fluid with other parameters. The third tireless process is the change of the parameters of the fluid during its movement along the channels. If the material of the channels has a lower temperature than the temperature of the fluid coming from the pump, then the cooling of the fluid occurs during its movement. The fourth tensile process is the change in the temperature of the material of the channels during the movement of the fluid with a different temperature. For connecting pipelines, the mass of the channels is insignificant and the change in temperature, taking into account the thermal conductivity and heat capacity of the material, occurs quickly. For channels passing through body elements, the time and amount of heat absorption by the material is large. The temperature change time exceeds the time that the drive actuates, and the amount of heat changes the temperature of the fluid along the channels. The influence of the above processes on the operation of the hydraulic drive determines changes in the viscosity in the order of hundreds of times, the hydraulic resistance ten times, and useful efforts several times, which determines the transition process and driveability of the drive.

Hydraulic systems of mobile equipment are unable to stabilize the temperature, or maintain the most favorable temperature indicators, of their hydraulic fluids (Kozlov et al. 2019, Sokol et al. 2018). These circumstances lead to a discrepancy in the calculated parameters of the hydraulic system compared to the real values of these indicators under operating conditions (Murashchenko et al. 2013, Sokol et al. 2018). The most significant consequences of such inconsistencies are in aviation and ship hydraulics, hydropower systems, and hydraulics of mobile machinery of the mining industry and agricultural purposes (Skorek 2014). The impossibility of waiting for thermal balance to be achieved, compressed timing of the operation of the drive signal control, limited energy resources for thermal stabilization of the machine, and a wide range of changes in conditions and operating modes are common for such systems. That is, the performance of the hydraulic drive at the design stage cannot be ensured without taking into account such transient changes. Taking into account the mutual influence of these processes in an unstable environment and changes in operating modes adds further complexity. For hydrodynamic systems of the specified class, it is necessary to take into account undisturbed heat fluxes in the calculation of the operational parameters, such as speed, effort, and time of processing the input signal. One of the most important problems with mobile technology is the reliability of fail-safe operation and functional completeness in conditions of rapid temperature fluctuations. An additional opportunity to improve the efficiency of the drive is to take into account nontypical temperature regimes in the control algorithms, for example, biased switching in the case of deceleration of the drive, and changes in the settings of pressure valves. The initial step for the practical application of this approach is to calculate the expected performance of the drive, taking into account tensile processes.

Calculations of unsteady processes 151

The purpose of the research is the development of a method for taking into account transient processes in the calculation of a hydraulic drive (effort, speed, flow, and pressure) working in nontypical temperature mode. Application methods which take into account the instability and the influence of temperature on the action of the hydraulic drive distributed over the length of the channels can be used during the modeling and design of hydraulic drives. Such a simulation will determine the functional-vulnerable elements of the drive which result in reduced performance. The next step is to define the rational parameters or change the structure or design of the drive.

13.2 ANALYSIS OF FEATURES AND VARIETIES OF CHANNELS OF HYDRAULIC DRIVE TYPES

The basis of this study was the typical structure of the hydraulic drive provided by means of functional modules and elements that combine the description of thermal, informational, energy, and hydraulic processes in the form of an operating cycle (Figure 13.1) (Skorek 2014, Kozlov et al. 2018, Sokol et al. 2018, Peña et al. 2015, Del' et al. 1975).

Figure 13.1 shows that the diameter of the channel varies, the thickness of the walls of the design itself is also different, and accordingly, it leads to the uneven distribution of temperature of the working fluid and the design itself. Therefore, the heat capacity at each site will be different and depends on the thickness. This feature should be taken

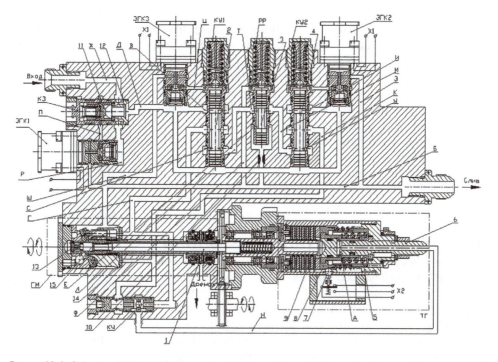

Figure 13.1 Scheme КПМ-148 flaps / preflight of an airplane (main positions: control calve 1 (KK1), electrohydraulic valve 3 (ЕГК3), flow controller (РП), electrohydraulic valve 2 (ЕГК2), shuttle valve (КЧ), bypass valve (КП), and control valve 2 (KK2)).

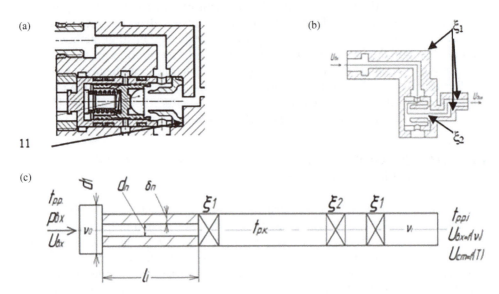

Figure 13.2 Image of the calculated positions of the trajectory of the motion of the fluid in a section of the beginning of the trajectory of the motion of the fluid (Figure 13.1, respectively).

into account when pre-design calculations of mobile drives with tireless processes of work are being performed. Figure 13.2 presents a step-by-step approach to considering the design with the further representation of its parts in the calculation model. The mode of power supply to the system before the inclusion of control elements are as follows: Figure 13.2a shows the basic structural scheme of the trajectory; Figure 13.2b shows the scheme of the trajectory of the motion of the fluid; and Figure 13.2c shows the calculation scheme of the hydraulic model.

In addition to the geometric differences, channels in the hydraulic drive have very different features of their inactivity: some channels are used often, others are rarely used; some are flow channels, others are dead-end channels. All of this results in both complexity of the calculations and complications of the dynamic processes in the work of the drive.

13.3 MATERIALS AND METHODS OF STUDYING THE NONSTATIONARY PROCESSES IN CHANNELS OF THE HYDRAULIC DRIVE

To construct a calculation model of pressure and drain lines, they are divided into areas of equal volume. Mathematical calculations of the sections of the channel of the hydraulic drive are proposed for a simplified calculation of the hydraulic model with a fixed diameter (section) of the channel in each section, a constant value of the temperature of the walls along the channel, a constant density along the channel, and a fixed value of the pressure difference between the pressure and drain lines (Figure 13.3a and b).

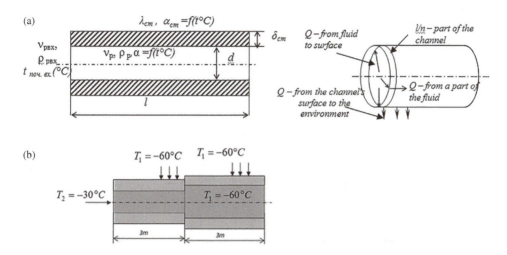

Figure 13.3 (a) Estimation model of the hydraulic channel distributed over a fixed diameter object; (b) Estimated model of hydraulic variable channel diameter divided into sections.

Mobile drives often work under volatile temperature conditions, making it necessary to take into account the Newton-Richman law (Mykheev et al. 1977, Petukhov et al. 1983): $dQ = \alpha(T_c - T_{p.ж.})$, respectively, under unstable temperature conditions, we have $\alpha \to \mu \to Nu \to Pr \to Re \to \Delta p$.

It is important to determine the coefficient of heat transfer (Mykheev et al. 1977, Petukhov et al. 1983) α, which affects Nusselt's criteria $Nu = (\alpha \cdot L) / \lambda$, and the Prandtl number $Pr = \upsilon/\chi = \mu c_p/\lambda$, which are related to the Reynolds number $Re = \rho UL/\mu = UL/\upsilon$ and have the following reciprocal link:

$$Nu = \frac{0.023 \, Re_D^4}{Pr^n}$$

where the index n when the temperature rises is $T° \uparrow \Rightarrow n=0.4$; and when lowered, $T° \downarrow \Rightarrow n=0.3$.

A modified method of "step-by-step" analysis is used, which is similar to method proposed by Thorner (Murashchenko et al. 2013, Mykheev et al. 1977, Petukhov et al. 1983), when the drive operates in non-isothermal modes and the temperature of the liquid in its elements is significantly different (Figure 13.4). This improves the method of calculation and approximates its initial values in comparison with experimental data.

Simulation of the process of stabilizing the speed of the fluid for the cooling (heating) conditions, taking into account the changes in the temperature of the working fluid, is determined by the process of the gradual displacement of the cooled (heated) fluid along the sections.

To account for non-isothermal processes, an algorithm for the sequential calculation of parameters and characteristics of fluid flow sections in a hydraulic channel at

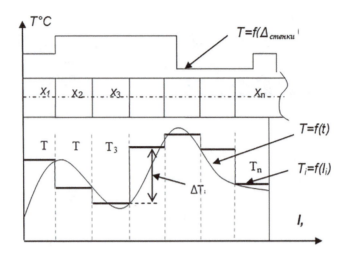

Figure 13.4 Scheme of taking into account the temperature distribution in the channel under the condition of distribution to the parts of the conditionally constant temperature of the fluid and the absence of changes in its density.

different temperature values in each section is proposed (Murashchenko et al. 2013), (Figure 13.5).

During calculations, it was also determined that one of the main factors influencing the operation of the hydraulic drive is the dependence of the viscosity of the working fluid on the temperature: $\mu = f(T°C)$. This dependence was determined on the basis of the Sterling equation $\mu_c/\mu_{p.p.} = \left(T/T_0\right)^{3/2} (T_0 + S)/(T + S)$, where $S = const$, and in

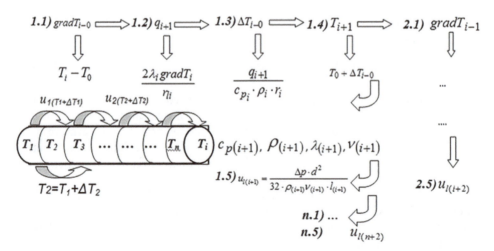

Figure 13.5 Scheme of the algorithm for finding the speed at a variable temperature process along the channel.

Calculations of unsteady processes 155

general terms: $\mu_c/\mu_{p.p.} = (T/T_0)^{n\mu}$. Dependence $T = f(t)$ in the calculation algorithm is used in a polynomial form (Mykheev et al. 1977, Petukhov et al. 1983):

$$T = a_0 + a_1 \cdot t + a_2 \cdot t^2 + \cdots + a_n \cdot t^n$$

The result is the expression: $\dfrac{\mu_0}{\mu} = 1 + \beta_{\mu 1}(T - T_0) + \beta_{\mu 2}(T - T_0)^2 + \ldots$ or $\mu_0/\mu = (T/T_0)^b$ (Mykheev et al. 1977, Petukhov et al. 1983). In the case of parameter value $b = 1$, we will get $\mu_0/\mu = (T/T_0)^1$, or:

$$\frac{\mu_c}{\mu_{p.p.}} = \left(\frac{a_0 + a_1 \cdot t + a_2 \cdot t^2 + \cdots + a_n \cdot t^n}{T_0} \right)^1$$

The dependences (Figure 13.6) of changes in viscosity and other parameters along the trajectory of the motion of the fluid with the consideration of temperature are $\mu_i = f(t_i, l_i)$, $\rho_i = f(t_i, l_i)$, $U_i = f(t_i, l_i)$, and $\Delta p_i = f(t_i, l_i)$. This makes it possible to predict the time of the hydraulic actuation during unsteady operating modes (Skorek 2014, Kozlov et al. 2018, 2019, Lure et al. 2013, Murashchenko et al. 2013, Mykheev et al. 1977, Petukhov et al. 1983, Sokol et al. 2018, Peña et al. 2015, Guana et al. 2015). The value of the temperature of the corresponding distribution of viscosity is found as:

$$t = \frac{\lambda^3 \cdot T^3 \cdot 2\pi \cdot (l+r) \cdot (t_c - t_0)}{r^2 \cdot l^2 \cdot c \cdot \rho \cdot \sqrt{\dfrac{\lambda}{c \cdot \rho} \cdot \pi}} \cdot (t_0 - t_c) + t_c$$

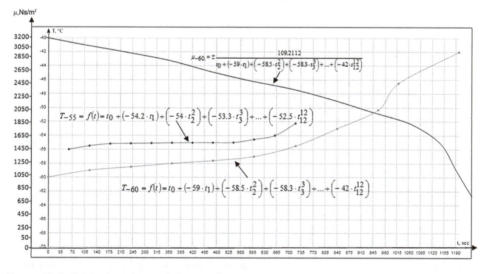

Figure 13.6 Critical analysis of the results.

Taking into account the value of temperature along the channel (with variable diameter), the velocity of fluid flow is (Guana et al. 2015):

$$U_i(1,2) = \frac{-32 \cdot \rho \cdot \left(\left(A1/d_1^2\right) + \left(A2/d_2^2\right)\right) \pm \left(\left(32 \cdot \rho \cdot \left(A1/d_1^2 + A2/d_2^2\right)\right)^2 + 2 \cdot \xi_{M1} \cdot \rho \cdot \Delta p\right)^{1/2}}{\xi_{M1} \cdot \rho},$$

where $A1 = v_1 \cdot l_1 + \sum_{0}^{i}(l_1 - l_{1i}) \cdot v_{1i}$ and $A2 = v_2 \cdot l_2 + \sum_{0}^{i}(l_2 - l_{2i}) \cdot v_{2i}$.

According to the results of the calculation, the dependences of the stabilization time and the stabilized value of the flow velocity of the fluid in the hydraulic channel during unstretched temperature regimes was determined (Figure 13.7, corresponds to the fixed diameter channel model from Figure 13.3). The obtained results, using the proposed model, are different from those obtained by conventional hydraulic calculations (Murashchenko et al. 2013). The graphs show the velocities in the channel in the corresponding section n for identical initial conditions on the channel geometry (length, diameter, number of sections) and at different Δt between the fluid that is in the channel $t_{p \cdot k.} = (-60°C$ and $-40°C) = var$ and the fluid, which is $t_{вх./поч.} = (-30°C) = const$. The time to stabilize at $t_{p \cdot k.} = (-40°C)$ is $T_{cm1} = 227$ (sec), which is about 100 seconds faster than with $t_{p \cdot k.} = (-60°C)$.

The process of stabilizing the speed at different temperature conditions at the beginning in the channel and the supply (–50°C and 0°C) was simulated. How the process of displacement of warmer liquid proceeds is shown, and that which is cooled in the channel (parts of the channel $l = 0.65/0.8$ m), is demonstrated in Figure 13.8.

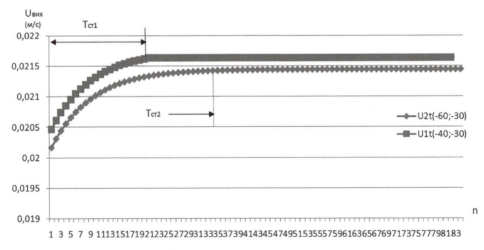

Figure 13.7 Function of the stability of the speed of the working fluid for different temperature values, where n is the number of sections of the channel.

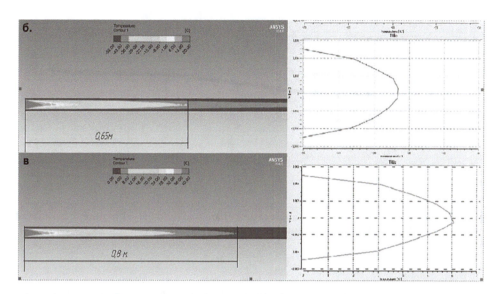

Figure 13.8 Simulation of the process transition from cold to warm in the channel.

In the second stage, considering the problem of practical direction, the pressure channel of a mobile drive with a total length of 12 m and a diameter of 0.06 m was simulated at ambient temperature and an initial liquid temperature of −60°C. It was determined that the element with the greatest hydraulic resistance is filled with a liquid with a minimum temperature for the maximum time. By repositioning the drive (transferring the element along the low temperature channel), the drive time was reduced to a stabilized speed (transient time) from almost 6 to 2.3 minutes.

In order to confirm the application of the proposed theoretical calculations of hydraulic channels in hydraulic drives under unstable operating conditions of the system, model and experimental determination of the parameters of the mobile multi-mode drive КПМ-148, placed in a thermobaric chamber, was carried out (the design scheme of the drive is shown in Figure 13.1) (Kozlov et al. 2019, Murashchenko et al. 2013). As a result of a comparison of the calculated and experimental data (Figure 13.9), it was established that the error of calculating the time of stabilization of speed is 14% $\left(\Delta t = \left(\left(t_{po3p} - t_{ekcn}\right) / t_{ekcn}\right) * 100\% = 14\%\right)$ and the relative error at the temperature setting was 8.1% $\left(\Delta T° = \left(\left(T°_{po3p} - T°_{ekcn}\right) / T°_{ekcn}\right) * 100\% = 8.1\%\right)$. Differences in the obtained values can be considered to be the result of not taking into account the change in liquid density along the channel and the conditions of convective cooling of the channel, and their elimination requires improvement of the method of calculating the unstable modes of operation of the hydraulic drive.

An analysis of the obtained graph of comparisons of the experimental data with an overlay of the theoretical calculations in determining the time of stabilization of speed follows. At points in the time line from 30 to 55 seconds, and in the area of 257 seconds, as shown in Figure 13.9, the calculations deviate from the points obtained

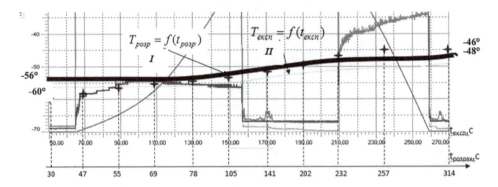

Figure 13.9 Graph of dependence of temperature change of fluid FH-51 in time: I, estimated temperature values and II, experimental temperature values.

experimentally, which can be justified by the fact that the parameters of the heat capacity of the material in the case of the drive were not taken into account. Since the range of wall thickness of the channels in the design varies from 0.6 to 1.6 mm, the process of heat transfer will be different; the heat from the liquid entering the cold case is unevenly distributed.

The next stage of the study is consideration of the convective heat transfer and different geometric parameters of the channels (Murashchenko et al. 2013). When taking into account the parameters that depend on the change in temperature in most sources, the heat transfer coefficient κ_i takes somewhat the shortened form to simplify the calculations, which in the case of transient processes produces a greater calculation error. The heat transfer coefficient k_i (W/(m^2·°C)), which is taken into account when the temperature changes (Mykheev et al. 1977, Petukhov et al. 1983), is as follows:

$$\Delta t = T_p - T_c = \frac{Q_c}{\left(\sum S_i \cdot k_i\right)}, \text{defined as } k_i = \frac{1}{\frac{1}{\alpha_{cm}} + \frac{\delta_{cm}}{\lambda_{cm}} + \frac{1}{\alpha_{H.c.}}}$$

where λ_{cm} is the coefficient of thermal conductivity of the wall material W/(m·°C); δ_{cm} is the wall thickness; and α is the appropriate coefficients of heat conductivity of the material from the wall to the fluid, and from the wall and the environment.

The results of this type of calculations can significantly improve the resulting dynamic performance of systems with transient nonstationary processes under different operating temperature conditions.

An example of a calculation program in Excel is shown in Figure 13.10, with the ability to select a fluid, different channel parameters, different initial temperature conditions, followed by the calculation of temperature changes, speed along the channel, and with the values of time stabilization of temperature and speed under transient processes.

(a)	Channel material			(b)	Parameters of the channel from influence of temperature change				
Geometric channel parameters				Parameters					
marking				in the beginning	marking		variables		marking
Wall thickness, m	δ		0.003	wall temperature	t	-20			
Channel Length, m	l		1.5	Heat capacity of walls, J / kg * deg	$c_p=$	0.92	Heat capacity of walls, J / kg * deg	$c_p=$	
Diameter inner, m	d		0.06	Thermal conductivity of the fluid, W	$\lambda=$	57	Thermal conductivity of the liquid	$\lambda=$	
				Heat transfer of walls, W / m² * hail	$\alpha=$	768	Heat transfer of walls, W / m² * hail	$\alpha=$	
				The density of the heat flux of the w	q	757	The density of the heat flux of the w	q	
				Heat transfer coefficient	κ=	646	Heat transfer coefficient	κ=	

(c)			Type of fluid				
Parameters of the working fluid in the channel at the beginning of the calculation and before entering the channel							
in the channel		marking		before entering the channel		marking	
temperature fluid		t	-50	temperature fluid		t	30
viscosity fluids, m²/s		v=	1.25	viscosity fluids, m²/s		v=	0,02
Density of fluid, kg/m³		ρ =	860	Density of fluid, kg/m³		ρ =	835,6
capacity of liquid, J / kg * deg		$c_p=$	1844	capacity of liquid, J / kg * deg		$c_p=$	1900
Thermal conductivity of the liquid,		$\lambda=$	0.1192	Thermal conductivity of the liquid,		$\lambda=$	
Heat transfer fluid, W / m² * hail, α		$\alpha=$	6	Heat transfer fluid, W / m² * hail, α		$\alpha=$	
The heat flux density, q, W / m², for		q	300	-1192 The heat flux density, q, W / m², for		q	

Figure 13.10 Calculation via Excel.

13.4 CONCLUSION

Taking into account the transients in the hydraulic channels of the hydraulic drive allows using the dependences of the distribution of viscosity, temperature, and density along the channel when calculating the force, acceleration, and speed of the output link of the drive. The error in the hydraulic calculations of the speed and effort of the hydraulic drive, caused by ignoring changes in the fluid temperature values, can range from several percent to several times. The greatest error occurs when the drive time is reduced during the transient process. The transition process in the drive motion has the longest time among the processes of temperature stabilization, heat transfer, and inertial load. The effect of stabilizing the viscosity and heat exchange can be from (30% to 50%) slow processes in channels of short length and up to 100% and more for high-speed drives. The calculation of rational values of pipeline diameters and the determination of the parts of the desired thermal insulation for the hydraulic cycle operation requires consideration of the distribution in time of repetition of the operating modes of the drive and the operation of the drive in unstated temperature conditions. The calculations performed in the example offer a layout of the arrangement of channels and local resistances and obtain the following calculation data (Murashchenko et al. 2013, Lure et al. 2013): reduction of pressure losses in nominal mode of operation by 16%; increase in average weighted energy efficiency by 14%; and reducing the time of working out the drive in unsteady conditions by 8%–12%, depending on the temperature state of the fluid.

REFERENCES

Balova, T. Kriulko, R.N. Kotyra, A. 2012. System of knowledge extraction from ontology of electronic educational resource. *Informatyka, Automatyka, Pomiary w Gospodarce i Ochronie Środowiska* 3: 3–4.

Del', G.D. Ogorodnikov, V.A. Nakhaichuk, V.G. 1975. Criterion of deformability of pressure shaped metals. *Izv. Vyssh. Uchebn. Zaved. Mashinostr* 4: 135–140.

Guana, L. Chenb, G. 2015. Pumping systems: Design and energy efficiency. Encyclopedia of Energy Engineering and Technology. CRC Press, United States of America, pp. 1–8.

Hanzha, A.M. Marchenko, N.A. Pidkopai, V.M. Niemtsev, E.M. 2017. Modeliuvannia protsesiv peredachi teploty vid kotelni do zhytlovoho masyvu na osnovi hidravlichnykh rozrakhunkiv skladnoi teplovoi merezhi. *Hidravlichni mashyny ta hidroahrehaty* 22(1244): 83–87.

Hraniak, V.F. Kukharchuk, V. Bogachuk, V.V. Vedmitskyi, Y.G., Vishtak, I.V., Popiel, P. Yerkeldessova, G. 2018. Phase noncontact method and procedure for measurement of axial displacement of electric machine's rotor. *Proceedings of SPIE* 1080866: 1825–1831.

Kotyra, A. Wójcik, W. Gromaszek, K. Popiel, P. Ławicki, T. Jagiełło, K. 2013. Detection of biomass-coal unstable combustion using frequency analysis of image series. *Przegląd Elektrotechniczny* 89(3b): 279–281.

Kozlov, L.G. Bogachuk, V.V. Bilichenko, V.V. Tovkach, A.O. Gromaszek, K. Sundetov, S. 2018. Determining of the optimal parameters for a mechatronic hydraulic drive. *Proceedings of SPIE* 10808: 1080861.

Kozlov, L.G. Polishchuk, L.K. Korinenko, M.P. Horbatiuk, R.M. Ussatova, O.S. 2019. Experimental research characteristics of counterbalance valve for hydraulic drive control system of mobile machine. *Przegląd Elektrotechniczny* 4: 104–109.

Lure, Z.Ya. Nykolenko, Y.V. Rыzhakov, A.N. 2013. Uravnenye sostoianyia y fyzyko-mekhanycheskye kharakterystyky rabochei zhydkosty pry modelyrovanyy perekhodnыkh protsessov v hydropryvode. *Promыshlennaia hydravlyka y pnevmatyka* 41(3): 49–58.

Malrai, F. et al. 2017. Power conversion optimization for hydraulic systems controlled by variable speed drives. *Journal of Process Control* 59: 67–71.

Murashchenko, A. et al. 2013. Simplified calculation of lines for hydraulic drive considering the change temperature of fluid. *Motrol. Motoryzacja i Energetyka Rolnictwa* 15(5): 173–179.

Mykheev, M.A. et al. 1977. *Osnovy teploperedachy*. Moscow. Enerhyia.

Peña, O.R. Leamy, M.J. 2015. An efficient architecture for energy recovery in hydraulic elevators. *International Journal of Fluid Power* 16(2): 83–98.

Petukhov, B.S. et al. 1983. Teploobmen y hydravlycheskoe soprotyvlenye v trubakh pry turbulentnom techenyy zhydkosty okolokrytycheskykh parametrov sostoianyia. *Hidravlichni mashyny ta hidroahrehaty* 1(21): 71–79.

Skorek, G. 2014. Sprawność energetyczna napędu hydrostatycznego. *Hydraulika i Pneumatyka* 6: 7–10.

Sokol, Y. Cherkashenko, M. 2018. *Synthesis of Control Schemes for Hydroficated Automation Objects*. Riga: LAP LAMBERT Academic Publishing.

Chapter 14

Analysis, development, and modeling of new automation system for production of permeable materials from machining waste

Oleksandr Povstyanoy, Oleg Zabolotnyi, Olena Kovalchuk, Dmytro Somov, Taras Chetverzhuk, Konrad Gromaszek, Saltanat Amirgaliyeva, and Nataliya Denissova

CONTENTS

14.1 Introduction 161
14.2 Research and analysis of the method of producing PPMs 162
14.3 Development of processing technology with use of machining waste 164
    14.3.1 Main powder properties 164
    14.3.2 Dry radial isostatic pressing (DRIP) of PPMs 165
    14.3.3 Sintering of the PPM 167
    14.3.4 Properties of PPM 168
14.4 Results and discussion 169
14.5 Conclusion 171
References 171

14.1 INTRODUCTION

The modern trend of sustainability of industrial development is increasing requirements for the quality of all types of manufactured products. The existence of traditional powder metallurgy technology makes it possible to produce new porous permeable materials (PPMs). Nevertheless, it is necessary to predict and control the parameters of their structure in the process of manufacturing, which include the granulometric composition of the coke charge, the shape of the particles, the density of the molded workpiece, the quality of the contacts, shaping, porosity, density, and their volume of distribution. Today, modeling using computer information technologies and the automation of production enables us to increase the efficiency of traditional technologies and to introduce non-waste production of common products, save energy, reduce labor costs, and control the parameters of the structure of porous powder materials in the process of their manufacture (Reut et al. 1998, Boginsky et al. 2001, Wójcik et al. 2012, 2014).

DOI: 10.1201/9781003224136-14

Current problems of creating porous technologies are solved due to the use of computer modeling, automation, and modern computer-aided drafting (CAD) systems, which optimize the technology for obtaining PPMs.

14.2 RESEARCH AND ANALYSIS OF THE METHOD OF PRODUCING PPMs

To obtain porous materials with high permeability, it is necessary to use powders with a large particle size, while at the same time, it is necessary to use powders of small particle size to obtain high fineness of cleaning. These contradictions lead to the need to find new technological techniques, automated tools, and computer modeling methods that will allow the creation of such structures of porous materials that provide the best possible combination of operational characteristics (Rud et al. 2014).

A literature review of the processing of metal-containing waste products showed that there is a sufficiently large number of technologies for obtaining powder from the sludge waste of tool and bearing steels (Povstyanoy et al. 2014, Rud et al. 2005).

To obtain filtering PPMs, a new installation was designed and manufactured. This device is used for pressing the sealing materials of various kinds: metallic and ceramic powders, graphite, fibers, wire, wire mesh, etc. (Figure 14.1).

To expand the range of porous powder products, save raw materials, and reduce expenditures on production, a solid elastic insert 19 (Figure 14.1) can be made by forming a set of inner folded inserts (Figure 14.2) (Sarantsev et al. 2005, Zabolotnyi et al. 2019). This allows us to get products with a wider range of sizes as well as to improve the technology and the culture of pressing.

PPMs made with the help of this installation (Figure 14.1) meet modern requirements as to the quality of products of this kind.

In the developed object-oriented CAD, the parameterization mechanism is implemented with the use of the parametric drawing and modeling system T-FLEX CAD and Pro / ENGINEER, which became the basis for the development of a system for modeling the parametric design of installations for dry radial isostatic pressing (Figures 14.3 and 14.4).

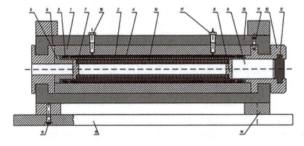

Figure 14.1 A device for the pressing of sealing materials of various kinds: metallic, ceramic powders, graphite, fibers, wire, wire mesh, etc.: 1, case; 2, reinforced elastic shell with cuffs; 3, inserts; 4, working camera; 5–8, intermediate stops; 9, ring; 10, cover; 11, screw; 12, slider; 13, rigid cylindrical frame; 14, ring; 15, work bench; 16, screws; 17, fittings; 18, stub; and 19, elastic liner.

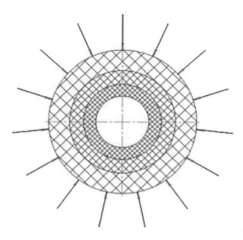

Figure 14.2 Scheme of pressing using an elastic insert formed in a set of inner folded inserts.

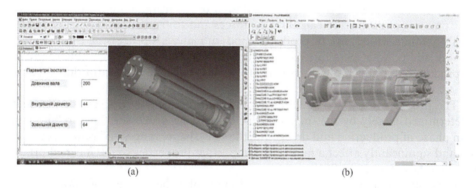

Figure 14.3 System of modeling of parametric design of installations for dry radial isostatic pressing of porous permeable materials based on (a) T-FLEX CAD (b) and Pro / ENGINEER.

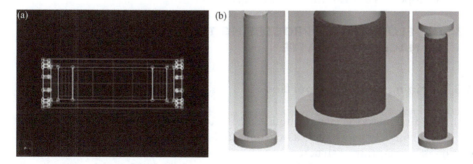

Figure 14.4 System of calculation of forces and voltage in detail of the design of the installation model for dry radial isostatic pressing of porous permeable materials (a) and modeling of powder filling process in the form of radial isostatic pressing (b) by using Pro / ENGINEER.

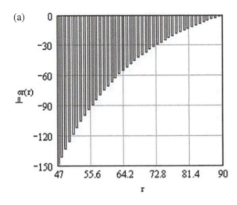 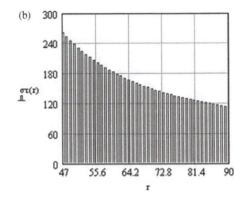

Figure 14.5 The curves of radial and tangential stress variation on the thickness of the isostat cylinder for pressing.

One of the important aspects of the design of the installation for dry radial isostatic pressing of PPMs is to calculate the forces and voltage in detail (Figure 14.5).

To obtain a wide range of products, modern materials processing technology for pressure and powder metallurgy is required. In many cases, they are able to provide the production of articles with special properties, especially those products that are difficult or impossible to produce by other methods.

The progress in powder metallurgy and the treatment of materials by pressure is determined largely by the improvement of pressing processes. They relate to the main phase of production and determine the size, shape, range, and power consumption and significantly affect a number of important properties of the finished product (Ogorodnikov et al. 2004, 2018).

Developed and existing technologies do not solve the problem of manufacturing PPM products with the optimal combination of structural characteristics and physical and chemical properties. It is important to control the quality of the products and to mechanize and automate the processes of pressing, equipment, and tools, by predicting their properties at the initial stage of formation.

14.3 DEVELOPMENT OF PROCESSING TECHNOLOGY WITH USE OF MACHINING WASTE

14.3.1 Main powder properties

Steel BBS15 refers to the class of high-carbon steels and is characterized by the presence of alloying elements such as chrome and manganese, which contribute to the creation of a fine-grained structure.

The starting material for PPM filtration is powder BBS15, which is obtained according to the technology described by Rud et al. (2005). The received metallic powder proved to be of high quality, containing particles of regular shape and size, and exhibited high technological properties. The conducted phase-structure analysis of

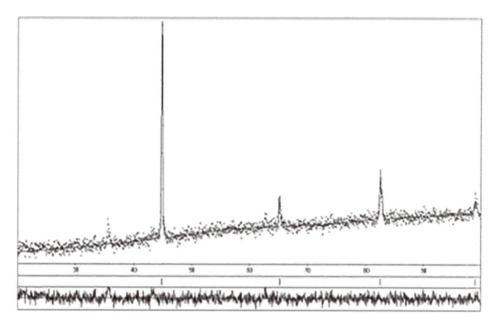

Figure 14.6 Phase analysis of the first sample of powder BBS15.

the starting material (Figure 14.6) by using a DRON-3.0 X-ray diffractometer showed that the phase with the structure of α-Fe comprises about 93%–95%, which indicates a rather high iron content in the powder.

14.3.2 Dry radial isostatic pressing (DRIP) of PPMs

There is one option possible to change the pore size of the filter material, which is adding a pore maker such as carbide powder $(CO(NH_2)_2)$ to the pit. The analysis of such research results (Rud et al. 2004) showed that the minimum content of pore maker that significantly increases the form of the powders is 5%–10%.

Figure 14.7 shows the distribution of densities along the radius of cylindrical billets based on powder BBS15 (Ø40, $L = 220$ mm) of several batches. According to the calculations, there is a slight change in density within 2% according to both the estimated and experimental data. The use of a radial pressing scheme produces filter membranes with an even density distribution across a wide range of pressures.

The pressing of powder BBS15 is carried out for 1.5–2.5 minutes at a pressure of 80–90 MPa. The dependence of the porosity of the filter PPM on the compression pressure is presented in Figure 14.8. It is evident from Figure 14.8 that under the same pressure, the porosity of the samples from the steel powder BBS15 is higher than the porosity of the samples of alloyed steel 12Cr18Ni10. Moreover, if the difference in porosity at small pressures is 5%–10%, then at pressures up to 900 MPa, it increases by 1.5 times. This suggests that as the pressure used to create the workpieces increases, the porosity is reduced, which is not desirable for filtering materials.

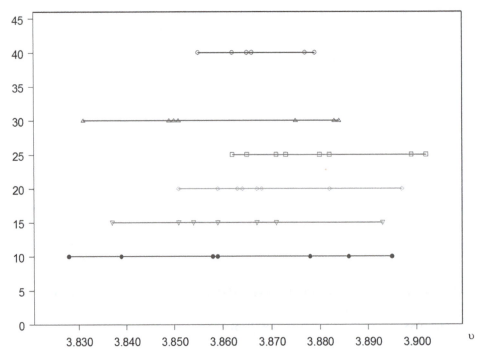

Figure 14.7 Distribution of densities along the radius of cylindrical workpieces based on powderBBS15 (Ø40, L = 220 mm) of 6 batches.

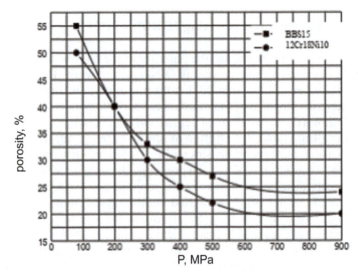

Figure 14.8 Experimental dependences of porosity of filtration PPM made of powder BBS15 and powder of the brand 12Cr18Ni10 on the pressure of pressing.

14.3.3 Sintering of the PPM

An important technological operation is the sintering of the preformed workpieces that are used to manufacture filtering materials (FM). The purpose of sintering is to provide sintered products of a specific porous structure and the corresponding physical, mechanical, hydraulic, and chemical properties. The main technological parameters of the process are influenced by the temperature regime, the medium of sintering, etc. Moreover, the pressure of the outgoing workpieces affects the sintering process. This means that by increasing the density of the presses, the workpiece shrinkage decreases, because the relative increase in density during sintering occurs predominantly in workpieces which are pressed at lower pressures.

Sintering of filter PPM made of powder BBS15 is carried out in a vertical vacuum furnace in three intervals (Krokhin 1980):

I. 1.5 hours at a temperature of 200°C–1,000°C;
II. 2 hours at a temperature of 1,000°C–1,050°C;
III. 2.5 hours at a temperature of 1,050°C–100°C.

The dependence of the porosity of the sintered products on the temperature of sintering is shown in Figure 14.9.

The dependence of the porosity of the sintered samples on the temperature of sintering confirms the conclusion that doping elements significantly affect the porosity, and the filter PPM with BBS15 has almost the same porosity during the sintering process (Hraniak et al. 2018).

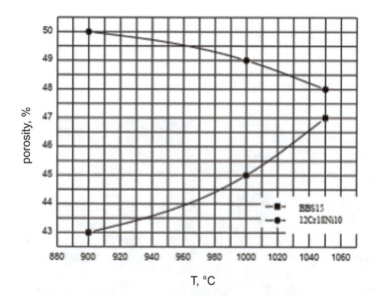

Figure 14.9 Experimental dependences of porosity of sintered filter PPM made of powder BBS15 and 12Cr18Ni10 on the temperature of sintering.

14.3.4 Properties of PPM

Filtering PPM Ø40×220 mm, obtained by DRIP from powder BBS15, is depicted in Figure 14.10. Figure 14.11 shows the structure of the filtering material. One can see the volumetric porosity and microporosity of the particles. The latter increases the specific surface of the filter material and, as a result, its adsorption properties, which is very important for filters used for cleaning liquids and gases.

Figure 14.12 shows the results of experimental studies obtained from a semi-industrial installation for DRIP. The performed experimental research was performed according to theoretical studies and are compared with the porosity distribution curve for a spherical powder brand 12X18H10 (Stepanenko et al. 1984). The results confirm that the DRIP method provides for obtaining powder-permeable materials with a more even distribution of pores.

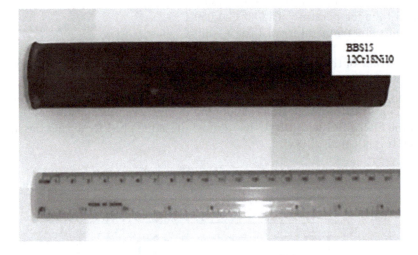

Figure 14.10 Sample of filter material made of powder BBS15.

Figure 14.11 Structure of filter material made of powder BBS15.

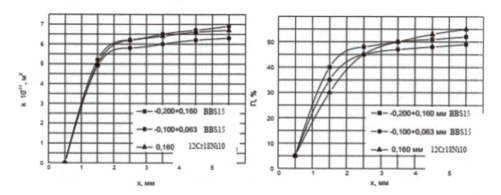

Figure 14.12 Distribution of the coefficient of sensitivity and porosity by filtering.

The presented experimental results show that the following basic characteristics of the BBS15 powder-based FM, such as the uniform porosity distribution by volume, the distribution of the coefficient of filter instability, and the volume porosity and microporosity of the particles, conform to the characteristic results of powder FM based on powder 12Cr18Ni10. In fact, this means that FM with BBS15 has similar characteristics to FM with 12Cr18Ni10, but the former is less expensive to produce and is much more competitive with similar filtering materials.

14.4 RESULTS AND DISCUSSION

The modeling of the porous structure of a specific multilayered PPM from a steel powder BBS15 was carried out in the MATLAB application package. According to the described methodology, software was developed in the C++ programming language ("FiltrN" program), which modeled the process of radial isostatic pressing with a given porosity of the PPM (Povstyanoy and Kuzmov 2016). Since the filter has the form of an elongated circular cylinder (Figure 14.13), the modeling was carried out for

Figure 14.13 Multilayer PPM made of steel powder BBS15 Ø40 × 220 mm by using the method of radial isostatic pressing.

the following parameters of the technological process: the diameter of the mandrel is 40 mm and the diameter of the reinforced elastic shell is 80 mm.

The powder falls into the space between the mandrel and the elastic shell. The relative bulk density of powder BBS15 is 20%, that is, the initial porosity is 0.8.

Based on our calculations, the distribution of porosity and radial velocity in the modeling of the radial isostatic pressing of a multilayered PPM is shown in Figure 14.14.

The developed method of computer modeling determined the distribution of porosity and other characteristics of the powder-permeable material and established the relationship between the technological modes of obtaining them and the operational characteristics.

The radial velocity makes it possible to control the distribution of porosity in a multilayered PPM and allows the factors that contribute to the density of distribution heterogeneity to be analyzed (Figure 14.15).

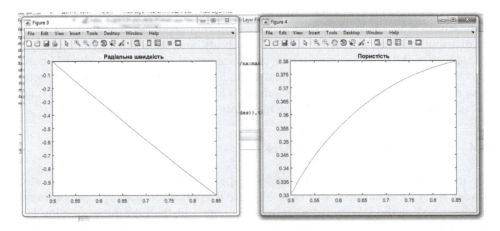

Figure 14.14 Display of graphical dependencies depicting the influence of radial velocity and porosity distribution in the modeling of radial isostatic pressing of multilayer PPM.

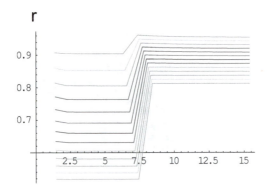

Figure 14.15 Generalized distribution of density along the radius of each layer of PPM at different phases of deformation.

The uneven distribution of porosity by cross section of the PPM was proved by computer modeling in accordance with the corresponding calculated data. The performed modeling confirmed the uneven distribution of porosity and qualitatively corresponds to the process of sealing and then to the process of accumulation of deformations. At the sites of PPM, which are located closer to the outer surface (the last layers), there is an increase of porosity. As a result, the influence of density on the distribution of porosity of the PPM was studied. In the areas located near the axis of the rod capsule, the magnitude of the accumulated deformation is lower. The existence of a counterweight increases the overall level of accumulated deformation and leads to a more even distribution. This increases the level of accumulated deformation in areas located near the axis of the rod of the capsule.

14.5 CONCLUSION

An analysis of the existing, traditional methods of pressing of PPMs showed that there is no technological process including the main positive features of traditional pressing which are non-defective. It is proven that the radial scheme of pressing can be the basis for creating rational equipment and technology for the production of filter materials including metals, ceramics, graphite, and waste industrial production, as it implements the main positive features. In this context, an installation for the pressing of a new generation of PPM is proposed. The installation comprises technology for the production of filter materials based on metals, ceramics, graphite, and industrial production wastes. PPMs produced with the help of this installation contain a whole set of properties which are required by this type of products.

Importantly, the method of computer modeling of PPMs was developed, which takes into account the peculiarities of the distribution of porosity and radial velocity in radial isostatic compression. As a result, a new modern automation system for manufacturing permeable materials from wastes of machining was proposed. The method of computer modeling not only determines the distribution of porosity and other characteristics of the powdered porous units but also predicts their impact on the operational properties of the PPM.

The new automation system for the production of permeable materials has allowed, first, to determine the distribution of porosity and other characteristics of powder-permeable material and, second, to establish the relationship between the technological modes of their obtaining the operational characteristics.

REFERENCES

Boginsky, L., Reut, O., Piatsiushyk, Y., Zabolotny, O. and Kupryianov, I. 2001. The development of processes of pressing of articles from powders on the bases of metals, ceramics and graphite. *15 International Plansee Seminar* 3: 197–209.

Hraniak, V.F., Kukharchuk, V., Bogachuk, V.V., Vedmitskyi, Y.G., Vishtak, I.V., Popiel, P. and Yerkeldessova, G. 2018. Phase noncontact method and procedure for measurement of axial displacement of electric machine's rotor. *Proc. SPIE* 1080866: 1825–1831.

Krokhin, N. 1980. Sintered material from bearing waste for parts of rolling bearings. *Practice of Application and Prospects for the Development of Parts Manufacturing by Powder Metallurgy*. Chelyabinsk, 33–35.

Ogorodnikov, V.A., Dereven'ko, I.A. and Sivak, R.I. 2018. On the influence of curvature of the trajectories of deformation of a volume of the material by pressing on its plasticity under the conditions of complex loading. *Materials Science* 54(3): 326–332.

Ogorodnikov, V.A., Savchinskij, I.G. and Nakhajchuk, O.V. 2004. Stressed-strained state during forming the internal slot section by mandrel reduction. *Tyazheloe Mashinostroenie* 12: 31–33.

Povstyanoy, O. and Kuzmov, A. 2016. Modeling of the porous structure in multilayer filtering powder materials. *III International Scientific and Practical Conference (Topical Problems of Modern Science and Possible Solutions)* 1: 5–9.

Povstyanoy, O. and Rud, V. 2014. Ecological and economic efficiency of using industrial waste for the manufacture of materials for structural purposes. *International Journal of Sustainable Development* 19: 89–94.

Reut, O. Boginskyi, L. and Petiushik, Y. 1998. *Dry Isostatic Pressing of Compactable Materials*. Minsk: Debor.

Rud, V. 2005. *Physical and Mechanical Principles of Complex Pressure Treatment and Vibration Processes in Powder Forming Technologies*. Diss. D.Sc. Kyiv.

Rud, V. Galchuk, T. and Povstyanoy, O. 2004. *Method of Obtaining a Metal Powder from Sludge Waste Bearing Production*. Patent of Ukraine 63558 A MPK 7 B22F9/04.

Rud, V. Galchuk, T. and Povstyanoy, O. 2005. Use of waste bearing production in powder metallurgy. *Powder Metallurgy* 1–2: 106–112.

Rud, V. et al. 2014. Structural characteristics of the blanks when filling molds with particles irregular forms. *New Technologies and Materials, Production Automation*: 89–95.

Sarantsev, V. et al. 2005. *Device for Pressing Products from Powders*. Patent of the Republic of Belarus 2252U MPK B 22F 3/00.

Stepanenko, A. Lozhechnikov, E. and Lozhechnikov, E. 1984. Production of steel powder from bearing production slimes. The influence of processing modes on the composition and properties of the powder. *Powder Metallurgy* 12: 82–85.

Wójcik, W. Gromaszek, K. Kotyra, A. and Ławicki, T. 2012. Pulverized coal combustion boiler efficient control. *Przegląd Elektrotechniczny* 88(11b): 316–319.

Wójcik, W. Shegebayeva, Z. and Koshymbaev, S. 2014. Definition of the objects of multivariable control of technological process of smelting industry on the basis of optimization model. *Informatyka, Automatyka, Pomiary w Gospodarce i Ochronie Środowiska* 1: 18–20.

Zabolotnyi, O. Sychuk, V. and Somov, D. 2019. Obtaining of porous powder materials by radial pressing method. *Advances in Design, Simulation and Manufacturing. DSMIE 2018. Lecture Notes in Mechanical Engineering*, 123–131.

Chapter 15

Study of effect of motor vehicle braking system design on emergency braking efficiency

Andrii Kashkanov, Victor Bilichenko, Tamara Makarova, Olexii Saraiev, Serhii Reiko, Andrzej Kotyra, Mukhtar Junisbekov, Orken Mamyrbaev, and Mergul Kozhamberdiyeva

CONTENTS

15.1 Introduction 173
15.2 Analysis of motor vehicle braking system design specifics and of theoretical premises for assessment of functionality of such braking systems 175
15.3 Results of field research on motor vehicle braking system efficiency in case of emergency braking 177
15.4 Recommendations for upgrading existing approaches to assessment of motor vehicle braking efficiency in the course of accident-related automotive equipment expert examination 180
15.5 Conclusions 182
References.. 182

15.1 INTRODUCTION

The safety of motor road traffic is a pressing problem in many countries. Each year, in the aftermath of road accidents, about 1.35 million people die, 20–50 million people suffer from bodily injuries and the amount of road accident losses and damages accounts for 3% of the GNP in most counties (WHO 2020). This gives rise to a number of tasks to be accomplished to reduce the traffic accident rate. The successful fulfilment of such tasks largely depends on the accuracy and credibility of road accident analysis procedures and on the ability to discover the cause-and-effect sequence.

The main road-accident-preventive method is the braking process (Kashkanov et al. 2010, Turenko & Saraiev 2015, Struble 2013). The efficiency of such process depends on the braking system design specifics and on the functionality of the motor vehicle (availability of an anti-lock braking system, emergency braking system, preventive safety systems, etc.) and is limited by the amount of the frictional force in the contact between the tyres and the road (Reif et al. 2014, Pacejka 2012, Saraiev & Gorb 2018).

There are a lot of different methods, tools and technologies used to assess the braking properties of a motor vehicle to ensure the functionality of the active safety

DOI: 10.1201/9781003224136-15

systems (Singh et al. 2013, Breuer et al. 2007, Singh & Taheri 2015), proper road traffic management (Laugier et al. 2011, Zhang et al. 2017), improvement of motor road design (Mannering & Bhat 2014, AASHTO 2018) and road accident analysis (Kashkanov 2018, Franck & Franck 2009, Danez & Saraiev 2018). The existing requirements (Reif et al. 2014, AASHTO 2018) define such braking efficiency assessment criteria, to be applied in the course of road tests, as the braking distance/length and the rate of steady deceleration. In investigating accident situations, an expert should determine the rate of steady deceleration by way of investigatory experimentation at the scene of the road accident or at a similar place, and based on the data obtained, he/she should calculate the stopping length thereafter (ENFSI 2015). In the case of damage to a motor vehicle, which precludes any road testing, an expert has to use either statistical average deceleration data or resort to an outdated calculating procedure (Puchkin 2010). This would result in increased inaccuracy and/or reduced credibility of the data which are used to prepare an expert opinion. The problem of the availability of a recorded vehicle motion video at the time of the occurrence of a road accident may be resolved in the future on a global level by way of widespread implementation of automated event-logging systems and road accident warning systems (DaSilva 2008, Hynd & McCarthy 2014); however, such systems are not widely used in the world.

So, an assessment of motor vehicle braking efficiency in the course of automotive equipment examination/testing for road accident reporting is associated with calculations based on the measurement results provided by an investigator or by court or based on the typical reference data (parameters and rates) found by an investigator in certain reference or technical books. For instance, in investigating the braking efficiency of a motor vehicle, such reference parameters may be as follows: the driver's reaction time, the braking system initial response time, the deceleration build-up time, the tyre-to-surface adhesion (friction) factor, the motor vehicle's steady deceleration, any road slopes and turn-of-the-road radii, speed of pedestrians, etc. Therefore, the objectivity of the expert's findings directly depends on the reasonable choice of the input data and on the calculation methods used by such expert.

The progress in the design of braking systems and in their functioning algorithms has entailed a significant difference in the braking efficiency between the present-day motor vehicles and outdated ones, which is the subject of this research work. The purpose of the research work is to upgrade the existing approaches to assessment of motor vehicle braking efficiency in the course of accident-related automotive equipment expert examination by way of considering the braking system design specifics while relying on an inertial analysis of tyre-to-surface engagement processes.

To achieve this purpose, the following points have been addressed:

- analysis of motor vehicle braking system design specifics and of the theoretical premises for the assessment of the functionality of such braking systems;
- field research on motor vehicle braking system efficiency in the case of emergency braking;
- drawing up the recommendations for upgrading the existing approaches to the assessment of motor vehicle braking efficiency in the course of accident-related automotive equipment expert examination.

15.2 ANALYSIS OF MOTOR VEHICLE BRAKING SYSTEM DESIGN SPECIFICS AND OF THEORETICAL PREMISES FOR ASSESSMENT OF FUNCTIONALITY OF SUCH BRAKING SYSTEMS

The history of braking-length-reducing systems starts with the anti-lock braking system (ABS), which is intended to prevent any blocking of the wheels when the brakes are applied and to maintain dynamic stability and steerability of the motor vehicle. Since 2004, ABS has been listed among the standard features of all new motor vehicles which are on sale in the European Union (Turenko & Saraiev 2015, Reif et al. 2014). Normally, in present-day motor vehicles, ABS is part of a more sophisticated electronic braking system, which may also include an electronic brake-force distribution (EBD) system, an electronic stability programme and an emergency braking system.

The electronic EBD system is primarily intended to prevent any rear wheel blocking via rear axle brake-force control. Its functioning is based on the ABS components, and it ceases to function at the moment that the drive wheels are being blocked.

The electronic stability programme (ESP) is a high-level active safety system, which is intended for vehicle dynamic stability control in a critical driving situation. The ESP includes an ABS, an EBD, an electronic differential system and an automatic slip regulation system. Since 2011, ESP has been listed among the standard features of all new cars sold in the European Union, the USA and Canada (Reif et al. 2014, AASHTO 2018).

The emergency braking system is an adaptive system intended to increase the efficiency of use, by a driver, of the primary brake system in emergency. The application of such an emergency braking system may be a decisive factor for preventing a road accident or, at least, for diminishing the consequences of a road accident by way of reducing the braking length by 15%–45% (Breuer et al. 2007, Rieger et al. 2005). There are two types of emergency braking systems: an emergency brake assist (BA) system, which allows maximizing the brake power when a driver presses a brake lever, and an automatic emergency braking system, which generates, in part or in full, the braking power without a driver's attendance. Such automatic emergency braking systems perform, besides their primary function, some other functions via combining the functioning of the vehicle's active and passive safety systems, thus becoming the so-called preventive safety systems (Ogorodnikov et al. 2004, 2018, Dragobetskii et al. 2105, Vorobyov et al. 2017).

However, no matter how effective the braking systems in modern motor vehicles might be, the capability of such systems are restricted to the rate (quality) of engagement of the tyres with the road surface. The rate of engagement of motor vehicle tyres with the road surface is determined by the nondimensional value, viz. by the coefficient of adhesion (μ) (Reif et al. 2014, AASHTO 2018, Puchkin 2010). The limit of the maximum implementation of the adhesion forces (Figure 15.1) affects the driving physics and is defined by the coefficient of adhesion and normal load.

Figure 15.1 shows that the coefficient of adhesion can be found from the wheel load value and latitudinal and longitudinal adhesion forces:

$$\mu = \sqrt{F_x^2 + F_y^2} / F_z. \tag{15.1}$$

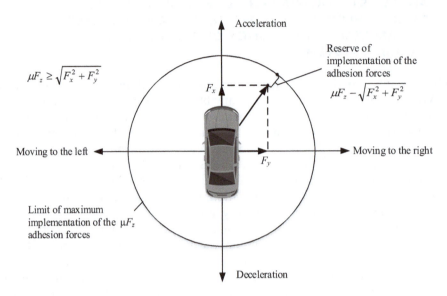

Figure 15.1 Tyre-and-road adhesion circle.

For simulation of the wheel motion dynamics for motor vehicles equipped with an electronic braking system, the coefficient of adhesion can be derived using different methods, among which the best known method is the one suggested by Pacejka (2012), Burckhardt (2005) and Denny (2005), i.e. the method of parabolas (Turenko et al. 2010).

The mechanical movement laws determine the rate of deceleration upon brake application. Such rate is limited by the cohesive properties of tyres on road surface.

$$j = \left(v_0^2 - v_k^2\right)/(2S_b) \le g \cdot (\mu \cdot \cos\alpha \pm \sin\alpha), \tag{15.2}$$

where S_b is the braking length; v_0 and v_k are the initial and final speed of the motor vehicle; α is the road gradient; g is the free-fall (gravity) acceleration; the symbol "–" is the downward movement and the symbol "+" is the upward movement.

In practice, while investigating the circumstances of a road accident, the European Network of Forensic Science Institutes (DaSilva 2008) determine the braking length in the following way:

$$S_b = \left(t_a + 0.5 \cdot t_{gd}\right) \cdot v_0 + v_0^2/(2 \cdot j), \tag{15.3}$$

where t_a is the brake response time and t_{gd} is the deceleration build-up time.

The analysis of the data provided by JATO Dynamics and Bosch experts as to the use of the driver's assist systems in new motor vehicles in Germany, France, Spain, UK, Italy, Belgium, the Netherlands and Russia as well as our own theoretical findings show that, in terms of the level of availability of braking length reduction systems, all M1-category motor vehicles in operation are to be classified for further research, on account of such features, as those without ABS, those with ABS and those with ABS and BA.

15.3 RESULTS OF FIELD RESEARCH ON MOTOR VEHICLE BRAKING SYSTEM EFFICIENCY IN CASE OF EMERGENCY BRAKING

The main parameter used in the assessment of motor vehicle braking efficiency is the rate of steady deceleration, which enables an expert to determine a braking or stopping length of a motor vehicle and the speed of such vehicle in the course of braking. The field investigations into the braking dynamics of M1-category motor vehicles were carried out from 2009, mainly at the scenes of road accidents or under conditions similar to the road accidents (Kashkanov et al. 2010, Turenko & Saraiev 2015). In the course of such investigations, we have used special, certified devices, such as MAHA VZM-100, MAHA VZM-300 and AMX 520 (Figures 15.2 and 15.3). The main

Figure 15.2 Devices used for measuring motor vehicle braking dynamics.

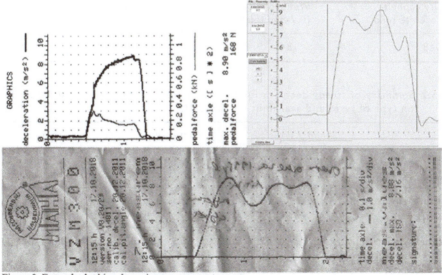

Figure 15.3 Examples braking dynamics measurement reports.

attention has focused on the technical state of the motor vehicle in question (its conformity to the existing technical requirements), on the motor vehicle braking system design (availability of ABS and BA) and on the type of the tyres used.

The following M1-category motor vehicles, manufactured in 1977 through 2015, have been tested on different types of road surface: 51 units without ABS, 63 units with ABS and 48 units with ABS and BA. The total number of steady deceleration measurements is shown in Table 15.1.

A summary of the 98% credible data is given in Table 15.2 and is shown in Figures 15.4 and 15.5. After processing with MS Excel tools, the results demonstrate that the braking process is well described by the normal law and is a stochastic process.

Table 15.2 demonstrates that, in terms of rate of deceleration in the course of braking, the summer tyres perform 3% better than the winter tyres on a dry asphalt-concrete pavement. On a wet asphalt-concrete surface (0.2-mm thick film of water), the winter tyres perform 3% better than the summer ones in terms of steady deceleration of a motor vehicle. Such tendency will increase in line with increasing thickness of the water film (Reif et al. 2014). The summarized steady deceleration test values differ by 9.5%–43.2% from the recommended statistical average data used for road accident reporting. This is essential for drawing up an expert opinion for the road accident in question. Therefore, to enhance the objectivity of the accident-related automotive equipment expert examination, it is necessary to take into account the braking system design specifics of the motor vehicle in question (functioning of ABS, BA and so on) and the type of the tyres.

It is evident from Figure 15.4 that the rate of steady deceleration of a motor vehicle without ABS decreases as the initial braking speed increases, which corresponds well

Table 15.1 Total number of tests

Dry asphalt-concrete road	Wet asphalt-concrete (0.2-mm thick film)	Packed snow	Icy
648	184	96	52

Table 15.2 Summary of steady deceleration figures in m/s^2 for 40 km/h speed depending on type of tyres and availability of ABS, BA

Type of road surface	Summer tyres			Winter tyres			Recom. value Puchkin (2010)
	Without ABS	With ABS	And BA	Without ABS	With ABS	And BA	
Dry asphalt-concrete	7.76	8.29	9.36	7.55	8.03	9.18	7.5
Wet asphalt-concrete[a]	5.86	6.71	7.16	6.14	6.83	7.21	5
Packed snow	–	–	–	2.98	3.11	3.28	2.5
Icy road	–	–	–	1.65	1.8	1.9	1.5

[a] 0.2-mm thick film.

Motor vehicle braking system design   179

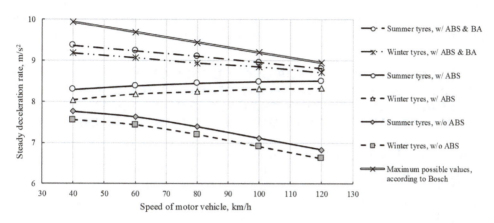

Figure 15.4 Steady deceleration of M1-category motor vehicle for dry asphalt-concrete road surface.

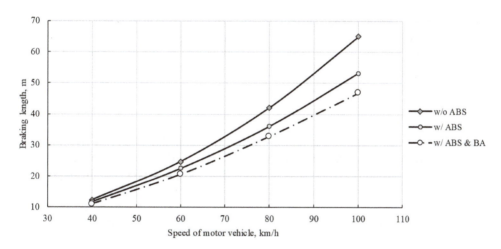

Figure 15.5 Summary of braking length values of M1-category motor vehicle for dry asphalt-concrete road surface.

with the test results (Kashkanov et al. 2010, Puchkin 2010). The motor vehicles with ABS demonstrate an opposite tendency, which is explicable on the basis of driving psychology in the case of braking (Greibe 2008, Pariota et al. 2017). The data obtained for the ABS-equipped motor vehicles correlate well with the results given in Turenko and Saraiev (2015) and Kudarauskas (2007). The motor vehicles provided with ABS and BA exhibit parameters which are close to maximum possible values, as found out by the Bosch experts (Reif et al. 2014) for motor vehicles with ordinary tyres under real operating conditions, which is explained by the functional features of such braking systems (Breuer et al. 2007, Rieger et al. 2005).

It is evident from Figure 15.5 that the braking length of motor vehicles with advanced braking systems is less than that of the motor vehicles without such systems. For instance, at the speed of 40 km/h, the difference between the motor vehicles equipped with ABS and BA and those without ABS is ca. 11% (between the vehicles with ABS and those without ABS, it is ca. 4%), whereas at the speed of 100 km/h, such difference is about 28% (18%). This trend grows as the initial braking speed increases.

15.4 RECOMMENDATIONS FOR UPGRADING EXISTING APPROACHES TO ASSESSMENT OF MOTOR VEHICLE BRAKING EFFICIENCY IN THE COURSE OF ACCIDENT-RELATED AUTOMOTIVE EQUIPMENT EXPERT EXAMINATION

Analysis of the field test results has proven the hypothesis that it is necessary to take into account the braking system design (functioning of ABS, BA) and the type of the tyres in the course of assessment of the emergency braking efficiency of a motor vehicle to minimize the probability of a wrong outcome of an automotive equipment expert examination. Therefore, the formula (15.3) which is recommended for determining the braking length of a motor vehicle and which, however, does not account for the above-said parameters needs to be refined. Proceeding from the functional features of the braking systems used in modern motor vehicles and from our test results, we suggest a number of improvements as follows.

For motor vehicles with an ABS (BA), we propose the following formula to be used for determining their braking length:

$$S_b = (t_a + 0.5 \cdot t_{gd}) \cdot v_0 + \frac{v_0^2 - v_s^2}{2 \cdot j_{ABS(BA)}} + \frac{v_s^2}{2 \cdot j}, \tag{15.4}$$

where $j_{ABS(BA)}$ is the steady deceleration rate with enabled/active ABS (BA); j is the steady deceleration rate with disabled/inactive ABS; v_s is the threshold lower speed of a motor vehicle before the ABS is disabled and $v_s = 15$ km/h (Reif et al. 2014).

For motor vehicles with an ABS (BA) using summer tyres:

$$j_{ABS} = j \cdot (1 + (0.0145 \cdot v_0 - 0.086) / j), \; j_{ABS+BA} = j \cdot (1 + 1.3917 \cdot e^{0.0028 \cdot v_0} / j). \tag{15.5}$$

For motor vehicles with an ABS (BA) using winter tyres:

$$j_{ABS} = j \cdot (1 + (0.0156 \cdot v_0 - 0.167) / j), \; j_{ABS+BA} = j \cdot (1 + 1.379 \cdot e^{0.0033 \cdot v_0} / j). \tag{15.6}$$

For motor vehicles without an ABS (BA), we propose the following formula to be used for determining their braking length:

$$S_b = (t_a + 0.5 \cdot t_{gd}) \cdot v_0 + \frac{v_0^2}{2 \cdot j_v}, \tag{15.7}$$

where j_v is the steady deceleration rate adjusted to the speed of the motor vehicle,

$$j_v = j \cdot (1 + (0.54 - 0.0119 \cdot v_a) / j). \tag{15.8}$$

Motor vehicle braking system design   181

The results of braking length assessment of M1-category motor vehicles with summer tyres derived with reliance on the models suggested (15.4)–(15.8) and according to the existing method (15.3) are shown in Figures 15.6 and 15.7.

Figures 15.6 and 15.7 demonstrate that it is advisable to use expressions (15.4) through (15.6) for assessment of the braking efficiency parameters of M1-category motor vehicles equipped with ABS (BA) on a dry asphalt-concrete surface. Expressions (15.7) and (15.8) are to be used for assessment of the braking efficiency parameters of M1-category motor vehicles without ABS on a dry asphalt-concrete surface when the initial speed is below 45 km/h and over 85 km/h. In other cases, the existing practices and formula (15.3) should be used.

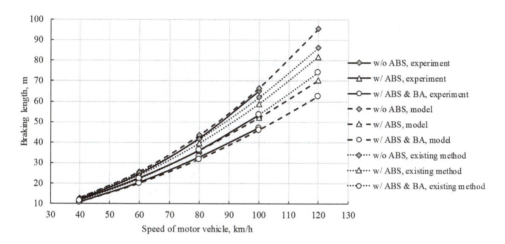

Figure 15.6 Results of assessment of braking distance of M1-category motor vehicles on dry asphalt-concrete surface.

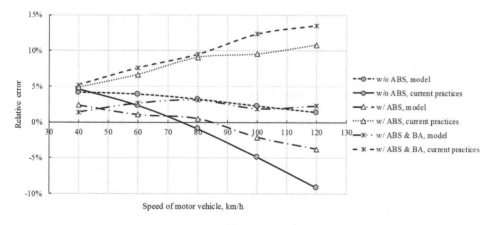

Figure 15.7 Error of braking length assessment for M1-category motor vehicles on dry asphalt-concrete surface.

15.5 CONCLUSIONS

The existing regulations and the current practices used for the assessment of emergency braking efficiency in the course of road accident reporting/documenting need to take account of the functionality of the state-of-the-art motor preventive/active safety electronic systems, the new achievements in the tyre design, as well as the psychological aspects of a driver's behaviour in an emergency.

The analysis of the theoretical grounds of the road-and-tyre interplay in the course of emergency braking as well as of the results of our field studies has brought us to the following conclusions:

1. Motor vehicle emergency braking is a complex dynamic process, which depends on many random factors such as the driver's physiological makeup, the design of the motor vehicle and the driving conditions and environment. The braking process is well described by the normal law and is a stochastic process.
2. The use of the approach suggested hereby minimizes the inaccuracy of simulation of the emergency braking efficiency for modern motor vehicles. The existing method appears to be more accurate only for motor vehicles without ABS within the speed range of 45 km/h to 85 km/h.
3. Verification of the effectiveness of the innovations suggested hereby has demonstrated why such things are to be used in practical work and has proven the need to extend such an approach to other categories of motor vehicles and other types and conditions of the road surface. Such innovations would increase the precision and quality of automotive equipment examinations/tests in road accident reporting.

REFERENCES

AASHTO Green Book. 2018. *A Policy on Geometric Design of Highways and Streets*, 7th Edition.

Breuer, J., Faulhaber, A., Frank, P. & Gleissner, S. 2007. Real world safety benefits of brake assistance systems. *Proceedings of the 20th International Technical Conference on the Enhanced Safety of Vehicles*; Lyon, France 18.06.2007.

Burckhardt, M. 2005. *Fahrwerktechnik: Radschlupf-Regelsysteme*. Wurzburg: Vogel.

Danez, S. & Saraiev, O. 2018. Mathematical modeling of speed change of vehicles at emergency braking. *Technology Audit and Production Reserves* 3(41): 22–28.

DaSilva, M.P.. 2008. Analysis of event data recorder data for vehicle safety improvement (http://www.nhtsa.gov/DOT/NHTSA/NRD/Multimedia/PDFs/EDR/Research/811015.pdf, accessed: 11.05.2020).

Denny, M. 2005. The dynamics of antilock brake systems. *European Journal of Physics* 26(6): 1007–1016. Doi: 10.1088/0143–0807/26/6/008.

Dragobetskii, V., Shapoval, A., Mos'pan, D., Trotsko, O. & Lotous, V. 2015. Excavator bucket teeth strengthening using a plastic explosive deformation. *Metallurgical and Mining Industry* 4: 363–368.

European Network of Forensic Science Institutes. 2015. *Best Practice Manual for Road Accident Reconstruction, ENFSI, ENFSI-BPM-RAA-01. Version 01*. (http://enfsi.eu/wp-content/uploads/2016/09/4._road_accident_reconstruction_0.pdf, accessed: 11.05.2020).

Franck, H. & Franck, D. 2009. *Mathematical Methods for Accident Reconstruction: A Forensic Engineering Perspective*. Boca Raton, FL: CRC Press.

Greibe, P. 2008. Determination of Braking Distance and Driver Behaviour based on Braking Trials. *87th Annual Meeting of the Transportation Research Board*, 13.01.2008 – 17.01.2008, Washington DC.

Hynd, D. & McCarthy, M. 2014. Study on the benefits resulting from the installation of Event Data Recorders (https://ec.europa.eu/transport/sites/transport/files/docs/study_edr_2014.pdf, accessed: 11.05.2020).

Kashkanov, A. 2018. Mathematical methods of decision making in autotechnical expertise of traffic accidents. *Automobile Transport* 43: 78–89. Doi: 10.30977/AT.2219-8342.2018.43.0.78.

Kashkanov, A. A., Rebedailo, V. M. & Kashkanov, V. A. 2010. *Estimation of Operational Braking Properties of Cars in Conditions of Inaccuracy of Initial Data*. Vinnytsia: VNTU.

Kudarauskas, N. 2007. Analysis of emergency braking of a vehicle. *Transport* 22(3): 154–159. Doi: 10.1080/16484142.2007.9638118.

Laugier, C., Paromtchik, I.E., Perrollaz, M., Yoder, J.-D., Tay, C., Yong, M., et al. 2011. Probabilistic analysis of dynamic scenes and collision risks assessment to improve driving safety. *IEEE Intelligent Transportation Systems Magazine* 3: 4–19.

Mannering, F.L. & Bhat, C.R. 2014. Analytic methods in accident research: Methodological frontier and future directions. *Analytic Methods in Accident Research* 1: 1–22.

Ogorodnikov, V.A., Dereven'ko, I.A. & Sivak, R.I. 2018. On the influence of curvature of the trajectories of deformation of a volume of the material by pressing on its plasticity under the conditions of complex loading. *Materials Science* 54(3): 326–332. Doi: 10.1007/s11003-018-0188-x.

Ogorodnikov, V.A., Savchinskij, I.G. & Nakhajchuk, O.V. 2004. Stressed-strained state during forming the internal slot section by mandrel reduction. *Tyazheloe Mashinostroenie* 12: 31–33.

Pacejka, H.B. 2012. *Tyre and Vehicle Dynamics*. 3rd ed. Butterworth-Heinemann: Elsevier.

Pariota, L., Bifulco, G.N., Markkula, G. & Romano, R. 2017. Validation of driving behavior as a step towards the investigation of Connected and Automated Vehicles by means of driving simulators. *5th IEEE International Conference on Models and Technologies for Intelligent Transportation Systems (MT-ITS)*, 26–28 June 2017, Naples, Italy.

Puchkin, V.A. 2010. *Basics of Expert Analysis of Road Accidents: Database. Expert Equipment. Solution Methods*. Rostov n/D: IPO PI YUFU.

Reif, K. & Dietsche, K.-H. 2014. *Bosch Automotive Handbook*. 9th ed. Karlsruhe: Robert Bosch GmbH.

Rieger, L., Scheef, J., Becker, H., Stanzel, M. & Zobel, R. 2005. Active safety systems change accident environment of vehicles significantly – a challenge for vehicle design. *19th International Technical Conference on the Enhanced Safety of Vehicles*, 06.06.2005 – 09.06.2005. Washington DC.

Saraiev, O. & Gorb, Y.A. 2018. *Mathematical Model of the Braking Dynamics of a Car*. SAE Technical Paper 2018-01-1893.

Singh, K.B. & Taheri, S. 2015. Estimation of tire-road friction coefficient and its application in chassis control systems. *Systems Science & Control Engineering* 3(1): 39–61. Doi: 10.1080/21642583.2014.985804.

Singh, K.B., Ali, A.M & Taheri, S. 2013. An intelligent tire based tire-road friction estimation technique and adaptive wheel slip controller for antilock brake system. *ASME Journal of Dynamic Systems, Measurement, and Control* 135(3). DOI: 10.1115/1.4007704.

Struble, D. 2013. *Automotive Accident Reconstruction: Practices and Principles*. Boca Raton, FL: CRC Press.

Turenko, A. Lomaka, S. Ryzyh, L. Leontiev, D. & Bykadorov, A. 2010. Calculation methods of realized coefficient of cohesion at wheel rolling in braking mode. *Automobile Transport* 27: 7–12.

Turenko, A. M. & Saraiev, O. V. 2015. *Evaluating the Effectiveness of Vehicle Braking in the Study Traffic Accident*. Kharkiv: KhNAHU.

Vorobyov, V., Pomazan, M., Vorobyova, L. & Shlyk, S. 2017. Simulation of dynamic fracture of the borehole bottom taking into consideration stress concentrator. *Eastern-European Journal of Enterprise Technologies* 3/1(87): 53–62.

World Health Organization. 2020. Road traffic injuries (http://www.who.int/mediacentre/factsheets/fs358/en/, accessed: 11.05.2020).

Zhang, R., Cao, L., Bao, S., Tan, J. 2017. A method for connected vehicle trajectory prediction and collision warning algorithm based on V2V communication. *International Journal of Crashworthiness* 22(1): 15–25.

Chapter 16

Essential aspects of regional motor transport system development

Volodymyr Makarov, Tamara Makarova, Sergey Korobov, Valentine Kontseva, Piotr Kisala, Paweł Droździel, Saule Smailova, Kanat Mussabekov, and Yelena Kulakova

CONTENTS

16.1 Introduction 185
16.2 Functional analysis of regional MT systems 186
16.3 Aspect of regional MT system development associated with the
      use of intelligent tyres 190
16.4 Specifics of intelligent tyres 192
16.5 Conclusion 194
References 195

16.1 INTRODUCTION

In their research work, German philosophers conclude that modern machinery and science are *colossal* factors, which have fundamentally changed the nature of our planetary system (PS), and the intensity of their effect (positive and/or negative) continues to grow. One peculiar feature of the present-day world is a rapid incorporation, in the lifecycles of our planet, of a significant part of the existing machinery, viz. motor vehicles. Vehicle transport flows run along the global network of expressways, which interconnect villages, cities, regions, countries and continents. There is no man on Earth who has never used a motor vehicle, in one way or another. Transport flows support the development of the global economy and migration of people to economically developed regions, where a motor vehicle has become an extension of the living space of residents, as much as it is has become a means of conveyance from home to a work place, school, place of recreation, etc. and back.

Nowadays, the most significant innovative solution in the field of motor transport (MT) is the creation of intelligent transport systems (ITS), which include such intelligent components as motor vehicles, roads and personnel involved in traffic and infrastructure management.

An intelligent motor vehicle is one containing an artificial intelligence system which allows it to not only transport, in an appropriate manner, all of the required material and people but also to enhance such an important feature as traffic safety. Each year, comfortable, human-driven but non-intelligent motor vehicles take the lives of over one million people (a magnitude comparable with the most hazardous diseases of today or with the plague disasters of Medieval Europe). At the Davos 2018 Meeting,

DOI: 10.1201/9781003224136-16

the utilization of artificial intellect to control a self-driving motor vehicle was declared a first-priority task of today (Davos 2018).

However, to create an ITS, it is necessary to build out widespread intelligent road infrastructure, on a global scale, which is a sophisticated system of resource-intensive facilities with immense computer-based support. Over the next decades, such ITSs will be developed and their functioning will be based on certain data flows. The ultimate goal of this research is to let nature function in its own way according to its own laws and to ensure the balance between the man-made world and virgin nature.

16.2 FUNCTIONAL ANALYSIS OF REGIONAL MT SYSTEMS

The purpose of this chapter is to select the most reasonable approaches to the development of an intelligent MT system for regions of countries, whose economy will benefit from such a system.

A regional motor transport systems (RMTS) should be developed to ensure efficient and accident-free traffic of motor vehicles (MVs). In particular:

- it should carry out certain transportation tasks within regions which are vital for the economy of the country, of the continent and of the whole world;
- it should provide for cost-efficient and reliable conveyance of cargo and passengers within the logistic supply chains;
- the most intensive expressway routes should be ascertained and sufficient funds should be invested in construction and renovation of the most promising expressways;
- it should ensure low-accident-rate motor traffic;
- it should actively use the available data flow;
- it should provide psychological, technical and computer-based support for drivers.

An initial survey of the most vital developmental aspects for a RMTS should start with an analysis of historic facts and continue to an analysis of the real achievements and functionality of similar systems in economically developed countries. Such historic observation would reveal the reasons why the flow of goods and people had to cover long distances. There was a stable transport corridor, viz. the Great Silk Way from China to Rome. It was characterized by intensive movement of goods, which required well-organized traffic along the roads in China. The roads in the Roman Empire were well-planned and well-built. Later, rapid marches of the Napoleonic troops, when Napoleon was successfully 're-arranging' the map of Europe, required new real-time transport and logistic solutions. In the Middle Ages, such circumstances were almost absent. So, the above examples testify to the fact that there were always strong reasons to implement large-scale transport system projects which contained such important components as traffic management, construction of roads and logistics.

However, the most imperative point now is to analyze the up-to-date solutions to the problem of creating an efficient MT system. One of the most obvious parameters which characterize the level of development of a MT system is the motor traffic flow (MTF) intensity. So, it is necessary to be able to assess such MTF intensity as an indicator of such flow efficiency. Efficient MTF would allow a RMTS to be gradually upgraded to an intelligent level of MT infrastructure functionality. In the process of assessment, the cost efficiency of the system functioning is to be studied. The MTF intensity is a main parameter of an expressway which designates the road to a certain weighted-average category and affects its transporting capacity, and, consequently, the input values used to calculate the funding resources required for road construction/renovation projects.

The motor flow assessment methods used in economically developed countries are noteworthy. If we consider the world practices used to assess the intensity of traffic in transport systems, we can see that the instrumental electronic approach was used to assess the intensity of MTF in the countries where the economy was on the rise and the MT infrastructure was developing fast. Historically, each country has gone, and is going, its own way in developing its transport system and economy, which influence each other. Indeed, based on the structure and the scope of transportation, we can estimate the level of the economy of a country or region, and based on the road network configuration, we can judge the appropriateness of locating industrial facilities there. Multiple factors affect the formation of certain types of MTF defined according to the types of transported matter; such flows can certainly have a bearing on the economic development. For instance, transit haulage is an essential item in the income of most European countries. Receipts from such transit-traffic and transit-related operations account for ca. 6% of their gross national product. In Holland, a transit-traffic share in the total bulk of the revenue derived from export services exceeds 40%, and in the Baltic countries, it amounts to 30%. Ukraine fails to use its transit-traffic potential in full, so it loses annually ca. USD 2.5 billion (Makarova 2013).

Hereinafter, we will consider the methods of motor flow intensity assessment used in Germany. These methods are valuable to analyze because they are suitable for regions where the economy is advancing. A survey of the MT system in Germany (a highly developed modern country) could be very useful in terms of planning how to use in practice the German experience of developing the Saxon MT System after the reunification of Germany.

The analysis shows that to ensure sustainable development of the economy of a region, it is necessary to increase two major elements which influence the intensity of the MTF: the number of up-to-date MVs to be used for passenger and cargo transport operations, and the length and quality of the motorways on which the MVs travel, making up a traffic flow.

Another important factor to be analyzed is the expressway density in the German Land of Saxony (km/km^2). It is 6–10 times higher than that in the regions of Ukraine, and the number of MVs is also much higher. However, the gap between the two said parameters is currently narrowing, thanks to the increasing stock of MVs in the regions of Ukraine. Therefore, in a region with a developing MT system, it is highly necessary to apply, in a reasonable manner, innovative methods for the assessment of the economic effect of the growing MTF to forecast the most promising expressways to be taken into account when planning and locating the motor depots and other similar facilities, as well as the road network in the region.

Below, we will consider, in more detail, the major routes of MTF running through the territory of Saxony. It is very important to note that, within the European segment, such routes run across Saxony and Ukrainian regions. The European E-40 route starts in France (at the seaport of Calais) and passes through Belgium, Germany, Poland and Ukraine. Another European route, E-55, which crosses Saxony, is a 3,305-km-long route which includes three ferry crossings. It starts in Sweden, crosses Denmark, Germany, Czechia, Austria and Italy and ends in Greece. The annual average rate of MTF along the major ways in Saxony is 40,000 units per day. Considering the whole scope of the MT flow, it is assumed that every second vehicle is in transit.

It is worth pointing out that assessment and further steady development of the Saxon transport routes, which are to satisfy the demand of the economy, are the major tasks and concerns of the Saxon Land Transport Ministry. Moreover, Saxony's geographic location in the centre of Europe plays an important role in global economic plans and estimates, as well as in the investing climate existing in said region. To enhance such a profitable function of Saxony as a transportation hub and transit-traffic territory, the Ministry is binding the Saxon routes to the Pan-European rail and expressway networks and is incorporating the existing Saxon routes into the European trade and transport flows.

To evaluate and monitor the MTF, the Saxon highways are furnished with over thirty automatic surveillance and metering stations. Every five years, a general federal survey of the motor traffic is performed with subsequent adjustment of the pattern of MTF. A peculiar feature of general assessment-and-analysis systems is the continuous process of their elaboration. Earlier in Saxony, there existed some transport telemetric solutions which were later modified (Makarova 2013).

The safety of the Saxon MTF, as a whole system, has been continuously evaluated, for several decades, by VUFO GmbH existing under the auspices of the Technical University of Dresden (Dresden 2020). VUFO studies the risk of road accidents around Dresden. The European E-55 and E-40 highways cross that territory.

The 2015 data, which characterize the merits of such an approach in Germany, in terms of two parameters, are as follows:

- number of road accident fatalities per 100,000 people is lowest in Germany – 4.4, followed by Finland – 4.7 and the Republic of Belarus – 7;
- number of deaths per 100 people involved in road accidents is also lowest in Germany – 0.9 followed by the USA – 1.6 and Finland – 3.9.

It should be noted that the Dresden scientists cooperate fruitfully with specialists of tyre factories and motor works, as well as with the road maintenance service, road police and health institutions. There are only two such road accident research centres in Germany.

To attract the interest of transport operators from the developed European countries (France, Belgium, Germany and Poland) in using the route which crosses the Ukrainian region with a developing MT system, it is necessary to ensure certain traffic conditions, including a low risk of road accidents.

Motor traffic is possible, thanks to the complicated and continuous work of drivers. Each driver has to analyze, often within the shortest time, such things as the traffic situation within the vicinity of his/her motor vehicle (including the intent of other drivers), the capability of his/her motor vehicle and, finally, to choose a quick response action. The limited personal perception of the reality enables the driver to draw only a relative and partial conclusion about the actual situation (Siegert et al. 1998). At the next moment, the driver has to use all of his/her driving skills to drive his/her vehicle. The driver bears a major portion of responsibility for the positive or negative outcome of a difficult traffic situation. Even the best traffic safety systems of a motor vehicle, however, cannot avert the driver's sudden wrong movements which contradict his/her motor vehicle's functional capability and the laws of physics (Siegert et al. 1998). In Germany, there are strict requirements on the duration of a driver's working day, and a breach of such requirements entails a penalty. To comply with such working time requirements, all MVs are equipped with state-of-the-art devices which monitor the working hours of a driver, who has to have, at least, 11 hours of rest after each 8 hours of work. To provide rest for drivers, German roads are provided with drive-in pockets (normally of a 10–15-vehicle capacity), spaced at a distance of 5–7 km which accommodate various services, such as catering, fuel-filling facilities, etc. Unfortunately, the Ukrainian expressway network lacks such services which can provide proper rest for drivers. This is a pressing problem to be resolved.

The most essential problem of the present-day logistic chains is appropriate conveyance (in terms of the length of the run) of cargo and passengers, which can be resolved only by way of creation of an ITS. Such a system will cause significant changes in the physical life of the planet and in the capacity of the atmosphere of Earth. Having said that, it is necessary to ensure environmental efficiency, in line with transportation cost efficiency, and in particular, a reduction in the toxic emissions of traffic products, hazardous for people and nature, generated by hundreds of millions of MVs and by the numerous infrastructure facilities. The latter should contribute to global sensory perception and intellectual growth. The visual image of the road is created by the engineering facilities and architectural design, with priority given to the functional aesthetic means (the availability of information screens, wireless Internet, solar batteries, air-conditioners, etc.), which together convert such a parking area into a stand-alone facility. In such a way, we can implement the humanistic and educating trend (Lebedeva 2013, Sitkevich 2016) for the sake of the cultural and spiritual growth of the society.

The requirements to the occupational fitness, professional skills and labour conditions of a driver must be selective, personalized and sufficient. A road accident, and the severe consequences it may cause, can strongly affect the health of a driver and of those involved in such accident. When a driver feels safe and confident, he/she can make and carry out free, fast and sound driving decisions. In the case of a road accident, Allgemeiner Deutscher Automobilclub (ADAC – German Automobile Club) (ADAC 2020) renders prompt technical and legal aid to its members, thereby providing them with efficient psychological support. Within 15 minutes of a road accident, an ADAC helicopter may land at the scene of the accident (within the German territory) delivering a lawyer and mechanic with the required equipment and spare parts on board. The members of the Club call such helicopters "the yellow angels". ADAC has existed for more than 100 years and its membership exceeds 15 million people. The above example demonstrates the private–public partnership (PPP) in developing the RMTS.

We have considered the following essential aspects of development of a regional MT system which can mature to an intelligent one:

- creation of the fundamentals of the motor-road infrastructure, which is very expensive and takes a significant amount of time to establish based on a PPP;
- setting up a regional system to be used to monitor the rate of MTF, including the centres/stations for road traffic management and evaluation of the motor flow rate;

- organizing computer-based working and psychological support to ensure proper labour conditions for drivers;
- 'incorporation', by the regional managerial staff, of the regional MT system components into the logistic network of other regions, rendering transportation support to major economic initiatives.

As of May 2019, in the Vinnytsia Region, there are high-quality roads running in the direction of the major MTF, in particular, to Zhytomyr and Khmelnitsky. The Vinnytsia-Nemyriv segment of another highway has been renovated, yet further on, up to the city of Uman, the quality of the road pavement does not conform to the existing requirements. Therefore, road renovation projects in the Vinnytsia Region are still underway; they are being carried out by Autostrada JSC, which efficiently performs roadworks at the European level. For the time being, in 2019, there are reasonable expectations that the most essential scope of roadwork on the European routes which cross Vinnytsia Region will be accomplished (Freeway 2020). The completion of the transport infrastructure construction projects from the Baltic Sea to the Black Sea can contribute to the growth of the economy in the Vinnytsia Region. PPP has facilitated the fulfilment of a considerable scope of work. Other above-mentioned aspects of the transport system development are to be further elaborated.

16.3 ASPECT OF REGIONAL MT SYSTEM DEVELOPMENT ASSOCIATED WITH THE USE OF INTELLIGENT TYRES

The authors have considered the aspect of the regional MT system development associated with the use of intelligent tyres for the subsequent creation of an ITS. The undercarriage of MVs are reliably supported by pneumatic tyres. The pressure inside the elastic envelope (tyre or tyre tube) maintains the dynamic movement of MVs in the proper manner. The weight of the unsprung wheels is comparatively low, thanks to the high air pressure inside the tyres. However, fast motion on compressed-air accumulators is disastrous when the tyre reinforcing elements are accidentally destroyed because of a breakage of the wheel elastic liner.

The second half of the twentieth century was the time when many motor vehicle components (units) were computerized. However, the wheels of MVs remained almost unchanged. The leading tyre manufactures publicized the creation of intelligent tyres only at the end of the twentieth century.

This delay was caused by the following factors:

- it would be hard to build an intelligent system into the elastic tyre structure, because a tyre is a well-balanced and advanced composite product: in fact, it is a mini-laboratory which turns the torque it receives into a motor-and-road interaction field of forces;
- it would be necessary to replace the air in the tyre, which is cheap and effective, with another type of lightweight and durable material which can withstand the adverse road environment.

The factors which have contributed to the creation of intelligent elastic tyres are as follows:

- the availability of on-line and continuous dataflow regarding the force field which dynamically varies on the wheel contact area of a running vehicle;
- it is possible to obtain a bifurcation diagram which would describe the motor vehicle's major road-holding parameters, in particular, the speed and the wheel turning angle;
- it is possible to transfer the data concerning the force field at the wheel contact area and the motor vehicle's road-holding parameters to the control unit of the self-driving automobile's artificial intelligence system, such data are to be used for traffic monitoring and control in the ITS.

Therefore, the on-line and permanent dataflow transferred from individual MVs and from the whole bulk of the MTF over to the artificial intelligence unit of an ITS contains the data which are required for drawing up an appropriate pattern of cargo and passenger conveyance. The latter is intended to maintain the balance between the maximum allowable motor traffic intensity and the acceptable level of road accidents in the MTF (Figure 16.1).

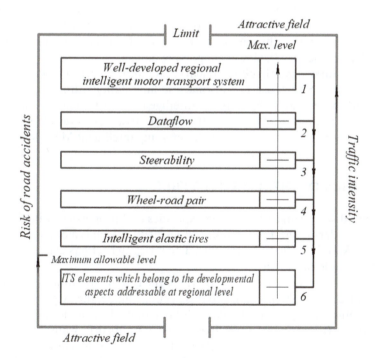

Figure 16.1 Mnemonic diagram which visualizes the model of regional MT system development in terms of an appropriate balance between the motor traffic intensity and the level of accidents on the roads in the region.

The mnemonic diagram shows two opposite attractive fields: the accident-risk axis on the left and the traffic intensity axis on the right. It is always desirable to raise the traffic intensity of the MTF in the regional MT system to the maximum possible rate, but within the limits of the given road traffic capacity (increasing the traffic intensity to the maximum level). However, in this case, the risk of road accidents increases too, and it may exceed the allowable limit. The minimum traffic intensity value is zero (a complete absence of traffic leads to a complete absence of road accidents). So, to keep the road accident risk within the allowable limits, it would be necessary to resort to the VUFO practices (Dresden 2020).

The first block of the mnemonic diagram explains the purpose of the research, viz. the creation of a well-developed intelligent RMTS. To achieve this high-level purpose, it is necessary to accomplish all lower consecutive steps, from 6 to 2.

The sixth block of the mnemonic diagram highlights the sequence of fulfilment of the major aspects required for the development of a RMTS. Block 5 contains the analysis of all possible types of elastic tyres. Block 4 initiates the elaboration of the *wheel-road* system analyzing the elastic parameters of intelligent tyres and the relation of such parameters with the forces existing in the *tyre-and-surface* contact. Block 3 analyses the road-holding stability and its relation with the bifurcation set diagrams. Block 2 deals with shape of the dataflow based on the parameters of the existing RMTS, which has the initial elements of an ITS (Block 6); this is followed by the selection of a certain intelligent tyre (Block 5). The modelling procedure goes through Block 4 and Block 3 where the specific numerical values are acquired; then a dataflow is formed (Block 2) and goes over to the intelligent control unit, which selects an algorithm to ensure the risk of accidents within the allowable limits. There is also a feedback whereby the higher blocks influence the lower supporting blocks.

Only tyres, unlike any other components of a motor vehicle, can give the artificial intelligence of the ITS such useful real-time and final data (within the *wheel-tyre* pair system) which would enable the artificial intelligence unit to achieve the balance, as demonstrated in the model. Therefore, the recent advances in developing intelligent elastic tyres for MVs are very opportune for creating intelligent MT systems.

16.4 SPECIFICS OF INTELLIGENT TYRES

Below, we will discuss some of the design specifics and properties of intelligent tyres. We suggest considering the Tweel model (Tweel 2020), which is a one-piece wheel consisting of a disk and elastic tyre elements. The airless wheel has elastic polyurethane spokes and a double-layer rim.

Now, we will consider an elastic tyre where the tyre tread imitates the brain coral structure (Lüders et al. 1996). Thanks to this peculiar shape, the external surface of the tyre which contacts the road surface can fulfil two opposite functions. It becomes stiff on a road surface (like a sponge) and, in contrast, also absorbs water from a wet road.

An innovative concept design developed by Goodyear, which was demonstrated by the company at the Geneva International Motor Show 2018 (Geneva 2018), is noteworthy. The most prominent feature of the *Oxygene* model proposed in Geneva is its ability to convert carbon dioxide into oxygen. Due to the immense contribution of MVs to the growing CO_2 concentration in the ambient air, the European Union adopted a directive as to the prospective reduction of CO_2 emissions. The requirements for motor manufacturers adopted in 2013 prescribe the following ultimate permissible limits for CO_2 emissions: 95 g/km in Germany; 105 g/km in Japan; 117 g/km in China and 121 g/km in USA. The requirements are binding on all MVs to be sold until 2020 (Karle 2018).

An airless elastic does not require any maintenance operations which are typical of pneumatic tyres (such as air pressure checks) nor does it require the driver's momentary response to any serious damage to the tyre liner in the course of driving. Thanks to the perforated structure of the wheel in question, its weight has been reduced. A car which runs on the *Oxygene* tyres holds the road well and is characterized by high manoeuvrability; its tyres quickly absorb water, if any, from the road surface.

In 2018, at the Geneva International Motor Show, Goodyear presented a new tyre concept for self-driving MVs (Goodyear 2020). The model had a spherical shape. The main concern was given to ensuring the driving safety, to improve the response-to-road function and to enhanced steerability.

In the future, drivers will have less freedom in driving a motor vehicle, therefore the role of intelligent tyres will increase. This is because they will be required to ensure a reliable force field in the contact of the motor vehicle with the road surface.

Below, we can see how advances in science and engineering facilitate the development of the intelligent MT system. Such a system must ensure high safety of the MTF. In addition, a motor vehicle itself should be very steerable to reduce the possibility of road accidents. Such steerability is affected by the lateral force which emerges when the elastic wheel moves with lateral displacement.

The first type of dependence of the lateral force ($\overline{Y}$) on the wheel lateral displacement (δ) characterizes the running of an ordinary pneumatic tyre. The dependence is described by the following formula:

$$\overline{Y} = \frac{\overline{k}\delta}{\sqrt{1+\left(\frac{\overline{k}\delta}{\varphi}\right)^2}} \qquad (16.1)$$

where $\overline{k}$. is the tangent of the angle of displacement (swing) of the linear segment of the curve and φ is the traction ratio of the elastic tyre against the road surface.

To create an intelligent tyre, we propose to alter the structure and the material of the components of the composite product in such a way that would cause the appearance of minor convexities and bulges, as well as non-linear segments in the dependency graph (curve).

Another type of dependence is a non-monotonic dependence which is characterized by the two parameters "*a*" and "*b*":

$$\overline{Y} = \frac{a\, tg\delta}{\sqrt{1+\frac{(tg|\delta|-b)^2}{b^2}}} \qquad (16.2)$$

where $a = \overline{k}\sqrt{2}$. is the coefficient which responds to the non-monotonic dependency maximum.

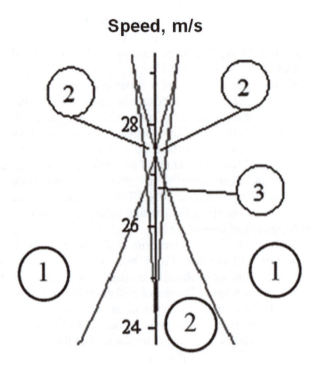

Figure 16.2 Part of diagram of bifurcation set in critical point domain.

Typical steerability values have been obtained for a motor vehicle equipped with tyres which provide for monotonic lateral force variation. The availability of the breakpoints on the curve which has been drawn up according to the formula (16.2) has increased the area of the lower fields 2 on the bifurcation diagram, where a steady-state mode of motion exists, and give rise to new upper fields (2 and 3) with steady-state modes of motion. The said bifurcation set diagram is shown in Figure 16.2 below.

The above diagram shows three zones with different levels of steerability: a zone with one steady yet unstable mode of motion (1); a zone with three steady modes of motion, of which one is stable (2); and a zone where five steady modes of motion can exist, of which two are stable and three are unstable (3).

The method developed by the authors hereof achieves the predetermined lateral forces in the wheel-road contact area, based on the elastic tyre stiffening behaviour, and thereby enhances the car steerability. The data flow 2 (Figure 16.1) will deliver the data array describing the linear speed and wheel-turn angle to the artificial intelligence unit of the ITS which will ensure steady movement of the motor vehicle.

16.5 CONCLUSION

The authors have addressed some important aspects which are to be considered while developing a regional MT system based on an analysis of successful existing practices. The authors have defined the engineering science-based scenario to facilitate motor vehicle flow accident-risk mitigation by way of using intelligent tyres, which are designed for self-driving MVs to be used in ITS.

The authors have elaborated a more detailed approach to shaping the dataflow transmitted to the intelligent MT system based on "wheel-road" input data. The above-mentioned will help to obtain new fields in the bifurcation diagram, which will ensure steady traffic with a low road accident risk. The authors intend to continue the research in two fields: surveying of MT system evolution in the developed countries and formation of possible wheel-road contact data flow to be used for intelligent transport system control. The above information will ensure free functioning of the Earth PS.

REFERENCES

ADAC. 2020. Allgemeiner Deutscher Automobil-Club (ADAC) (http://www.adac.de/, accessed: 11.05.2020).
Davos. 2018. Worldeconomicforum (https://www.weforum.org, accessed: 11.05.2020).
Dresden. 2020. Verkehrsunfallforschung an der TU Dresden (http://www.vufo.de/, accessed: 11.05.2020).
Freeway. 2020. Current state of roads (https://vinrada.gov.ua/main.htm, accessed: 11.05.2020).
Geneva. 2018. Geneva 2018: Live moss appeared in Goodyear tires (http://autonews.autoua.net/novosti/18497-zheneva-2018-v-shinah-goodyear-poyavilsya-zhivoj-moh.html, accessed: 11.05.2020).
Goodyear. 2020. Goodyear Eagle-360 tires are recognized as one of the best inventions of 2016 (https://www.4tochki.ru/news/novosti-rynka-shin-i-avtokomponentov/2016/noyabr/21/shiny-goodyear-eagle-360-priznany-odnim-iz-luchshih-izobreteniy-2016-goda.html, accessed: 11.05.2020).
Karle, A. 2018. Elektromobilität – Grundlagen und Praxis 3. München: Carl Hanser Verlag GmbH & Co. KG.
Lebedeva, N.N. 2013. Ministry of Education of the Republic of Belarus. BNTU. Science, education, production, economics. *Materials of the Eleventh International Scientific and Technical Conference* 2: 132.
Lüders, A., Hofmann, O. & Brinkmann, H. 1996. Beitrag zum Problem der Laufunruhe von Fahrzeugrädern. ATZ 73(1): 1–8.
Makarova, T.V. 2013. *Assessment of Socio-Economic Efficiency of Traffic Intensification of the Region's Vehicles and the Mechanism of its Provision*. Kyiv: Economics and Enterprise Management.
Michelin. 2020. Tweel: Airless tire (https://tweel.michelinman.com, accessed: 11.05.2020).
Siegert, E., Geisler, H, van Zanten, A., Becker, R., et al. 1998. *Fahrsicherheitssysteme*. BOSCH. Braunschweig, Wiesbaden: Vieweg.
Sitkevich, A.M. 2016. Improvement of the road safety system in the Republic of Belarus on the basis of planning and implementation of a complex of preventive measures. Prospects for the development of the transport complex. *Materials of the 2nd International behind Scientific - Practice Conference* October 4–6, 2016, Minsk, Transtechnika, 169–173.

Chapter 17

Improvement of logistics of agricultural machinery transportation technologies

Ievgen Medvediev, Iryna Lebid, Nataliia Luzhanska, Volodymyr Pasichnyk, Zbigniew Omiotek, Paweł Droździel, Aisha Mussabekova, and Doszhon Baitussupov

CONTENTS

17.1 Introduction 197
17.2 Literature review 198
17.3 Research objectives 199
17.4 Statement of basic materials 200
17.5 Conclusion 206
References 207

17.1 INTRODUCTION

One of the main challenges for sustainable human development is the growth of the general population and global changes in climate conditions which are increasing the shortage of agricultural food products year after year. Therefore, agricultural products are gradually turning into a category of resources, the production of which is of strategic importance for the successful economic development of most countries in the world. This fact accounts for the serious attention of scientists to the development of new approaches, and modernization and improvement of agro-industrial production technology at all stages: selection, cultivation, harvesting, transportation, and storage of agricultural products. It should be noted that the successful development of the agrarian sector directly depends on the efficiency of agricultural production transportation support.

As evidenced by the results of the conducted research, transport processes account for up to 35% of all costs in the total amount of agricultural work, and when considering energy consumption, they account for up to 40%. Transportation expenses form about 25% of expenditures affecting the net cost of the most important types of agricultural products. The task of agrarian production transport services is the timely harvesting and transportation of products from fields to processing enterprises and storage facilities. It is noted that the main part of the agrarian enterprises' industrial infrastructure is formed by road transport ensuring continuity and a rhythmical agricultural production flow (Prydiuk 2015).

Since the classification of agricultural cargo is broad, farms need to have a wide range of specialized vehicles with seasonal application leading to significant idle time

DOI: 10.1201/9781003224136-17

and low efficiency in the use of expensive equipment. At present, there are significant costs of a financial and time nature involved in the harvesting of grain crops due to expenditures for transporting grain harvesting equipment to the points of its use, so the problem of improving the organization and transportation technologies with the use of special vehicles is extremely relevant. Its solution will contribute to boosting the efficiency of agricultural machinery and reduce the amount of such equipment which is required.

17.2 LITERATURE REVIEW

There are a great number of scientific publications focused on the problems of the further development and efficient use of transport, presenting analyses of the current state of transport systems and proposing directions for their long-term development. Thus, in Engström (2016), it is noted that by 2050, the total volume of freight transport in European countries will have increased by 80%, and international cargo shipping will have risen fourfold. The given data indicate the need to find new technical and technological solutions for the modernization of the international transport system for the timely provision of necessary freight traffic volumes. It should be noted that equally tough problems need to be resolved in the field of passenger transportation. Thus, the problems of passenger and freight transport development have been studied within the scope of the 7th Framework Program of the European Commission where the need to bridge certain gaps between numerous studies on the future of the European transport system is identified. The report on the program implementation states that one of the promising directions of innovation development is conducting research and the realization of design and engineering developments aimed at transport infrastructure improvement in the field of land use (FUTRE 2014). In view of this, the problems of further improvement and optimization of transport services provided to agrarian enterprises for the production of agricultural products are subject to numerous scientific studies and developments.

The directions of an innovative increase in the application efficiency of heavy road transport using "multi-lift," the technological system of container transportation, and in the organization of seasonal transportation of agricultural goods are presented by Kernychnyi (2016). The application of innovative management methods of an agricultural enterprise's management on the basis of transport logistics is considered in research by Glukhova (2011). The general characteristics of features and competitive conditions for the operation of the Ukrainian transport service market are given by Kirian (2014), the ways of increasing the efficiency of goods transportation by road on the basis of modern technologies are determined by Tomliak (2014), and the problems of ensuring the transport process efficiency in transporting agricultural products are considered in a study by Pepa and Cherniuk (2012). An analysis of the organizational peculiarities of agricultural enterprises' logistic systems was carried out by Vishnevskaya et al. (2015). The prospects for the further development of agrarian logistics for the market of agricultural crops are presented in a publication by Potapova (2017). The directions and possibilities of increasing the transportation support efficiency of the export of Ukrainian goods to the European Union relying on modern cargo transportation technologies are analyzed in an article by Kozachenko et al. (2014). In a paper by Kolodiichuk (2016), a conceptual model of logistic system optimization of the grain production subcomplex as part of the Ukrainian agro-industrial complex is proposed, and the question of the influence of the logistics management system in the grain production subcomplex of the agro-industrial complex on an increase in added value is considered in an article by Svitovyi (2017). The problems of management and the evaluation of the development prospects of Ukraine's transport and logistics systems in the context of European integration are the subject of research in a paper by Maselko and Shevchenko (2020). In a publication by Ohota (2014), considerable attention is paid to the study of ways to improve the efficiency of international traffic management, and the issues of improving the quality management system of transport services are discussed by Panchuk (2017).

The development of transport and logistics infrastructure is extremely important for solving transport problems in both passenger transportation and the carriage of all cargo types. Therefore, various aspects of transport and logistics infrastructure improvement and modernization are considered in a number of scientific works. So, Pasichnyk et al. (2013) perform an analysis of the factors determining the state and ensuring the formation of transport and logistics infrastructure. The problems of functioning and optimization of the infrastructure and technologies of cargo handling in a container terminal are presented by Pasichnyk and Kushenko (2016a, b). The application of grain cargo container transportation technologies and the directions of the corresponding infrastructure improvement are given in a paper by Pasichnyk et al. (2018). Results of an analysis of the approaches and the corresponding methodology of the Ukrainian transport and customs infrastructure formation are presented in an article by Pasichnyk (2016).

The results of a study on the construction of an effective supply logistics strategy and the provision of spare parts for equipment maintenance are considered by Wagner et al. (2012). The optimization method of cargo transportation routes by road based on ant colony optimization is given in a publication by Khalipova et al. (2018).

The results of the systematization of prerequisites for bringing Ukraine's transport policy to the norms of the European Union are given by Syryichyk et al. (2015). Spivakovskyi (2015) analyzes Ukraine's position in the international market for forwarding services, and Fedorko (2015) addresses the European benchmarks for ensuring the quality of providing transport and logistics services.

The problems of the influence of the developmental level of the Ukrainian transport system on the country's international economic image are considered in a paper by Harsun (2013).

At the same time, there are currently a number of unresolved problems which require further analysis consisting in the development of efficient transport and logistics technologies for the rapid delivery of agricultural machinery to the places of its intended use.

17.3 RESEARCH OBJECTIVES

The purpose of this chapter is to analyze and improve technologies and means of large-sized and heavy agricultural equipment transportation to reduce the time spent on delivery and preparation for harvesting. To achieve this goal, the following objectives have been formulated:

- to carry out an analysis of possible transport and technological schemes and transportation means for large and heavy goods;
- to determine appropriate domestic vehicles used for the transportation of heavy goods;
- to formulate the requirements for vehicles that can provide appropriate carriage on the basis of the initial parameters of the agricultural machinery chosen for transportation;
- to consider the identification peculiarities of auto trailers for the carriage of large and heavy goods and the rules for their customs clearance.

The solution of these tasks will contribute to improved efficiency of using agricultural machinery for its intended purpose on the basis of the optimization of the transport processes involved in their movement (STP 2015).

17.4 STATEMENT OF BASIC MATERIALS

Reduction in the time of agricultural machinery transportation affects not only the period of harvesting but also crop yields in general. The development and application of efficient transportation technologies can reduce equipment idle time and improve production output and product quality which may result in lowering agricultural production costs.

The delivery of agricultural machinery, primarily grain harvesters, as oversized and heavy cargoes of special purpose has always held a specific place in the services provided by transport companies. This is rather a complicated work which requires cargo carriers of the highest qualification and with a good command of the relevant modern professional technologies. The complexity of the transportation of harvesters lies in the large size of this machinery. The slow machine travel speed and its large dimensions do not allow harvesters to independently move over long distances.

Among combines, the transportation of which may cause some difficulties, several models can be distinguished. These include (Troitska & Shylymov 2010) grain and potato harvesters, beet harvesters, silage machines, forage harvesters, and flax harvesters. Organization of the transportation of the specified agricultural machinery falls under the category of the carriage of oversized heavy goods.

The issue of improving the technologies and means of oversized and heavy goods transportation is a rather complex problem. A significant number of scientific studies by famous scientists (T.K. Amirov, A.M. Kotenko, V. K.Mironenko, Yu.M. Nerush, D.K Preiher., Pravdin, M.Ya. Postan, N.A. Troitska, P.S. Shylaev) have made a considerable contribution to the development of this subject.

Oversized and heavy goods transportation implies, first of all, the development of special conditions and an individual transportation plan specific for each type of oversized cargo. Technical agricultural means also belong to the category of oversized cargo. So, the conditions of transportation must guarantee the integrity of all mechanisms of this machinery in its carriage.

The following technological procedures are included in the technological scheme of oversized and heavy goods transportation (Kotenko et al. 2014):

- calculation and drawing up of a logistic transportation plan;
- development of loading and fastening schemes on a vehicle and their coordination;
- obtaining a transportation permit;
- selection and preparation of the route (strengthening of bridges, dismantling of aerial networks, etc.);
- creation and development of special transport solutions specifically for each cargo;
- installation of special equipment for transportation, if necessary;
- escorting of cargo transportation by representatives of the State Automobile Inspection;
- the railing distance;
- customs clearing and preparation of all necessary documents for customs formalities (if necessary).

All types of transport of oversized and heavy cargo includes goods that exceed the size and load of the modern rolling stock and the existing dimensions of limiting devices and structures. In automobile transport, large-sized, long, and heavy goods, the sizes of which together with the vehicle exceed the following parameters, are categorized as oversized and heavy cargo: a height of 4 m, a length of 22 m, and a width of 2.6 m, according to the weight of goods with a vehicle of more than 38 tons.

The use of road transport is quite efficient for the transportation of large and heavy cargoes to clearly specified destinations and at a distance of up to 300 km. Motorways have more opportunities to transport oversized and heavy goods as compared to railway transportation and considerably lower restrictions on the dimensions and other transportation parameters. The main advantage of motor vehicles is that this mode of transport provides door-to-door cargo delivery. The parameters of the functioning of motor transport to a large extent also facilitate the timely delivery of goods to their destination. Therefore, at present, the transportation of oversized cargo by road is one of the most common types of oversized transportation. The carriage of such cargo types by road is carried out with the help of specially equipped towing trucks of different load-carrying capacities. In the case of oversized cargo transportation and handling, special stairs with a different angle of inclination are used for the oversized cargo of different kinds.

For the transportation of harvesters, it is recommended to use specialized low-bed semitrailers with a rear load of up to 53 tons and a front load of up to 45 tons. The cargo on such semitrailers is fastened with chain binders and special straps for fixing and safe movement of cargo. The main technological scheme involves fixing a harvester on the platform in its assembled state. It is possible to deliver a machine in a more simple way with the removal of the combine wheels, which are transported separately. But in this case, the time spent on preparation for and the transportation of such equipment, as well as for preliminary starting procedures, is significantly increased (Beliaev et al. 2011).

The technical condition of large and heavy vehicles and their equipment has to comply with the Road Traffic Regulations and the instructions of production plants. A towing truck must be equipped with devices which will make the automobile come to a stop using the emergency braking system in the event of a failure of the brake lines connected with the trailer.

Large and heavy vehicles must be equipped with the following parts:

- at least two truck chocks for additional fixation of the wheels of the towing truck and each of the trailers in the event of a forced stop;
- a "keep left" and a "keep right" sign with diameters of 600 mm each, made of reflecting material in accordance with the standard requirements;
- eight traffic cones with alternating horizontal reflective bands of white and red colors (cone height is 600 mm and the width of the white and red bands is 150 mm each);
- a towing bar;
- a red flashing flashlight or a triangular safety reflector;
- a set of snow chains (from October 1 to April 1);
- an orange vest with reflective elements;
- at least one self-powered rotating warning light of an orange color, the use of which is agreed with the division of the National Police.

A license issued to the carrier by the authorized division of the National Police also serves as a permit to install and use an orange-colored rotating warning light on large and heavy vehicles and on a pilot car with an indication of its model, state registration number and the permit validity. Nowadays, the problem of transporting agricultural machinery is becoming even more urgent given that most regions of Ukraine today do not have their own regional agricultural centers. Therefore, agricultural enterprises in these parts of the country are forced to organize the transportation of such equipment from other regions of Ukraine.

In particular, agricultural enterprises in the Dnipropetrovsk region have to rent combines from agricultural enterprises located in the southern regions of Ukraine, and this raises the question of determining methods of delivering such agricultural machinery as harvesters. The problems that arise in the case of self-propelled combines are primarily due to the fact that such equipment can move on general-purpose roads only in certain cases and provided that there are special permits from the state authorities. In addition, due to the weight of such equipment (over 10 tons) and their low speed (up to 25 km/h), self-propelled combine harvesters are too inefficient, given the time needed for such movement, fuel consumption, and especially taking into account its cost.

However, among the necessary conditions for the self-movement of such agricultural machinery as a combines, the availability of regular refueling and the possibility of maintenance and repair should be mentioned. Therefore, this type of transfer can be used for self-transporting at a distance of up to 25 km. Today, the most commonly used combine harvester models in Ukraine are John Deere harvesters (USA), namely the John Deere 9500 model. The parameters of this combine harvester are given in Table 17.1.

According to the size of a combine during transportation, it belongs to oversized and heavy goods. Thus, its carriage has to be done following certain rules and in compliance with the corresponding requirements. For its transportation, it is recommended to use the products of the enterprise "VARZ" LLC, which manufactures special vehicles for carrying out oversized and heavy goods transportation. Among the wide assortment of this factory's products, special attention is paid to semitrailers with a capacity of 20 tons.

Table 17.1 John Deere 9500 combine specifications

Parameters	Value	Unit of measurement
Engine type	Proprietary solution (John Deere 6076 model)	–
Engine power	228	HP
Engine capacity	8	L
Fuel tank capacity	530	L
Fuel consumption under normal load	62	L/h
Maximum productivity	9	t/h
Maximum achievable speed	25	km/h
Maximum unloading rate	4,500	L/min
Minimum turning radius	3.2	m
Length (without attachments)	4	m
Width (without attachments)	9	m
Weight (without attachments)	11.5	t

Let us consider the technical characteristics of the most common specialized equipment models offered by the manufacturer: VARZ NPV-3811, VARZ NPV-4711, and VARZ NPV-6011. The VARZ NPV-4711 semitrailer is intended for the transportation of agricultural, construction, road, and military equipment. Besides this, this semitrailer model can be used for non-divisible load transportation. The total weight of transported goods should not exceed 46.9 t. (VARZ 2020). As for the semitrailer type, it is a low-bed semitrailer. The semitrailer frame is equipped with additional extended wideners to allow broadening the bed to up to 3.0 m. Among the technical specifications of this model, it is possible to note the fact that the brake system is provided with the ABS system which includes the main filter, axle arm, cross valve, control cock, floor level crane, automatic brake controller, condensate drain valve, coupling heads, two-line cock for semitrailer brake release, main brake valve, and quick brake release cock. Its technical specifications are given in Table 17.2. The exterior of the VARZ NPV-4711 semitrailer model is shown in Figure 17.1.

A comparative analysis of the parameters of a semitrailer truck and a truck-trailer of identical load-carrying capacity shows that, in the case of the latter, the engine power is required to be 35%–40% more than for a semitrailer truck, which decreases the economic performance of the cargo transportation. A semitrailer truck also has advantages over a truck-trailer based on some other parameters, for example, the overall length, which matters in maneuvering, movement in a motor column, and placing them in parking lots.

Consequently, for the transport of such oversized and heavy goods as agricultural machinery, it is reasonable to use the VARZ NPV-4711 model with average load characteristics, which meets all the technical requirements set for the carriage of such cargoes. Let us now calculate the fuel consumption rate for moving a combine from the southern region of Kherson to Dnipro at a distance of 329 km using a Mercedes Actros 1846 LS L-Haus semitrailer truck. Fuel consumption with load is calculated using the formula:

$$Q_n = 0.01(H_{san} \cdot S + H_w W) \cdot (1 + 0.01D) \qquad (17.1)$$

Table 17.2 VARZ NPV-4711 technical specifications (VARZ 2020)

Parameter	Value
Cargo weight, kg	46,900
Semitrailer gross weight, kg	16,000
Semitrailer total weight, kg	62,900
Weight of semitrailer loading on truck wheels, kg	42,500
Total weight of semitrailer loading on weight-bearing swivel connection, kg	23,500
Dimensions of the semitrailer bed:	
Length, mm	11,013
Width, mm	2,550
Width with a widener, mm	3,000
Suspension	BPW Pneumatic Aggregate
Maximum achievable speed, km/h	70
First, second and third axles	Fixed
Third and fourth axles	Adjustable
Dovetails	Mechanically driven with a spring auxiliary mechanism (two-level)
Dovetail angle, degrees	15
Braking mechanism	Drum type brakes 300 × 200
Braking system	WABCO (with ABS)
Landing gear	JOST SIMOL
Tires (20 + 2)	235/75/R 17.5
Plate wheels	17.5 × 6.75
Electrical equipment	"VIGNAL", with retractable shields "Oversized load" and outline marker lamps
Assembled kingpin	CF 3.5-inch
Parking brakes	Mechanical (brake cable, nut and bolt)
Platform	Nondetachable, drop sided

Figure 17.1 Exterior of VARZ NPV-4711.

where Q_n is the standard fuel consumption, l; S is the kilometerage of a vehicle or a road train, km; H_{san} is the standard rate of fuel consumption in relation to the kilometerage of a vehicle or a road train with kerb weight, l/100 km; H_w is the standard rate of fuel consumption for transport work l/100 km; W is the volume of transport work, t/km; and D is the correction factor (a total relative increase or decrease), %.

The basic standard fuel consumption rate is 37 L/100 km and the standard rate of fuel consumption for transport work (H_w) is 1.3 l/100, t/km. Standard fuel consumption is determined using the formula:

$$W = H_{total} L \tag{17.2}$$

where H_{total} is the total weight of transported goods, t and L is the length of transportation, km:

$$W = (16 + 11.5) \cdot 329 = 9047.5 \text{ t/km} \tag{17.3}$$

Consequently, the volume of transport work performed by a vehicle on the route is 9,047.5, t/km. The standard fuel consumption with the load will be calculated using the formula (17.1):

$$Q_n = 0.01(37 \cdot 329 + 1.3 \cdot 9047.5) \cdot (1 + 0.01 \cdot 0) = 239.5 \text{ l} \tag{17.4}$$

Thus, the standard fuel consumption on the specified route using a Mercedes Actros 1846 LS L-Haus semitrailer truck and a VARZ NPV-4711 semitrailer for the transportation of a John Deere 9,500 combine along the Kherson-Dnipropetrovsk route, with a length of 329 km, will amount to 239.35 L. The calculations show that in the case of a self-propelled combine, its own consumption of fuel will be an order greater. The results of the comparative analysis aimed at the calculation of the dependence of the transportation costs on the length of the route are shown in Figure 17.2. The analysis of the data shows that the transportation costs in two different cases will be the same for a route of 25 km. That is, moving the combine under its own power is efficient at a distance of up to 25 km. For longer distances, it is expedient to use a semitrailer truck.

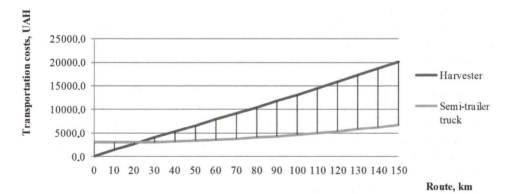

Figure 17.2 Dependence of the transportation costs on the length of the route for the two options.

Consequently, the most efficient method of transporting a harvester along the chosen route is its carriage by a Mercedes Actros 1846 LS L-Haus semitrailer truck and a VARZ NPV-4711 semitrailer. The fuel consumption amounts to 239.35 L in this case, according to the calculations carried out above, which is almost one tenth of that of a self-propelled combine. To reduce the cost of the combine transportation, a carrier company can plan the transportation in such a way that the Dnipropetrovsk-Kherson route is used for other transportation as well. For example, accompanying goods (building materials, road machinery, etc.) can be transported along with it.

It should also be noted that the use of locally produced vehicles reduces the expenses for their purchase and also stimulates the development of the domestic transport industry. This can be shown to be true because of the fact that when moving newly purchased vehicles through the customs border of Ukraine, according to Part 1., Article 51 of the Customs Code of Ukraine (CCU 2012), the declarant defines their customs value in accordance with the norms of this code and calculates the rates of customs duties and fees. For calculation of customs payments for a VARZ NPV-4711 semitrailer (code 8716 40 00 10 in the Ukrainian Commodity Coding System), all of the obtained data were used. For calculation purposes, it was also taken into account that the semitrailer was new, and the specified price was set by the manufacturer in the cost documents when it was sold to the carrier company as totaling USD 40,000.

In accordance with the current regulations available on the Internet resource (MDOS 2020), for the semitrailer customs clearance, the following customs payments need to be calculated:

1. Import duty in the amount of 10% of the customs value of a semitrailer.
2. Value-added tax of 20%.

Thus, the total amount of customs payments comprises, according to the calculation, UAH 329,272.65, which should be paid in the case that the specified vehicle is imported.

17.5 CONCLUSION

The results of the conducted research show that for the transportation of heavy goods (large agricultural machinery), it is necessary to use special vehicles of domestic production which will reduce the time and financial resources for its carriage. In particular, for the transportation of grain harvesters, it is expedient to use the VARZ NPV-4711 semitrailer with an average capacity which meets all of the technical transportation requirements.

The scientific novelty of the research results lies in the fact that the proposed approach to organizing the transportation of agricultural machinery using special vehicles will significantly improve the efficiency of the application of the agricultural machinery. The use of the proposed transport and logistics method raises the possibility of the transition to qualitatively new criteria for agricultural machinery utilization efficiency which is characterized by a high speed of its delivery to the places of harvesting along optimally effective transportation routes.

The practical value of the results of the study consists in the application of the proposed schemes and recommendations to reduce the time and financial resources for agricultural machinery transportation in the harvest season. The results of the investigation can be of use for further research in the direction of development of the theory of large and heavy agricultural cargo transportation efficiency.

REFERENCES

Beliaev, B.M., Mirotin, L.B., Nekrasov, A.H. & Pokrovskyi, A.K. 2011. *Management of Processes in Transport and Logistics Systems.* Moscow: MADI.

CCU. 2012. Customs Code of Ukraine: Law of Ukraine No. 2756-VI of March 13, 2012. *Official Bulletin of Ukraine* 32: 9.

Engström, R. 2016. The roads' role in the freight transport system. *Transportation Research Procedia* 14: 1443–1452. Doi: 10.1016/j.trpro.2016.05.217.

Fedorko, I.P. 2015. European guides for the quality assurance of transport and logistics services. *Development of Management Methods in Transport* 1: 49–62.

FUTRE. 2014. Future prospects on transport evolution and innovation challenges for the competitiveness of Europe Acronym: FUTRE Grant Agreement no: 314181 Programme. *7th Framework Programme Funding Scheme: Support Action Start date,* 1st October 2012 Duration: 24months.

Glukhova, I.Yu. 2011. Transport agrology as one of innovation management directions at agricultural enterprises. *Theoretical and Practical Aspects of Economics and Intellectual Property: A Collection of Scientific Works* 1: 167–172.

Harsun, L.H. 2013. Transport component of the international economic image of the country. *Culture of the Black Sea Region Nations* 264: 160–163.

Kernychnyi, B.Ya. 2016. Innovative ways to increase the efficiency of using heavy road transport (in the context of the organization of agricultural cargo seasonal transportation). *Proceedings of Dnepropetrovsk National University of Railway Transport named after Academician V. Lazaryan "Problems of Transport Economics"* 11: 31–36.

Khalipova, N., Pasichnyk, A., Lesnikova, I., Kuzmenko, A., Kokina, M., Kutirev, V., & Kushchenko, Y. 2018. Developing the method of rational trucking routing based on the modified ant algorithm. *Eastern-European Journal of Enterprise Technologies* 1/3(91): 68–76.

Kirian, O.I. 2014. General characteristics of the domestic competitive market for transport services. *Bulletin of National Technical University "KhPI". Series: Technical Progress and Production Efficiency* 33: 3–13.

Kolodiichuk, V.A. 2016. Conceptual model of the logistics system optimization in the grain production subcomplex of Ukraine's agro-industrial complex. *Economy of Agro-Industrial Complex* 5: 60–65.

Kotenko, A.N., Lavrukhin, O.V., Shylaev, P.S., Svitlychna, A.V., Shevchenko, V.I. & Pylypeiko, O.N. 2014. Transportation of oversized and heavy cargoes in transport systems. *UkrDazt: Collection of Scientific Works* 145: 50–59.

Kozachenko, D.M., Okorokov, A.M. & Hrevtsov, S.V. 2014. Transportation support for the export of Ukrainian goods to the European Union. *Bulletin of University of Customs and Finance. Series: Technical Sciences* 2: 141–148.

Maselko, T.Ye. & Shevchenko, S.H. 2020. Problems of management of Ukraine's transport-logistic systems and development prospects in the context of the European integration (http://www.nbuv.gov.ua/portal/chem_biol/nvnltu/17_2/301_Maselko_17_2.pdf, accessed: 12.05.2020).

MDOS. 2020. Internet resource of the MD Office Software (http://www.mdoffice.com.ua, accessed: 12.05.2020).

Ohota, B. 2014. Improving the efficiency of international traffic management. *Economic Bulletin* 1: 35–41.

Panchuk, O.V. 2017. Improving the quality management system for transport services. *Global and National Problems of the Economy* 19: 626–630.

Pasichnyk, A., & Kushenko, E. 2016a. Creating a model of functioning the objects container terminal. *Proceedings of the International Research and Practice Conference. "News of Science and Education"* 18(42): 91–96.

Pasichnyk, A., & Kushenko, E. 2016b. Analysis of the work of Ukrainian sea ports and prospects of their development based on the container terminal "Transport Investment Service". *Customs Scientific Journal, University of Customs and Finance of the World Customs Organization*, 9–21.

Pasichnyk, A., Lebed, I.H., Miroshnychenko, S.V. & Kushchenko, Ye.S. 2018. Improving the infrastructure and technologies of container transportation of grain cargoes. *Technical Service of Agro-Industrial, Forestry and Transport Complexes* 14: 174–183.

Pasichnyk, A., Vitruh, I. & Kutyrev, V. 2013. Factors that influence the formation of the transport-logistics networks. *Systemy i srodki transportu samohodowego*, 517–526.

Pasichnyk, A.M. 2016. *Methodology of Formation of Transport and Customs Infrastructure in Ukraine*. Dnipropetrovsk: UMSF.

Pepa, T.V. & Cherniuk, L.H. 2012. Quality assurance system of the transport process in the carriage of agricultural products. *Collection of Scientific Works of Vinnytsia National Agrarian University. Series: Economic Sciences* 4(70): 59–68.

Potapova, N.A. 2017. Prospects for the development of agro-logistics in the markets for agricultural crops (http://repository.vsau.org/getfile.php/13015.pdf, accessed: 12.05.2020).

Prydiuk, V.M. 2015. Organizational features of agricultural goods transportation by road. *Agricultural Machines* 28: 68–72.

Spivakovskyi, S. 2015. Position of Ukraine in the International Market for Freight Forwarding Services. *Economy of Ukraine* 1: 75–78.

STP. 2015. Program the concept of the state target program for the development of the agrarian sector for the period until 2020. The Cabinet of Ministers of Ukraine (http://zakon5.rada.gov.ua/laws/show/1437-2015-r, accessed: 12.05.2020).

Svitovyi, O.M. 2017. Improvement of logistics system management in grain production sub-complex of the agrarian and industrial complex as a factor of an increase in added value (http://www.vestnik-econom.mgu.od.ua/journal/2017/24-1-2017/14.pdf, accessed: 12.05.2020).

Syryichyk, T., et al. 2015. *Transport Policy of Ukraine and its Approximation to the Norms of the European Union*. Kyiv, Ukraine: Blue Ribbon Center.

Tomliak, S.I. 2014. Ways of increasing the efficiency of goods transportation by road. *Inter-University Collection "Scientific Notes"* 46: 529–537.

Troitska, N.A. & Shylymov, M.V. 2010. *Transport and Technological Schemes of Certain Types of Goods Transportation*. Moscow: KNORUS.

VARZ. 2020. Official site of the manufacturer "VARZ" LLC (http://www.varz.dp.ua/, accessed: 12.05.2020).

Vishnevskaya, O.M., Dvoinisiuk, T.V. & Shyhyda, S.V. 2015. Features of logistics systems of agricultural enterprises (http://global-national.in.ua/archive/7-2015/25.pdf, accessed: 12.05.2020)

Wagner, S.M., Jonke, R. & Eisingerich, A.B. 2012. A strategic framework for spare parts logistics (www.researchgate.net/publication/259729351, accessed: 12.05.2020).

Chapter 18

Optimization of public transport schedule on duplicating stretches

Iryna Lebid, Nataliia Luzhanska, Irina Kravchenya, Ievgen Medvediev, Andrzej Kotyra, Paweł Droździel, Aisha Mussabekova, and Gali Duskazaev

CONTENTS

18.1 Introduction 209
18.2 Review of scheduling methodology 210
18.3 Scheduling methodology of route vehicle on duplicating stretches 211
18.4 Application 214
18.5 Conclusion 216
References 218

18.1 INTRODUCTION

One of the quality indicators of public transport services is punctuality, which is directly related to a well-planned vehicle route schedule. The problem of the improvement in passenger service quality and the efficiency of urban public transport is to align the schedules of different routes on duplicating stretches, thereby contributing to more regular traffic intervals and vehicle occupancy (Lakhno & Sobchenko 2016, Dudnikov et al. 2010).

The presence of duplicating stretches is accompanied by the formation of transport queues at the transport stops, as well as the irregularity of traffic intervals and vehicle occupancy, thus increasing passengers' waiting time for vehicles and leading to a negative impact on comfort while traveling (Kochegurova et al. 2015).

The regularity of vehicle arrival at the transport stop, which is serviced by one route, is ensured by the observation of necessary traffic intervals. However, if there are several routes servicing a particular stretch in a transport network, it is necessary to coordinate the time schedules of different routes on duplicating stretches by adjusting the departure time for each to avoid the formation of queues at the transport stops. If there are several transport stops shared by duplicating stretches, it will be hard enough to completely eliminate queues because of the difference in length of these route stretches, their traffic speed, and passenger traffic flow at the transport stops (Yasenov et al. 2014).

It would appear to be possible to achieve the abovementioned coherence in vehicle schedules of different routes through the primary coordination of traveling time through "basic" transport stops with further calculation of the traveling time through the other transport stops of the route.

DOI: 10.1201/9781003224136-18

18.2 REVIEW OF SCHEDULING METHODOLOGY

A lot of articles are devoted to the scheduling methodology of vehicle routes (Kravchenya & Podkolzin 2019, Kazhaev 2020, Yurchenko et al. 2015, Currie & Bromley 2005, Subiono et al. 2018). The problem of urban public transport scheduling is a variant of the assignment problem and belongs to class NP-hard, whose complexity grows exponentially with the increase of the variable number and possible values. Furthermore, it is characterized by a large amount of reference information of different composition and a significant number of requirements which are difficult to formalize.

The reported difficulties impede automation of the urban public transport scheduling procedure, despite the wide range of integer programming techniques:

1. Total or partial enumeration of possibilities (e.g. branch-and-bound algorithm), their quantity analysis, and the determination of the best option (Taha 2007, Kravchenya et al. 2014).
2. Simulation of the actions of a person developing a schedule.
3. Step-by-step schedule designing according to determined optimization criteria.

The first type of algorithms are the precision (classical) methods (Taha 2007, Kravchenya et al. 2014) and are used for urban public transport scheduling that contain a small number of routes. However, to develop a schedule with a large number of transport stops, the application of these algorithms is not acceptable on the grounds of the exponential growth of the number of options.

The main disadvantage of an algorithm application based on partial enumeration or a branch-and-bound algorithm is amendment of earlier assignments and the repeating of some steps in the case of inapplicability of the obtained schedule version. The reason is that the developed schedule has an impact on the designing of a new one. Adjustment or complete changing of the earlier developed schedules is thus required. The application of total enumeration of all possibilities, under the condition of a high dimension amount to algorithm circularity, makes it unacceptable because of the enormous time expenditure.

Therefore, among the disadvantages of precision methods are awkwardness and complexity of the obtained mathematical model of the scheduling problem, a sharp increase in time expenditure in combination with the growth of the reference information, and finding a solution owing to NP-hard nature of the scheduling problem in its classical posing.

The second type of algorithms relate to heuristic and metaheuristic approaches (genetic algorithm, simulated annealing, and ant colony optimization) (Gorokhova 2015). The most important disadvantage of these techniques is the difficulty of assessing the threat of route vehicle assignment on the feasibility of implementing further assignments.

The most effective algorithms are the algorithms of the third type. They are based on an approach called "step-by-step method of designing" or "directed search method". The purpose of using this approach is to exclude or diminish the enumeration of possibilities and provide an acceptable quality of developed urban public transport schedule aligning time intervals among consecutive buses of different routes on duplicating stretches (Usov et al. 2016).

18.3 SCHEDULING METHODOLOGY OF ROUTE VEHICLE ON DUPLICATING STRETCHES

The scheduling methodology of vehicle routes on duplicating stretches includes the following steps:

Step 1. In the first step, urban network analysis is conducted. For different routes, it is necessary to determine:

- all duplicating stretches $D = \{D_1, D_2, ..., D_{Nd}\}$;
- their lengths $LD = \{LD_1, LD_2, ..., LD_{Nd}\}$;
- vehicle arrival/departure frequency of different routes at each transport stop of each duplicating stretch $SDT = \{SDT_1, SDT_2, ..., SDT_{Nst}\}$;
- social value of service area $KD = \{KD_1, KD_2, ..., KD_{Nd}\}$: passenger railroad and bus terminals, attractions, large-scale enterprises, educational establishments, etc.

Step 2. The second step is the alignment of time intervals among consecutive vehicles on duplicating stretches. The duplicating stretches $D = \{D_1, D_2, ..., D_{Nd}\}$ are ranked in descending order of the transport stop amount (length) and the social value of the service area. The routes $MD = \{MD_1, MD_2, ..., MD_{Nmd}\}$ with the largest duplicating stretches are determined with the following ranking in ascending order of number of vehicles.

The basic transport stop SB_k is assigned for determined routes. When choosing a basic transport stop, the stretch length in which routes are duplicated, the vehicle frequency of different routes along such stretches, and the social value of the service area should be taken into account. Compulsory arrival/departure frequencies (compulsory assignments) of vehicles are defined, pegged to:

- passenger service of railway and bus terminals;
- beginning or end of the enterprise working hours;
- beginning or end of the working shift and lunch break of vehicle drivers;
- beginning or end of the classes in the educational institutions.

The optimal time interval among arrivals of vehicles of all routes I_M^* and duplicating stretch routes I_D^* at the transport stop is calculated:

$$I_M^* = \frac{T}{\sum_{i=1}^{n} N_{Mi}} \qquad (18.1)$$

$$I_D^* = \frac{T}{\sum_{i=1}^{n} N_{Di}} \qquad (18.2)$$

where T is the planning period (in general, one hour of the day); ND_i is the number of runs carried out on the duplicating stretch at the time interval i; and NM_i is the number

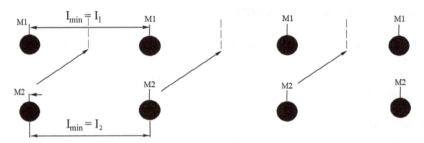

Figure 18.1 Scheme of bus traffic shift of two routes with the same minimum values of traffic intervals.

of runs carried out on the route at the time interval i. Shifting the route schedule at a basic transport stop is undertaken for a defined time interval $[\tau_i, \tau_{i+1}]$.

Figure 18.1 shows a scheme of shifting the bus traffic of two routes with the same minimum values of traffic intervals. Bus traffic routes are indicated with letters M_1 and M_2 and I_{min} is the minimum interval between consecutive buses of the same route.

Figure 18.2 shows a scheme of shifting the bus traffic of two routes with different numbers of vehicles on the routes and different traffic intervals. The arrival times of vehicles of the first and second routes at the transport stop by schedule are indicated through t_{11}, t_{12}, t_{21}, and t_{22}, and I_d is the calculated optimal time interval between bus arrivals of different routes.

Figure 18.3 shows a scheme of shifting the bus traffic of three routes with different numbers of vehicles on the routes and different traffic intervals. This phase is repeated until intervals $(I_1 \approx ... \approx I_i \approx ... \approx I_N)$ as equal as possible among vehicles arriving at the transport stop are obtained.

The problem, therefore, is limited to ensuring equal intervals among arrivals at the transport stop of vehicles where possible. In other words, it is necessary to minimize the mean difference deviation $D(I)$ between consecutive vehicles from the optimal value. The objective function can be written down in the following format:

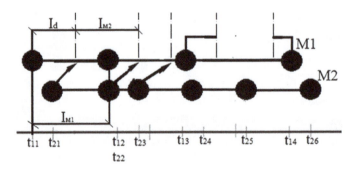

Figure 18.2 Scheme of bus traffic shift of two routes with different numbers of vehicles on the routes and different traffic intervals.

Optimization of public transport schedule 213

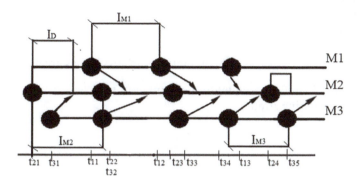

Figure 18.3 Scheme of bus traffic shift of three routes with different numbers of vehicles on the routes and different traffic intervals.

$$D(I) = \sum_{i=1}^{N_D}\left|I_D^* - I_i\right| + \sum_{i=1}^{N_{M_1}}\left|I_{M_1}^* - I_i\right| + \ldots + \sum_{i=1}^{N_{M_k}}\left|I_{M_k}^* - I_i\right| \to \min \qquad (18.3)$$

Next, calculation of the traveling time through the rest of the transport stops of the route regarding the basic stops and transition to the next duplicating stretch are performed.

Step 3. During the next step, synchronization of the adjusted vehicle schedules, taking into consideration the schedules at the stops which are used as transfer facilities for passengers, is executed. When it is impossible to synchronize the obtained vehicle schedules among each other, some steps must be repeated.

Step 4. The quality of the adjusted schedule for each transport stop is determined by transport stop load factor K_D, passengers' waiting time T_W for vehicles of all routes, and the lowest deviation value of the intervals among consecutive vehicles from the optimal value $D(I)$. A two-dimensional Boolean assignment matrix showing the assignment of route j to time interval i is formed for the basic transport stops (Table 18.1):

Table 18.1 Assignment matrix for basic transport stops

Arrival time	Duplicating stretch			K_D	$\|I_D^* - I_i\|$	T_W	N_P	$\|I_{M1}^* - I_i\|$	$\|I_{Mk}^* - I_i\|$
	No. M_1	No. M_2	No. M_k						
t_1	x_{11}	x_{12}	x_{1k}	K_{D1}	$\|I_D^* - I_1\|$	T_{W1}	N_{P1}	$\|I_{M1}^* - I_1\|$	$\|I_{Mk}^* - I_1\|$
t_2	x_{21}	x_{22}	x_{2k}	K_{D2}	$\|I_D^* - I_2\|$	T_{W2}	N_{P2}	$\|I_{M1}^* - I_2\|$	$\|I_{Mk}^* - I_2\|$
…	…	…	…	…	…	…	…	…	…
t_i	x_{i1}	x_{i2}	x_{ik}	K_{Di}	$\|I_D^* - I_i\|$	T_{Wi}	N_{Pi}	$\|I_{M1}^* - I_i\|$	$\|I_{Mk}^* - I_i\|$
…	…	…	…	…	…	…	…	…	…
t_n	x_{n1}	x_{n2}	x_{nk}	K_{Dn}	$\|I_D^* - I_n\|$	T_{Wn}	N_{Pn}	$\|I_{M1}^* - I_n\|$	$\|I_{Mk}^* - I_n\|$
Sum	$\sum_{i=1}^{n} x_{ij}$	$\sum_{i=1}^{n} x_{ij}$	$\sum_{i=1}^{n} x_{ij}$	$\sum_{i=1}^{n} K_{Di}$	$\sum_{i=1}^{n} I_D^*$	$\sum_{i=1}^{n} T_{Wi}$	$\sum_{i=1}^{n} N_{Pi}$	$\sum_{i=1}^{n}\left|I_{Mi}^* - I\right|$	$\sum_{i=1}^{n}\left|I_{Mk}^* - I_i\right|$

$x_{ij} = 1$, if route j is assigned to time interval i;
$x_{ij} = 0$, if not;
$x_{ij} = 1^*$ – compulsory assignment (step 2).

For each time interval of the assignment matrix, the next values are defined:

- transport stop load factor, equal to the number of vehicles arriving at a transport stop on a duplicating stretch

$$K_{D_i} = \sum_{j=1}^{k} x_{ij}; \qquad (18.4)$$

- $|I^*_D - I_i|$ – deviation value of intervals among consecutive vehicles from the optimal value of the duplicating stretch;
- $|I^*_{Mk} - I_i|$ – deviation value of intervals among consecutive vehicles from the optimal value of the route k;
- $T_W = I_i \lambda_i$ – passengers' waiting time, where λ_i is the density of incoming passenger traffic using the vehicles of a duplicating stretch.

Step 5. The last step is to implement the simulation model of urban public transport within the simulation modeling system, GPSS World (Shevchenko & Kravchenya 2007, Krause et al. 2012), to battle test the optimization technique of route vehicle scheduling.

18.4 APPLICATION

The described technique was tested in the existing transport network of Gomel, Belarus. Currently, transportation of passengers in Gomel is carried out on 81 regular bus routes. Six long duplicating stretches of bus movement of three or more routes were defined, and the main bus route schedules were optimized. One of the most important duplicating stretches of the transport network is the "Institute Gomel project – Ogorenko street" stretch for routes No. 17, 18, and 34 (Figure 18.4). It includes 13 public transport stops of 7.5 km length. This combined stretch passes through the streets with the heaviest traffic of the Centralniy District of the city (Sovetskaya and Internatsionalnaya streets). A huge amount of waiting passengers are at the transport stops, as well as delays of public transport caused by involuntary stopping to wait for an opportunity to enter a station, because of the simultaneous arrival of several vehicles along different routes.

Bus route No. 17 "Medgorodok – Microdistrict Klenkovsky" is the most demanded and overcrowded bus route in Gomel, as it passes through a large number of facilities of mass attractions, which include two hypermarkets, a department store, cinemas, parks, theaters, and a circus. It runs through the Sovietsky, Centralny, and Zheleznodorozhny districts and part of the route passes along the largest artery of the city – Sovetskaya Street. The route length in the forward direction is 18.75 km and the number of public transport stops is 35. The route length in the return direction is 18.73 km and the number of public transport stops: 33. Total traveling time is 1 hour 5 minutes.

Bus route No. 18 "Medical radiology center – Microdistrict Klenkovsky" passes through the Novobelitsky, Centralny, and Zheleznodorozhny districts of Gomel, has 18

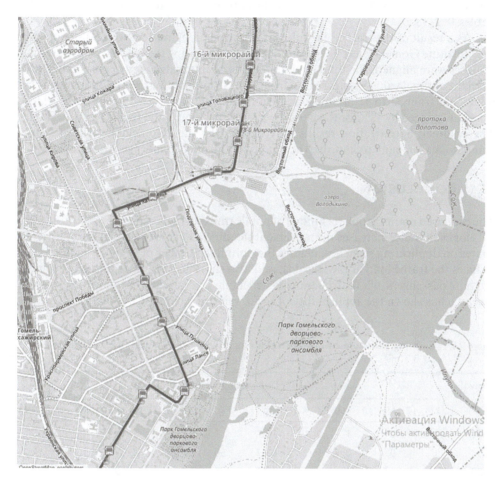

Figure 18.4 Scheme of bus routes No. 17, 18, and 34 on duplicating stretch.

common public transport stops in a row with route No. 17, and also 13 public transport stops with route No. 34, including the Central District stretch of the city, which is most crowded with passenger traffic, with its majority of cultural and recreational facilities, public recreation places, as well as shopping areas and enterprises. The route length in the forward direction is 19.56 km and the number of public transport stops is 33. The route length in the return direction is 19.64 km and the number of public transport stops is 33. Total traveling time is 1 hour 2 minutes.

Bus route No. 34 "Microdistrict Lyubenskiy – Old Volotova" passes through the Sovietsky, Centralny, and Zheleznodorozhny districts of Gomel, has 13 common public transport stops in a row with routes No. 17 and 18. The route length in the forward direction is 13.89 km and the number of public transport stops is 24. The route length in the return direction is 14.15 km and the number of public transport stops is 26. Total traveling time is 49 minutes.

The existing and adjusted schedules of bus arrival along routes No. 17, 18, and 34 at the "Institute Gomelproject" transport stop during the rush hour from 7 a.m. to

8 a.m. and the interval between the rush hour from 11 a.m. to 12 p.m. are presented in Tables 18.2–18.5.

As a result of the traffic schedule optimization along routes No. 17, 18, and 34 on the "Institute Gomelproject – Ogorenko street" duplicating stretch, the traffic intervals of buses were aligned, the total deviation value of intervals among consecutive buses from the optimal value has decreased from 33 to 15 minutes for the period between rush hours and from 32 to 11 minutes for rush hours. Herewith, total passengers' waiting time has reduced by 13% for the period between rush hours and by 30% for rush hours. The obtained optimization results may be used by the "Gomeloblavtotrans" Open Joint Stock Company to improve the quality of the public passenger transportation.

18.5 CONCLUSION

The proposed optimization technique of public transport schedules adjusts traffic intervals for each route, increases the movement steadiness of consecutive buses of different routes on duplicating stretches, defines the optimum number of vehicles on the routes, reduces the traffic load at the stations, and also shortens the waiting time for vehicles by those passengers, who can be transported along several route options. Experimental research has shown the applicability of the developed technique in practice.

Table 18.2 Existing schedule of bus arrival along routes No. 17, 18, and 34 from 7 a.m. to 8 a.m.

Arrival time	\multicolumn{10}{c}{Duplicating stretch}									
	No. 17	No. 18	No. 34	K_D	$|I^*_D - l_i|$	T_W	N_P	$|I^*_{M1} - l_i|$	$|I^*_{M2} - l_i|$	$|I^*_{M3} - l_i|$
7:00	0	0	1	1	–	–	–			1
7:04	1	0	0	1	0.67	50	20	0		
7:09	0	1	0	1	1.67	75	25		2	
7:11	1	0	0	1	1.33	15	10	0		
7:12	0	0	1	1	2.33	5	5			2
7:18	1	0	0	1	2.67	105	30	0		
7:22	0	1	0	1	0.67	50	20		2	
7:24	0	0	1	1	1.33	15	10			2
7:25	1	0	0	1	2.33	5	5	0		
7:35	1	0	0	1	6.67	275	50	3		
7:36	0	1	1	2	2.33	5	5		1	2
7:42	1	0	0	1	2.67	105	30	0		
7:47	0	0	1	1	1.67	75	25			1
7:49	1	1	0	2	1.33	15	10	0	2	
7:56	1	0	0	1	3.67	140	35	0		
7:59	0	0	1	1	0.33	30	15			2
Amount	8	4	6	18	31.67	965	295	3	7	10

Optimization of public transport schedule 217

Table 18.3 Adjusted schedule of bus arrival along routes No. 17, 18, and 34 from 7 a.m. to 8 a.m.

Arrival time	Duplicating stretch									
	No. 17	No. 18	No. 34	K_D	$\|I^*_D - l_i\|$	T_W	N_P	$\|I^*_{M1} - l_i\|$	$\|I^*_{M2} - l_i\|$	$\|I^*_{M3} - l_i\|$
7:00	0	0	1	1	–	–	0			1
7:03	1	0	0	1	0.33	30	15	0		
7:07	0	1	0	1	0.67	50	20		2	
7:10	1	0	0	1	0.33	30	15	0		
7:13	0	0	1	1	0.33	30	15			3
7:17	1	0	0	1	0.67	50	20	0		
7:20	0	1	0	1	0.33	30	15		2	
7:23	0	0	1	1	0.33	30	15			0
7:25	1	0	0	1	1.33	15	10	1		
7:31	1	0	0	1	2.67	105	30	1		
7:33	0	0	1	1	1.33	15	10			0
7:36	0	1	0	1	0.33	30	15		1	
7:39	1	0	0	1	0.33	30	15	1		
7:43	0	0	1	1	0.67	50	20			0
7:46	1	0	0	1	0.33	30	15	0		
7:49	0	1	0	1	0.33	30	15		2	
7:53	1	0	0	1	0.67	50	20	0		
7:56	0	0	1	1	0.33	30	30			3
Amount	8	4	6	18	11.31	635	295	3	7	7

Table 18.4 Existing schedule of bus arrival along routes No. 17, 18, and 34 from 11 a.m. to 12 p.m.

Arrival time	Duplicating stretch									
	No. 17	No. 18	No. 34	K_D	$\|I^*_D - l_i\|$	T_W	N_P	$\|I^*_{M1} - l_i\|$	$\|I^*_{M2} - l_i\|$	$\|I^*_{M3} - l_i\|$
11:05	1	0	0	1	–	45	15	2		
11:07	0	0	1	1	4	9	6			2
11:17	0	1	0	1	4	165	30		6	
11:25	0	0	1	1	2	108	24			2
11:26	1	0	0	1	5	3	3	6		
11:40	1	0	0	1	8	315	42	1		
11:44	0	1	1	2	2	30	12		7	1
11:54	1	0	0	1	4	165	30	1		
11:57	0	1	0	1	3	18	9		7	
Amount	4	3	3	10	33	858	171	10	20	5

Table 18.5 Adjusted schedule of bus arrival along routes No. 17, 18, and 34 from 11 a.m. to 12 p.m.

Arrival time	Duplicating stretch									
	No. 17	No. 18	No. 34	K_D	$|l^*_D - l_i|$	T_W	N_P	$|l^*_{M1} - l_i|$	$|l^*_{M2} - l_i|$	$|l^*_{M3} - l_i|$
11:04	0	0	1	1	—	30	12			0
11:10	1	0	0	1	0	63	18	2		
11:16	0	1	0	1	0	63	18		5	
11:21	0	0	1	1	1	45	15			2
11:28	1	0	0	1	1	84	21	3		
11:38	0	0	1	1	4	165	30			3
11:42	1	0	0	1	2	30	12	1		
11:46	0	1	0	1	2	30	12		10	
11:54	1	0	0	1	2	108	27	3		
12:03	0	1	0	1	3	135	6		3	
Amount	4	3	3	10	15	753	171	9	18	5

REFERENCES

Currie, G. & Bromley, L. 2005. Developing measures of public transport schedule coordination quality. *28th Australasian Transport Research Forum, 28 Sep – 30 Sep 2005*, Sydney, Australia.

Dudnikov, O.M., Vinogradov, M.S. & Vinogradov, I.M. 2010. Method of development of the buses schedule of different routes with consideration of a compatible part of their movement. *Vesti of Automobile and Road Institute* 2(11): 21–31.

Gorokhova, E.S. 2015. Formation of the schedule of passenger transport with the help of colony optimization (http://earchive.tpu.ru/bitstream/11683/17122/1/conference_tpu-2015-C04-v1-075.pdf, accessed: 13.05.2020).

Kazhaev, A.A. 2020. Simulation model of loading public transport stops (https://cyberleninka.ru/article/n/imitatsionnaya-model-zagruzki-ostanovochnyh-punktov-gorodskogo-marshrutnogo-transporta, accessed: 13.05.2020).

Kochegurova, E.A., Fadeev, A.S., Piletskya, A.Y. & Yurchenko, M.A. 2015. *Calculation of Performance Indicators for Passenger Transport Based on Telemetry Information. Engineering Technology, Engineering Education and Engineering Management*. London: Taylor & Francis Group, 847–85.

Krause, J., Spicker, M., Wörteler, L., Schäfer, M. & Zhang, L. 2012. Interactive visualization for real-time public transport journey planning. Proceedings of *Sigrad*, 95–98.

Kravchenya, I.N., Burduk, E. L. & Alymova, T.V. 2014. *Mathematical Modeling. Linear and Nonlinear Programming, Network Planning and Control*. Gomel: BelSUT.

Kravchenya, I.N. & Podkolzin, A.M. 2019. Optimization of public transport schedule of different routes on duplicating stretches. *Organization and Road Safety* 2: 54–61.

Lakhno, V.A. & Sobchenko, V.M. 2016. The automatic system's model of decision-making support for dispatching control of the city passenger traffic. *Proceedings of Dnepropetrovsk National University of Railway Transport "Science and Transport Progress"* 2(62): 61–69.

Shevchenko, D.N. & Kravchenya, I.N. 2007. *GPSS Simulation: A Learning Method*. Gomel: BelSUT.

Subiono, S., Fahim, K. & Adzkiya, D. 2018. Generalized public transportation scheduling using max-plus algebra. *Kybernetika* 54(2): 243–267.

Taha, H.A. 2007. *Operations Research: An Introduction*. 8th Ed. Upper Saddle River, NJ.

Usov, S.P., Obrezkova, V.E., Lipenkova, O.A. & Lipenkov, A.V. 2016. Improving the efficiency of public transport by adjusting the schedules of duplicating routes. *Problems of Quality and Operation of Vehicles. Materials of the 11th International Scientific and Technical Conference*, Penza, 15th March 2016, 382–390.

Yasenov, V.V., Eliseev, M.E., & Lipenkov, A.V. 2014. Analysis of problems in the work of public passenger transport in Nizhny Novgorod. *Trudy of NSTU Named after R. E. Alekseeva* 4(106): 249–254.

Yurchenko, M., Kochegurova, E., Fadeev, A. & Piletskya, A. 2015. Calculation of performance indicators for passenger transport based on telemetry information. *Engineering Technology, Engineering Education and Engineering Management*, 847–851.

Chapter 19

Dynamic properties of symmetric and asymmetric layered materials in a high-speed engine

Orest Horbay, Bohdan Diveyev, Andriy Poljakov,
Oleksandr Tereschenko, Piotr Kisala, Mukhtar Junisbekov,
Samat Sundetov, and Aisha Mussabekova

CONTENTS

19.1 Introduction 221
19.2 Beam modelling 222
19.3 Layered asymmetric plate 223
19.4 FGM beam 225
    19.4.1 Symmetric beam 226
    19.4.2 Asymmetric beam 227
    19.4.3 Comparison of functionally graded and three-layered plates 228
19.5 Discussion 229
19.6 Conclusion 229
References 230

19.1 INTRODUCTION

Plates from FGMs with properties continuously varying across their thickness attract the attention of many researchers. Such materials are used in many structures of mechanical engineering because of their high strength. For example, a homogeneous elastic layer of ceramic material glued to a metal structure serves as thermal protection in a high-temperature environment of the high flash high-speed engine. However, the thermoelastic properties of a ceramic or metallic material between two heterogeneous layers are resulting in an inevitable incompatibility of elastic fields in the contact zone. That leads to delamination and failure due to a sudden change in pressure and shear strains in this zone. Therefore, the origination of such a distinct interface between two dissimilar materials, as a rule, has to be avoided in the design of composite structures. This can be achieved by gradually changing the proportion of volumes of laminate constituents across its thickness and creating a transition zone in which the thermal properties vary continuously instead of sharply jumping between those of the ceramic material and metal layer. Therefore, FGMs have advantages in terms of strength in comparison to homogeneous and multi-layered materials, which is confirmed by data obtained with the help of numerous models and experimental studies.

    To model FGM plates, as well as laminated composite ones, an optional theory is required to accurately estimate the effect of shear stresses on their operation. There

DOI: 10.1201/9781003224136-19

are many studies devoted to the analysis of free and forced vibrations of functionally graded structures (Elishakoff & Guede 2004; Cheng & Batra 2000; CelikGul & Kurtulus 2016), but works on the optimization of their frequency spectrum and damping are few in number. At present, various approaches to modelling the structures containing heterogeneous thin-walled elements are being developed intensively. Results of comparative analyses of the theories of layered elements subjected to various loading conditions and the method of investigation are presented in Hu et al. (2008), Carrera (2003) and Vaer et al. (2017). In Diveyev et al. (2009, 2012, 2018), refined equations of dynamic bending of layered composite beams were obtained. On the basis of the same technique, the dynamic characteristics of FGM beams were determined in our work.

19.2 BEAM MODELLING

Various high-order displacement models have been developed in the literature by considering combinations of displacement fields for in-plane and transverse displacements inside a mathematical sub-layer. To obtain more accurate results for the local responses, another class of laminate theories commonly named the layer-wise theories approximate the kinematics of individual layers rather than a total laminate using the 2-D theories. These models have been used to investigate the phenomena of wave propagation as well as vibrations in laminated composite plates. Numerical evaluations obtained for wave propagation and vibrations in isotropic, orthotropic and composite laminated plates have been used to determine the efficient displacement field for economic analysis of wave propagation and vibration. The numerical method developed in this chapter follows a semi-analytical approach with an analytical field applied in the longitudinal direction and a layer-wise power series displacement field used in the transverse direction. The goal of this chapter is to develop a simple numerical technique, which can produce very accurate results compared with the available analytical solution. The goal is also to provide one with the ability to decide on the level of refinement in higher-order theory that is needed for accurate and efficient analysis.

Let us consider the stress-strain state of a beam, asymmetric with respect to its midplane, as the sum of symmetric (S) and antisymmetric (A) states. As in Carrera (2003), Sitek et al. (2006), Dobrzański et al. (2006) and Diveyev et al. (2009), we introduce the following kinematic approximations for the asymmetric beam as the sum of symmetric and antisymmetric components $U = U_s^e + U_a^e$

$$U_s^e - \begin{cases} u_s^e = \sum_{i,k} u_{iks}{}^e z^{(2i-1)} \sin k\pi x/L, \\ w_s^e = \sum_{i,k} w_{iks}{}^e z^{(2i-2)} \cos k\pi x/L, \end{cases} \quad (19.1)$$

$$U_a^e - \begin{cases} u_a^e = \sum_{i,k} u_{ika}{}^e z^{(2i-2)} \sin k\pi x/L, \\ w_a^e = \sum_{i,k} w_{ika}{}^e z^{(2i-1)} \cos k\pi x/L, \end{cases} \quad (19.2)$$

where $\phi_k(x)$ and $\gamma_k(x)$ are a priori known coordinate functions (for every beam clamp conditions) and u_{ik}^e, w_{ik}^e, u_{ik}^e, w_{ik}^e are the unknown set of parameters.

By substituting equations (19.1, and 19.2) into the Hamilton variation equation

$$\int_{t_1}^{t_2}\int_V \left(\sigma_{xx}\delta\varepsilon_{xx} + \sigma_{zz}\delta\varepsilon_{zz} + \tau_{xz}\delta\varepsilon_{xz} - \rho\frac{\partial u}{\partial t}\delta\frac{\partial u}{\partial t} - \rho\frac{\partial w}{\partial t}\delta\frac{\partial w}{\partial t}\right)dVdt = \int_{t_1}^{t_2}\int_S P\delta U \quad (19.3)$$

assuming single-frequency vibration

$$u_{ik}^e = \bar{u}_{ik}^e e^{i\omega t},\, w_{ik}^e = \bar{w}_{ik}^e e^{i\omega t},\, u_{ik}^d = \bar{u}_{ik}^d e^{i\omega t},\, w_{ik}^d = \bar{w}_{ik}^d e^{i\omega t}$$

we obtain the following system of linear algebraic equations for determining the amplitudes of displacements

$$-\omega^2[M]\bar{U} + i\omega[C]\bar{U} + [K]\bar{U} = [A]\bar{U} = \bar{f} \quad (19.4)$$

where ε_{ii} and ε_{ij} are the longitudinal and shear strains, respectively; σ_{ii} and τ_{ij} are the normal and shear stresses, respectively; V is the volume of the beam; S is surface of the elastic fixation; U and P are the vectors of amplitudes of displacements and external forces; $t_1 - t_2$ is time interval (Kukharchuk et al. 2016, 2017, Vedmitskyi et al. 2017); M is the matrix of masses; C is the matrix of viscoelastic damping; K is the stiffness matrix and ω is vibrations frequency. This solution scheme is rather bulky. However, it can be simplified by decomposition of the problem for the asymmetric beam into the sum of solutions for beams with mechanical properties symmetric and antisymmetric towards its midplane.

The operator of the problem (19.1–19.4) can be presented in the symbolic form $L = L_S + L_A$, where L_S and L_A are operators with symmetric and antisymmetric distributions of mechanical properties. Now, we obtain a simplified solution scheme

$$\begin{aligned}L(U)\delta U &= (L_S + L_A)(U_S + U_A)(\delta U_S + \delta U_A) \\ &= L_S(U_S)\delta U_S + L_A(U_S)\delta U_A + L_A(U_A)\delta U_S + L_S(U_A)\delta U_A\end{aligned} \quad (19.5)$$

The remaining terms are equal to zero because they are integrals of products of symmetric and antisymmetric functions (Ogorodnikov et al. 2018, Polishchuk et al. 2019, Kozlov et al. 2019).

19.3 LAYERED ASYMMETRIC PLATE

Imagine a plate as the sum of two plates: a plate with symmetric and asymmetric mechanical properties relative to the median plane (Figure 19.1).

Using the stiffness matrix K of the resolving system of equations (19.6), it is possible to calculate the values of damping η for beams of heterogeneous materials by the known formula

$$\eta = \frac{\eta_1[q_1]^T[K_1][q_1] + \ldots + \eta_N[q_N]^T[K_N][q_N]}{[q]^T[K][q]} \quad (19.6)$$

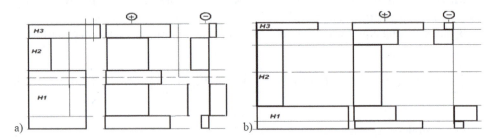

Figure 19.1 Construction of an asymmetric three-layer package to symmetric and anti-symmetric: (a) the thickness of the first layer exceeds the half-thickness of the package and (b) the thickness of the first layer is less than half the thickness of the package.

where $[K_i]$ and η_i are the stiffness matrix and damping of an interlayer, q is the vector of the solution and the superscript T indicates transposition. We assume, as usual, that the damping matrix is proportional to the stiffness matrix, i.e. $C_i = \eta_i [K_i]$

Additional approximations for the outer layers are (as in [19.7–19.9]).

$$U_s^N - \begin{cases} u_s^N = \sum_{i,k} u_{iks}{}^N z^{(i)} \sin k\pi x / L, \\ w_s^N = \sum_{i,k} w_{iks}{}^N z^{(i)} \cos k\pi x / L \end{cases}, U_a^N - \begin{cases} u_a^N = \sum_{i,k} u_{ika}{}^N z^{(i)} \sin(z) \sin k\pi x / L, \\ w_a^N = \sum_{i,k} w_{ika}{}^N z^{(i)} \sin(z) s \cos k\pi x / L \end{cases}, N = 1,2 \quad (19.7)$$

Figure 19.2 shows the frequency response function (FRF) and the ratio of damping in the sandwich as a whole to damping in its soft layers for an asymmetric three-layer beam with a mild damping core (Figure 19.1a) with a different thickness of a thin, rigid layer H_3

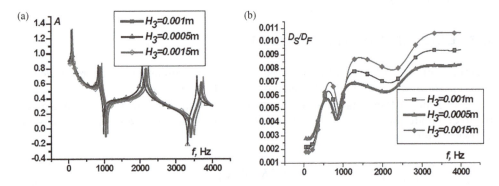

Figure 19.2 Dynamic characteristics of laminate with different thickness of the upper rigid layer H_3: (a) FRF and (b) relative damping.

We consider vibration testing of the beam for these geometrical parameters: length $L = 0.27$ m and thickness $H_1 = 0.01$ m and $H_2 = 0.0635$ m. Elastic modules are as follows: $C_{xx} = 240$ MPa, $C_{xz} = 150$ MPa, $C_{zz} = 200$ MPa MPa, $G = 58$ MPa and $\rho = 240$ kg/m^3. For the outer layers, $C_{xx} = 47$ GPa and $G = 0.6$ GPa (Ogorodnikov et al. 2004, 2018 Dragobetskii et al. 2015).

19.4 FGM BEAM

On the basis of the same technique, the dynamic characteristics of FGM beams were determined in our work. The problem can be simplified by decomposition of the problem for the asymmetric beam into the sum of solutions for beams with mechanical properties symmetric and antisymmetric to its midplane. An example of such a decomposition for Young's modulus is shown in Figure 19.3.

In contrast to the approach presented in the works of Diveyev et al. (Diveyev et al. 2009, 2011, 2012, 2017, 2018,; Kernytskyy et al. 2017), the stiffness coefficients C_{xx}, C_{xz}, C_{zz} and G_{xz} vary across the beam thickness, because the mechanical characteristics of FGM depend on its geometric parameters and are closely related to the concentration of a doping material. Very often, this is the percentage VZ of ceramic in metal. Predominantly, a power-law dependence for this concentration across the beam thickness of the beam congruent to the concentration of doping material is considered (Elishakoff & Guede 2004, Vasilevskyi 2013, 2014, 2015):

$$V_Z = V_0\left(\alpha_0 z^a + \beta_0 z^\beta\right), C_{xx} = C_{xx}{}^0\left(\alpha_0 z^a + \beta_0 z^\beta\right),...,G_{xz}{}^0\left(\alpha_0 z^a + \beta_0 z^\beta\right)S \tag{19.8}$$

For the case of elastic moduli and damping varying continuously across the beam thickness, according to (19.9)

$$\eta = \frac{\int \eta(z)[q]^T [K][q]dz}{[q]^T [K][q]} \tag{19.9}$$

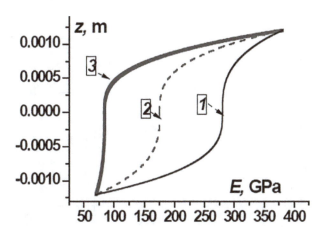

Figure 19.3 Across-the-thickness distributions of Young's modulus at $n = 1.5$ (1), 3 (2) and 6 (3).

19.4.1 Symmetric beam

Let us consider the cylindrical bending of a symmetric rectangular beam of length $L = 0.3$ m and thickness $H = 0.0024$ m simply supported and loaded by a force normal to its surface. The equations of dynamic equilibrium at such a kind of deformation do not depend on beam width. Let us consider the layer made of aluminium and covered with a ceramic alloy on both sides. The Young's modulus and density of aluminium (the core of the beam) are $E_A = 70$ GPa and $\_A = 2{,}700$ kg/m^3, respectively, and of the ceramic alloy (the top and bottom layers) are of $E_A = 370$ GPa and $A = 3{,}700$ kg/m^3 (Ogorodnikov et al. 2004, Dragobetskii et al. 2015, Vasilevskyi 2013).

Let us consider the case of an isotropic material with a concentration of the doped material varying according to a power law. In the same way, the elastic constants and density also vary across the beam thickness (Loy et al 1999, Cheng & Batra 2000, Elishakoff & Guede 2004):

$$E(z) = E_A + (E_C - E_A) z^n / H^n, \; G(z) = E(z)/(1+2\nu), \; \rho(z) = \rho_A + (\rho_C - \rho_A) z^n / H^n$$

At negative values of z, the elastic moduli are assumed to be symmetric $E(-z) = E(z)$. The Poisson ratio is assumed to be constant, $\nu = 0.3$.

We used approximations of different orders across the thickness. The longitudinal coordinate functions are given in the form $\phi_k(x) = \sin\left((2k-1)\pi x / 2L\right)$, $\gamma_k(x) = \cos\left((2k-1)\pi x / 2L\right)$

The frequency response of the symmetrical beam at different degrees of approximation n in equation (19.8) is shown in Figure 19.4.

To obtain the frequency response of the beams, we considered a concentrated harmonic force applied normally to a midplane point of the beam. To determine all

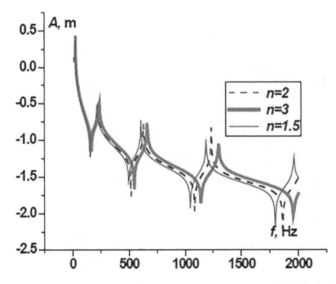

Figure 19.4 The frequency response of a symmetric beam at $n = 1.5$ (1), 3 (2) and 6 (3).

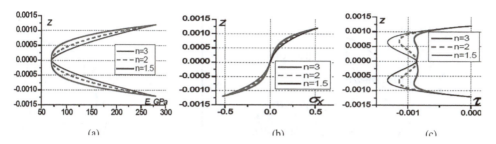

Figure 19.5 (a) Distributions of Young's modulus across the beam thickness, variations in the normal σ_x (b) and shear τ (c) stresses across the beam thickness.

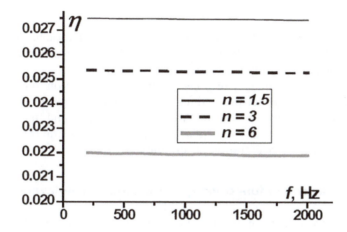

Figure 19.6 Relative damping.

natural frequencies without distortion, the force applied was close enough to the centre of the beam.

The data of Figure 19.5 show variations in Young's modulus and dimensionless stresses across the beam thickness at a frequency of 200 Hz for different values of n.

The damping η in the beam, calculated by formula (19.7) for different values of n, is illustrated in Figure 19.6. We assumed hypothetically the distribution of damping η to be the same as that of stiffness (19.8), namely

$$\eta(z) = \eta_A + (\eta_C - \eta_A) z^n / H^n \qquad (19.10)$$

Figure 19.6 presents the relative damping.

19.4.2 Asymmetric beam

Let us now consider the cylindrical bending of a simply supported and centrally loaded asymmetric beam of length $L = 0.3$ m and thickness $H = 0.012$ m. The beam is made of aluminium and has a top layer made of a ceramic alloy. The characteristics of

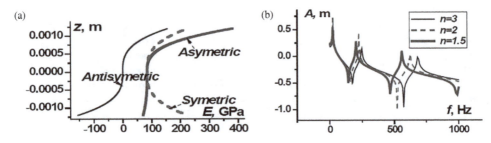

Figure 19.7 Variations in Young's modulus E across the beam thickness (a) and the corresponding frequency responses f (b) at n = 1.5 (1), 3 (2) and 6 (3).

aluminium and the ceramic alloy are the same as in the symmetric beam (see the previous section). The elastic moduli and density of the beam vary according to a power law (19.8), but the Poisson ratio is $\nu = 0.3$.

Let us consider the frequency response of the beam in relation to the distribution of its mechanical properties across the thickness. The data of Figure 19.7 illustrate variations in the elastic modulus across the thickness of the beam and the corresponding frequency response at different values of n. As expected, the natural frequencies of the beam increase with the concentration of ceramic.

19.4.3 Comparison of functionally graded and three-layered plates

Let us compare stress distributions in a symmetrical FGM beam ($n = 3$) and an equivalent three-layered beam (sandwich). The mechanical properties of face layers of the sandwich beam are the same as those of rigid layers of the FGM beam. We select the thicknesses of these layers in such a way that the frequency responses of the FGM and sandwich beams coincide (Figure 19.8). The sandwich beam was calculated using relations given in Carrera (2003) and Sitek et al. (2006), with the layer-by-layer approximation of the stress state. The data of Figure 19.8 illustrate the variation in Young's

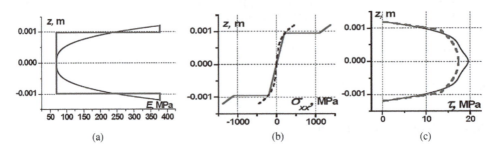

Figure 19.8 Variations in Young's modulus of FGM and sandwich across their thickness a, variation in the normal σ_{xx}-b and shear τ-c stresses across the thickness of FGM (– – –) and sandwich (——) plates.

modulus of FGM (dashed line) and sandwich (solid line) beams and the corresponding stress distributions at a frequency of 200 Hz.

19.5 DISCUSSION

Various displacement models have been developed by considering combinations of displacement fields for in-plane and transverse displacements inside a mathematical sub-layer to investigate the phenomenon of wave propagation as well as vibrations in laminated composite plates. Numerical evaluations obtained for wave propagation and vibrations in isotropic, orthotropic and composite laminated plates have been used to determine the efficient displacement field for economic analysis of wave propagation and vibrations in laminated composite and FGM plates. The numerical method developed follows a semi-analytical approach with analytical field applied in the longitudinal direction and layer-wise displacement field used in the transverse direction. This work aims at developing a simple numerical technique, which can produce very accurate results in comparison with the available analytical solution and also to decide on the level of refinement in higher-order theory that is needed for accurate and efficient analysis. Every theory has its limitations, including the finite elements method. It is well known that the exact solutions of elasticity theories are singular near the corners. Thus, the oscillation solutions found for stresses, for example, in (Huang 2003) by ABAQUES, are not obviously physically convenient to the case of study.

19.6 CONCLUSION

On the basis of refined relationships for laminates, the dynamic characteristics of layered and functionally graded beams in cylindrical bending were obtained. The calculation schemes of asymmetric beams are based on splitting the operator of the problem and the mechanical properties into symmetric and antisymmetric components. The values of damping for a FGM plate were determined. The distribution of damping across the thickness of the beam was assumed to obey the same power law as Young's modulus. Calculation schemes for both the asymmetric and symmetrical beams were developed, and the numerical results obtained are presented. The effect of the penetration depth of the stiff upper constituent on the frequency response of beam was investigated. The distributions of stresses across the plate thickness are presented at different degrees of concentration of the surface substance. These distributions are shown in comparison with those for a three-layered plate with rigid outer layers, whose thicknesses were selected in such a way as to obtain a frequency response identical to that of the FGM beam. A principal difference was obtained only for the longitudinal normal stresses, which were much higher in the three-layered plate. Contrary to the widespread opinion, the interlayer shear stresses in the FGM beam differed only minimally from those in the equivalent sandwich plate. A significant difference was observed only for the longitudinal stresses, which were less significant for the strength of the laminate. We could not find in the literature damping indices for a mixture of the base material (aluminium) and ceramic. To determine these parameters and the stiffness of such a mixture, the identification schemes developed earlier for the elastic moduli of composites can be proposed.

REFERENCES

Carrera, E. 2003. Historical review of zig-zag theories for multilayered plates and shells. *Applied Mechanics Reviews* 56: 287–308.

CelikGul, G. & Kurtuluş, F. 2016. Synthesis and characterization of double phase metal nickelates/borates. *Journal of Achievements in Materials and Manufacturing Engineering* 76(1): 5–8.

Cheng, Z.Q. & Batra, R.C. 2000. Exact correspondence between eigenvalues of membranes and functionally graded simply supported polygonal plates. *Journal of Sound and Vibration* 229: 879–895.

Diveyev, B. Butyter, I. & Shcherbyna, N. 2009. Combined evolutionary non-deterministic methods for layered plates mechanical properties identification. *Proceeding of 16th International Congress on Sound and Vibration (ICSV-16)*, July 5–9, 2009, Krakow, Poland 785.

Diveyev, B. Horbay, O. Kernytskyy, I. Pelekh, R. & Velhan, I. 2017. Dynamic properties and damping predictions for laminated micro-beams by different boundary conditions. *Proceedings of MEMSTECH*, 20–23 April, 30–34.

Diveyev, B. Horbay, O. Nykolyshyn, M. Smolskyy, A. & Vikovych I. 2011. Optimisation of anisotropic sandwich beams for higher sound transmission loss. *Proceedings of MEMSTECH*, 11–14 May, 33–37.

Diveyev, B. Horbay, O. Pelekh, R. & Smolskyy, A. Acoustical and vibration perfofmence of layered beams with dynamic vibration adsorbers. *Proceeding of 19th International Congress on Sound and Vibration (ICSV-12)*, July 8–12, 2012, Vilnius, Lithuania 1: 1494–1498.

Diveyev, B. Kohut, I. Butyter, I. & Cherchyk, G. 2012. Determination of the elastic moduli of layered beams based on the results of experimental investigations and numerical models. *Materials Science* 48(3): 281–288.

Diveyev, B, Konyk, S. & Crocker, M. 2018. Dynamic properties and damping predictions for laminated plates: High order theories e Timoshenko beam. *Journal of Sound and Vibration* 413: 173–190.

Dobrzański, L.A. Honysh, R. & Fassois S. 2006. On the identification of composite beam dynamics based upon experimental date. *Journal of Achivments in Materials and Manufacturing Engineering* 16(1–2–2): 114–123.

Dragobetskii, V. Shapoval, A. Mos'pan, D. Trotsko, O. & Lotous, V. 2015. Excavator bucket teeth strengthening using a plastic explosive deformation. *Metallurgical and Mining Industry* 4: 363–368.

Elishakoff, I. & Guede, Z. 2004. Analytical polynomial solutions for vibrating axially graded beams. *Mechanics of Advanced Materials and Structures* 11: 517–533.

Hu, H. Belouettar, S. Potier-Ferry, M. & Daya, M. 2008. Review and assessment of various theories for modeling sandwich composites. *Composite Structures* 84: 282–292.

Huang, S.J. 2003. An analytical method for calculating the stress and strain in adhesive layers in sandwich beams. *Composite Structures* 60: 447–454.

Kernytskyy, I. Diveyev, B. Horbaj, O. Hlobchak, M. Kopytko, M. & Zachek, O. 2017. Optimization of the impact multi-mass vibration absorbers, *Scientific Review, Engineering and Environmental, Sciences* 26(3): 394–400.

Kozlov, L.G. Polishchuk, L.K. Piontkevych, O.V. Korinenko, M.P. Horbatiuk, R.M. Komada, P. Orazalieva, S. & Ussatova, O. 2019. Experimental research characteristics of counter balance valve for hydraulic drive control system of mobile machine. Przeglad Elektrotechniczny *Przeglad Elektrotechniczny* 95(4): 104–109.

Kukharchuk, V.V. Bogachuk, V.V. Hraniak, V.F. Wójcik, W. Suleimenov, B. & Karnakova, G. 2017. Method of magneto-elastic control of mechanic rigidity in assemblies of hydropower units. *Proceedings of SPIE* 104456A.

Kukharchuk, V.V. Hraniak, V.F. Vedmitskyi, Y.G. & Bogachuk, V.V. 2016. Noncontact method of temperature measurement based on the phenomenon of the luminophor temperature decreasing. *Proc. SPIE* 100312F.

Loy, C.T. Lam, K.Y. & Reddy J.N. 1999. Vibration of functionally graded cylindrical shells. *International Journal of Mechanical Sciences* 41: 309–324.

Ogorodnikov, V.A. Dereven'ko, I.A. & Sivak, R.I. 2018. On the influence of curvature of the trajectories of deformation of a volume of the material by pressing on its plasticity under the conditions of complex loading. *Materials Science* 54(3): 326–332.

Ogorodnikov, V.A. Savchinskij, I.G. & Nakhajchuk, O.V. 2004. Stressed-strained state during forming the internal slot section by mandrel reduction. *Tyazheloe Mashinostroenie* 12: 31–33.

Ogorodnikov, V.A. Zyska, T. & Sundetov S. 2018. The physical model of motor vehicle destruction under shock loading for analysis of road traffic accident. *Proceedings of SPIE* 108086C.

Polishchuk, L.K. Kozlov, L.G. Piontkevych, O.V. Horbatiuk, R.M. Pinaiev, B. Wójcik, W. & Abdihanov, A. 2019. Study of the dynamic stability of the belt conveyor adaptive drive. *Przeglad Elektrotechniczny* 95(4): 98–103.

Sitek, W. Trzaska, J. & Dobrzański, L.A. 2006. An artificial intelligence approach in designing new materials. *Journal of Achivments in Materials and Manufacturing Engineering* 18(1–2–2): 114–123.

Vaer, K. Anton, J. Klauson, A. Eerme, M. Ounapuu, E. & Tšukrejev, P. 2017. Material characterization for laminated glass composite panel. *Journal of Achivments in Materials and Manufacturing Engineering* 81(1): 11–17.

Vasilevskyi, O.M. 2013. Advanced mathematical model of measuring the starting torque motors. *Technical Electrodynamics* 6: 76–81.

Vasilevskyi, O.M. 2014. Calibration method to assess the accuracy of measurement devices using the theory of uncertainty. *International Journal of Metrology and Quality Engineering* 5(4).

Vasilevskyi, O.M. 2015. A frequency method for dynamic uncertainty evaluation of measurement during modes of dynamic operation. *International Journal of Metrology and Quality Engineering* 6(2): 202.

Vedmitskyi, Y.G. Kukharchuk, V.V. & Hraniak, V.F. 2017. New non-system physical quantities for vibration monitoring of transient processes at hydropower facilities, integral vibratory accelerations. *Przeglad Elektrotechniczny* 93(3): 69–72.

Chapter 20

Selection and reasoning of the bus rapid transit component scheme of huge capacity

Volodymyr Sakhno, Victor Poliakov, Victor Bilichenko, Igor Murovany, Andrzej Kotyra, Gali Duskazaev, and Doszhon Baitussupov

CONTENTS

20.1 Introduction 233
20.2 Research results 234
20.3 Recommendations 238
20.4 Conclusion 240
References 241

20.1 INTRODUCTION

Over the past decade, bus rapid transit, a high-speed bus service, is being developed in the United States, the countries of South America, Turkey, Russia and some European countries as a new type of passenger transport. This type of passenger transport is cheaper than the subway, as it requires, with less investment, to establish passenger transportation on the routes of the largest passenger traffic. In Kyiv, last year, the KCSA announced a tender for a feasibility study for a transitway from Troyeschina to Sevastopol Square. It is planned to develop a transitway line from Darnytska Square to the Forest Solid. Metro lengths can reach 25–28 m and weigh up to 45 tons. The use of such vehicles should be confirmed by solving certain technical problems related to the choice of power plant, mass and layout parameters, which largely determine the characteristics of traction-speed properties, manoeuvrability, controllability and environmental friendliness of BRT (Omelnitcky 2018).

One of the directions of development of modern ecologically clean vehicles is considered hybrid power plants based on a combination of an internal combustion engine and an electric machine. Hybrid technologies have been known for more than 40 years. Various technologies are used in power plants of passenger cars: electric, hybrid and fuel elements (Genta et al. 2014). The first attempts to use hybrid power plants on trucks begin in the 80s of the twentieth century. Since 1981, HinoMotors Ltd. has been researching diesel–electric hybrid systems, which resulted in the commercial sales and operation of a large-scale hybrid diesel bus in 1991 (Ehsani, Gao & Emadi 2010).

New developments in the field of creating multi-level vehicles and optimization techniques for their designs are aimed at minimizing fuel and energy costs, improving manoeuvrability and manageability. Much theoretical data on the optimization

DOI: 10.1201/9781003224136-20

of complex mechanical systems and multi-purpose optimization methods are presented in Mastinu, Gobbi and Miano (2006). In the paper by Bazhinov et al. (2008), the scheme solutions and peculiarities of construction of motor vehicles with a hybrid power plant, electric systems and hybrid car complexes are considered. The analysis of constructions and the classification of multi-purpose vehicles of the traditional design, and the general laws of their movement are considered in the paper by Vysotsky, Kochetov and Kharitonchik (2011). In works by Sakhno et al. (2016), Sakhno, Poliakov et al. (2018) and Sakhno, Gerlici et al. (2018), the questions of stability of transitways of various layout schemes are considered. At the same time, one of the unresolved problems is the rational distribution of power between the two power plants, depending on the vehicle operating conditions. In this regard, the goal of the work is to determine the parameters of the hybrid power plant, the choice and justification of the link scheme of the BRT, which provide the necessary characteristics of the traction and speed properties and manoeuvrability.

20.2 RESEARCH RESULTS

Earlier studies found that the BRT can compete with the subway in the case of its passenger capacity of 180–250 passengers and the interval of the movement at 30–45 seconds. Such passenger capacity can be provided only by a three-axle transitway within the bus (as a tractor) and two trailers. The total mass of the transitway can be taken at the level of 38,000–40,000 kg (the mass of the three-way hinged-connected bus – 28,000 kg + trailer weight – 10,000–12,000 kg). The total length of such a transitway is 25,000 mm (Figure 20.1).

Determine the need for the power of the hybrid power plant, based on the condition that the diesel engine should provide a regular traffic transitway at a speed of 25 m/s, and diesel and electric motor should ensure the dispersal of the transitway on lower gears (1,2 in the gearbox) with an acceleration of 1.5–1.0 m/s². Let us record the equation of the power balance for these two BRT modes (Sakhno, Kostenko & Zagorodnov 2014). We will get:

$$N_{ev} = \frac{f_v \times M_a \times g \times v_{max} + k_B \times F \times v_{max}^3}{1,000 \times \eta_M} \quad (20.1)$$

$$N_{eP} = \frac{f_0 \times M_a \times g \times v_p + \delta_i \times M_a \times v_p \times \frac{dv_i}{dt}}{1,000 \times \eta_M} \quad (20.2)$$

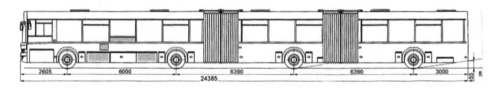

Figure 20.1 Compound diagram of a three axes BRT.

where N_{ev} and N_{eP} are the powers of a diesel engine and a hybrid power plant, respectively; f_0 and f_v are the coefficients of rolling resistance at a speed of 1 m/s and maximum, respectively, $f_0 = 0.02$ and $f_v = 0{,}024$; v_p and v_{max} are, respectively, the speed during acceleration and maximum speed of the metro bus, $v_p = 2.5$ m/s and $v_{max} = 25$ m/s; k_B denotes the air drag coefficient, $k_B = 0{,}6$ Ns^2/m^4; F is the cross-sectional area of the metro bus, $F = 7.5$ m^2; M_a is the total mass of the bus, $M_a = 40{,}000$ kg; g is the standard acceleration of free fall; η_M denotes the mechanical transmission coefficient of efficiency, $\eta_M = 0.9$ and dv_i/dt is the first gear acceleration of the metro bus, $dv_i/dt = 1.5$ m/s^2.

According to the selected data, the diesel power was 300 kW and the power of the hybrid power plant was 385 kW. As an electric motor, it is possible to use the engine BRASA HSM1-12/18/13, the speed characteristic of which is shown on Figure 20.2. This engine is used together with a gearbox, the gear ratio of which is 9.7. It is noticed that this engine can be installed on the first or second trailer. The final choice of location will be made after the analysis of the BRT manoeuvrability.

At the previous stage, the manoeuvrability of the transitway can be viewed on rigid wheels in the lateral direction (Zakin 1967, Farobin 1993, Polyakov &Sakhno 2013). The scheme of turning the transitway on the rigid sideways wheels for uncontrolled wheels of trailers is shown in Figure 20.3 and for controlled wheels of trailers in Figure 20.4.

Determine the overall radius of rotation and the overall traffic lane (OTL) for the transitway for different control circuits. At the same time, let us take into consideration the fact that the OTL of vehicles, including the transitway, is determined by their circular motion (Underdog et al. 2003, Boboshko 2006). Component parameters of the transitway are shown in Figure 20.1. As the prescriptive parameter, we take an external dimensional radius – the radius of the vector point Ar = 12.5 m. By sequentially defining from ΔOArBr, the segment OBr and then the segment OB, with ΔOBO$_1$ segment OO$_1$, with ΔOO$_2$C segment of the OC, with ΔOCE segment OE and further internal overall radius we get:(Smirnov 1990, Vedmitskyi, Kukharchuk & Hraniak 2017):

$$R_{io} = OE - 0.5 B_{wt}, \tag{20.3}$$

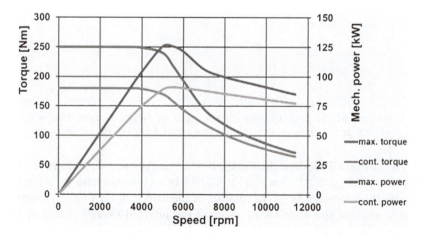

Figure 20.2 Speed characteristics of the electric motor BRASA HSM1-12/18/13.

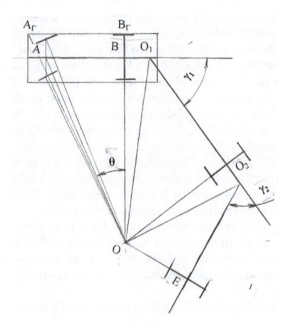

Figure 20.3 Scheme of BRT turning on the rigid sideways wheels.

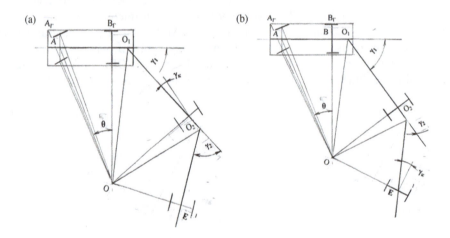

Figure 20.4 Scheme of the BRT turning on the rigid sideways wheels on the controlled wheels of the first trailer (a) and the second trailer (b).

where B_{wt} is the overall width of the trailer, $B_w = 2.5$ m. As a result, we have $R_{io} = 4.87$ m, OTL$=R_{dr}-R_{io}= 12.5-4.87=7.63$ m $>$[OTL$=7.2$ m]. In accordance with the given method, we determine the necessary angles of rotation of the wheels of the first or second trailer, which would provide the normalized value of the internal overall radius of rotation $R_{io} = 5.2$ m and the required value of the OTL. In the case of controlled wheels of the first trailer, the average turning angle of its wheels is 14.2° and the second trailer is 5.4°.

Further analysis of the options of the link scheme of the transitway will be carried out in two directions:

- the first – the wheels of the first trailer are controlled by an electric motor mounted on the trailer and the wheels of the second trailer are guided by the kinematic turning method;
- the second – the first trailer is a controlling wheel with which you can turn the trailer on the power circuit (different in size, or even in the direction of the moments on the inner and outer wheels) and the second trailer is uncontrolled.

The advantages of the first layout scheme are the possibility of its implementation with the wide application of standard components, and the main drawback is the complexity of the design. The advantage of the second scheme is the compact design solution, in which the controlled wheels and the trailer turning is carried out by the same motor wheels, that is, for managing the transitway a combined mode of rotation was used (Kukharchuk, Kazyv & Bykovsky 2017).

In Underdog et al. (2003) and Boboshko (2006), we obtain the equations of the dynamics of the circular motion of a machine with a combined turning control method

$$\frac{d\omega}{dt} = \frac{tg\bar{\alpha}}{1+\frac{b^2+i_z^2}{L^2}tg^2\bar{\alpha}} \times \begin{bmatrix} \frac{1}{m \cdot L}\left(\frac{M_{k1}}{r_{d1}}+\frac{M_{k2}}{r_{d2}}\right) - \frac{fg}{L} - \frac{fh}{L^3} \times \\ \times \left(V_{x1}^2 + b\frac{dV_{x1}}{dt}\right)tg^2\bar{\alpha} + \frac{B}{2L^2m}tg\bar{\alpha} \times \\ \times \left(\frac{M_{k1}''-M_{k1}'}{r_{d1}} - \frac{M_{k2}''-M_{k2}'}{r_{d2}}\right) \end{bmatrix} \quad (20.4)$$

where ω is the angular velocity of the car in the plane of the road; t is the time and $\bar{\alpha}$ is the average turning angle of the guiding wheels,

$$\bar{\alpha} = \frac{1}{2}(\alpha' + \alpha'') \quad (20.5)$$

α' and α'' are the angles of rotation of the internal (in relation to the centre of rotation) and external guiding wheels, respectively; b is the distance from the rear axle to the projection of the centre of the car's mass on a horizontal plane; m is the weight of the car; L is the longitudinal wheelbase of the car; M_{k1} and M_{k2} are the total torque on the wheels of the driving axle of the bus and the trailer, respectively; r_{d1} and r_{d2} are the dynamic radii of driving wheels of a bus and a trailer, respectively (accepted $r_{d1} = r_{d2} = r_d$); f is the coefficient of rolling resistance of the wheels of the BRT; h is the height of the centre of the car's mass (trailer); i_z is the radius of inertia of the car (trailer) relative to the vertical axis; V_{x1} is the linear speed of the metro bus in the direction of the longitudinal axis; M'_{K1} and M''_{K1} are the twisting moments on the inner and outer driving wheels of the bus, $M'_{K1} = M''_{K1}$; and M'_{K2} and M''_{K2} are the twisting moments on the inside and outside of the trailer controlled wheels.

From equation (20.4), it is seen that the difference in torque on the controlled wheels

$$\Delta M_{k1} = M''_{K1} - M'_{K1} \tag{20.6}$$

and

$$\Delta M_{k2} = M''_{K2} - M'_{K2} \tag{20.7}$$

increase the angular acceleration $d\omega/dt$ of the transitway in the plane of the road. If you change the sign of these differences to the opposite in equation (20.4), then you can achieve a situation in which $d\omega/dt = 0$.

We represent equation (20.4) in the form

$$\frac{d\omega}{dt} = a(\alpha)[b(M) - c - d(v,\alpha) + k(\alpha,M)] \tag{20.8}$$

where

$$a(\alpha) = \frac{tg\overline{\alpha}}{1 + \frac{b^2 + i_z^2}{L^2} tg^2\overline{\alpha}},$$

$$b(M) = \left[\frac{1}{m \cdot L}\left(\frac{M_{k1}}{r_{d1}} + \frac{M_{k2}}{r_{d2}}\right)\right],$$

$$c = \frac{fg}{L},$$

$$d(v,\alpha) = \left[\frac{fh}{L^3} \times \left(V_{x1}^2 + b\frac{dV_{x1}}{dt}\right) tg^2\overline{\alpha}\right],$$

$$k(\alpha, M) = \left[\frac{B}{2L^2 m} tg\overline{\alpha} \times \left(\frac{M''_{K1} - M'_{K1}}{r_{d1}} - \frac{M''_{K2} - M'_{K2}}{r_{d2}}\right)\right] \tag{20.9}$$

The solution of equation (20.8) makes it possible to determine the difference between the moments on the wheels of the trailer link, which can provide the metered indicators of the BRT manoeuvrability.

20.3 RECOMMENDATIONS

In determining the components of equation (20.8), it is taken into account that there is a dependence between the angle of rotation of the controlled wheels of the bus (trailer) and the speed of the transitway (Smirnov 1990)

$$v_{kr} = \sqrt{\left(\frac{\sqrt{\phi^2 - f^2}}{tg\alpha} - f\right)gL\cos\alpha}, \tag{20.10}$$

where v_{kr} is the critical handling speed; φ is the gear ratio of the wheels with the road; f is the rolling resistance coefficient of wheels; g is the acceleration of free fall; L is the base of the bus.

Since the second way of the transitway turning is considered by turning the trailer behind the power circuit (different in size or even direction of the moments on the inner and outer wheels), then the calculation of critical speed will be done for the trailer for the following output data: $b = 3.25$ m; $m = 11,500$ kg; $L = 6.5$ m; $r_{d1} = r_{d2} = r_d = 0.526$ m; $f = 0.02$; $h = 1.25$ m, height of the centre of the trailer mass; and $i_z^2 = 6.25$ m².

Figure 20.5 shows the dependence of the critical speed on handling for the bus as part of the considered transitway.

As there is a certain dependence between the speed on the controllability and the steering angle of the controlled wheels in the future, all the calculations used the corresponding values of the velocity v_x and the angle of rotation of the controlled wheels α trailer.

Indicators of manoeuvrability of vehicles, as noted earlier, determine the amount of traffic. In this case, the angular and circumferential acceleration is equal to zero, i.e. $d\omega/dt = 0$ and $dV_{x1}/dt = 0$. Since the coefficient $a(\alpha) \neq 0$, then

$$b(M) - c - d(v,\alpha) + k(\alpha, M) = 0 \tag{20.11}$$

Let us solve equation (20.11) with respect to α. At the same time, let us take into account that, at constant motion, the components of equation (20.11) $b(M)$ i c are equal (with the manoeuvrability of the transitway with a slight speed, the strength of the air resistance can be neglected), i.e.

$$b(M) = \left[\frac{1}{m \cdot L}\left(\frac{M_{k1}}{r_{d1}} + \frac{M_{k2}}{r_{d2}}\right)\right] = \frac{fg}{L} \tag{20.12}$$

Under this condition

$$d(v,\alpha) = k(\alpha, M) \tag{20.13}$$

Further calculations will be made for two conditions of the BRT movement. For the first condition, we set the speed of the steady equal to 5 m/s and determine the required

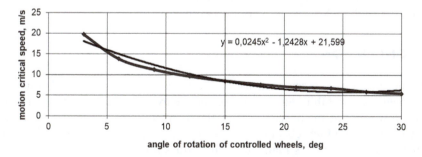

Figure 20.5 Dependence of the critical speed on controllability from the angle of rotation of the bus controlled wheels.

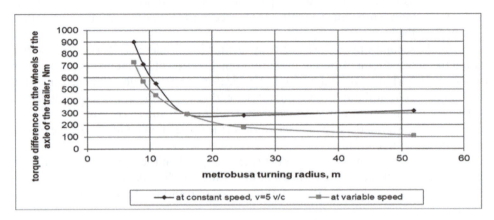

Figure 20.6 Dependence of the moment difference on the wheels of the axle of the trailer to provide BRT movement in a circle of given radius.

difference of torque on the wheels of the trailer, and for the second condition from Figure 20.5 choose the angle of rotation of the controlled wheels and accordingly the speed of the transitway and again determine the difference in moments on the trailer wheels.

Figure 20.6 shows the dependence of the torque difference on the wheels of the axle of the trailer, which corresponds to the angle of rotation of its controlled wheels for two driving conditions in motion along the curve of a given radius.

As it follows from Figure 20.6, at a fixed speed of rotation, $v = 5/s$, the difference in torque on the inner and outer wheels of the trailer while moving along the curves of the small radius is progressively increasing. This is explained by a non-linear change in the value of $t_{g\alpha}$ to provide the required radius of rotation of the transitway. At the same time, taking into account the speed of movement and its relationship with the radius of rotation (through the angle of rotation of controlled wheels) in the given dependence, Figure 20.6, leads to the fact that the difference in torque on the wheels of the trailer decreases with an increase in radius to 18 m (that corresponds to the required angle of rotation of the trailer wheels to achieve normalized manoeuvrability) and remains almost unchanged throughout the range of change in the radius of rotation.

Thus, to ensure the required level of the BRT manoeuvrability, the difference in torque on the trailer wheels should not be less than $\Delta M = 275$ Nm.

20.4 CONCLUSION

1. The necessary passenger capacity of the transitway, its mass and dimensional parameters are established. The power of the hybrid power plant is determined. It is shown that at the previous stage, the manoeuvrability of the transitway can be viewed on rigid wheels in the lateral direction. Due to the selected layout parameters of the BRT, its manoeuvrability does not correspond to the operating normative documents.

2. The kinematic and power modes of rotation of the trailer to improve the manoeuvrability of the transitway are considered. The necessary angles of rotation of the controlled wheels of the first or second trailer are set. The advantages of the power of rotation by the wheels of the trailer are shown. The required torque difference on the trailer wheels is determined, in which the radius of rotation of the metro bus should be the same as that of the first trailer controlled by the kinematic turning method. At the steady speed of rotation, $v = 5/s$, and variable speed and its connection with the radius of rotation to ensure the required level of the BRT manoeuvrability, the difference torque on the wheels of the trailer should not be less than $\Delta M = 275$ Nm.

New developments in the field of creating multi-level vehicles and optimization techniques for their designs are aimed at minimizing fuel and energy costs, improving manoeuvrability and manageability. In this case, one of the unresolved problems of hybrid power plants is a rational power distribution between the two power plants, depending on the operating conditions of the vehicle. In this chapter, the parameters of a hybrid power plant are determined, in which the necessary indicators of traction-speed properties and manoeuvrability are provided.

Indicators of manoeuvrability of the transitway, defined on the rigid sideways of the wheels, require an analytical and experimental verification, taking into account the lateral removal of the wheels of the transitway. Further research will be devoted to this.

REFERENCES

Bazhinov, O.V. Smirnov, O.P. Serikov, S.A. Gnatov, A.V. & Kolesnikov A.V. 2008. *Hybrid Cars*. Kharkiv: KhNADU.

Boboshko, A.A. 2006. *Unconventional Ways of Turning Wheel Cars*. Kharkiv: Publishing House of KhNADU, 172.

Ehsani, M. Gao, Y. & Emadi, A. 2010. *Modern Electric, Hybrid Electric, and Fuel Cell Vehicles: Fundamentals, Theory, and Design*. Boca Raton, FL: CRC Press.

Farobin, Ya.E. Yakobashvili, A.M. & Ivanov, A.M. 1993. Three-Way Trains. *Mechanical Engineering* 23: 17–22.

Genta, G. Morello, L. Cavallino, F. & Filtri, L. 2014. *The Motor Car: Past, Present and Future*. Dordrecht: Springer.

Kukharchuk, V.V. Kazyv, S.S. & Bykovsky, S.A. 2017. Discrete wavelet transformation in spectral analysis of vibration processes at hydropower units. *Przeglad Elektrotechniczny* 93(5): 65–68.

Mastinu, G. Gobbi, M. & Miano, C. 2006. *Optimal Design of Complex Mechanical Systems: With Applications to Vehicle Engineering*. New York: Springer.

Omelnitcky, O.E. 2018. Analysis of metrobus constructions. *Avtoshlyakhovyk Ukrayiny* 3: 7–11.

Polyakov, V.M. & Sakhno, V.P. 2013. Three-axle trains. Maneuverability. NTU: 200 s.

Sakhno, V.P. Gerlici, J. Poliakov, V. Kravchenko, A. Omelnitsky, O. & Lask, T. 2018. Road train motion stability in BRT system. *XXIII Polish-Slovak Scientific Conference Machine Modelling and Simulation. MMS 2018 - Book of abstracts, September 4–7, 2018*. Rydzyna: Book of abstracts.

Sakhno, V.P. Kostenko, A.P. & Zagorodnov, M.I. 2014. *Operational Properties of Motor Vehicles. At 3 hrs. 1. Dynamism and Fuel Efficiency of Motor Vehicles*. Donetsk: Noulig.

Sakhno, V.P. Poliakov, V. Murovanyi, I. Selezniov, V. & Vovk, Y. 2018. Analysis of transverse stability parameters of hybrid buses with active trailers. *Scientific Journal of Silesian University of Technology. Series Transport* 101: 185–201.

Sakhno, V.P. Poliakov, V. Timkov, O. & Kravchenko, O. 2016. Lorry convoy stability taking into account the skew of semitrailer axes. *Transport Problems* 11(3): 69–76.

Smirnov, G.A. 1990. *Theory of the Movement of Wheeled Machines.* Moscow: Mashinostroenie, 352.

Underdog, M.A. Volkov, V.P. Kirchaty, V.I. & Boboshka, A.A. 2003. *Maneuverability and Brake Properties of Wheeled Cars.* Kharkiv: Publishing house KhNADU, 403.

Vedmitskyi, Y.G. Kukharchuk, V.V. & Hraniak, V.F. 2017. New non-system physical quantities for vibration monitoring of transient processes at hydropower facilities, integral vibratory accelerations. *Przeglad Elektrotechniczny* 93(3): 69–72.

Vysotsky, M.S. Kochetov, S.I. & Kharitonchik, S.V. 2011. *Fundamentals of Designing Modular Highway Trains.* Minsk: Belarus.

Zakin, Y.H. 1967. *Applied Theory of the Movement of a Train.* Moscow: Transport, 225.

Chapter 21

Physical-mathematical modelling of the process of infrared drying of rape with vibration transport of products

Igor P. Palamarchuk, Mikhailo M. Mushtruk,
Vladislav I. Palamarchuk, Olena S. Deviatko,
Waldemar Wójcik, Aliya Kalizhanova, and Ainur Kozbakova

CONTENTS

21.1 Introduction 243
21.2 Literature review and problem statement 244
21.3 Aim, objective, material and methods 246
21.4 Results and discussion 246
21.5 Conclusion 251
References 251

21.1 INTRODUCTION

Infrared drying is carried out by transferring the heat of the product from the source of the radioactive energy, which either uses gas burners or special incandescent lamps (80%–90% of which energy is converted into infrared radiation energy) or special radiation surface slabs. In the process of heat exchange, the radiant flow penetrates partially into the capillary-porous bodies to a depth of 0.1–2 mm and is almost absorbed due to a series of reflections from the walls. In this case, the heat transfer coefficient has a significant length of the drying process in comparison with the convective or conductive method and decreases by 30–100 times (Sorochinsky 2002, Burdo 2010). High-frequency jet drying is based on the fact that the molecules located between two plate-like electrodes are driven by oscillatory motion under the influence of an alternating electric field. This motion of the molecules causes the material to heat over its entire thickness. The inner layers of the product are heated more than outer ones due to the contact of the latter with the environment. As a result, the movement of moisture from the centre to the periphery increases.

In the same direction when drying humidity changes, the gradients of temperature and humidity coincide on the sign, contributing to the migration of moisture from the centre to the surface. Therefore, the speed of such drying significantly exceeds the speed of convective drying, although its implementation requires 3–4 times more power consumption (Khir & Pan 2006, Sukhenko et al. 2017). In addition, the high thermal stresses acting on the surface layer of products when moving in the working area are also problematic. Constructive improvement of thermo-radiation dryers in

DOI: 10.1201/9781003224136-21

recent years has allowed equalising the energy consumption during their operation with convection machines. The usage of vibration and wave effects has made it possible to significantly reduce the heat load on the product layers. This, consequently caused the urgency of the development of promising schemes of processes and equipment for drying in the conditions of infrared irradiation.

21.2 LITERATURE REVIEW AND PROBLEM STATEMENT

Among the machines that accumulate all the given signs of efficient technological equipment there are undoubtedly conveyor machines. The word *conveyor* is derived from the English word *convey*, i.e. 'transport', and means a continuous machine for moving loose, lump and piece cargoes (Sorochinsky 2002).

With the increase in the mechanisation of production processes and the introduction of more efficient technological equipment into production, conveyor machines perform not only a pure transport function but are also used as transfer devices in technological automated lines for the manufacture and processing of parts or assembly of systems. Due to high productivity, directivity of action and automation of control, conveyors become a component and an integral part of the modern technological process; they establish and regulate the pace of production, ensure its rhythmicity, increase labour productivity and increase output (Derevenko 2004, Palamarchuk et al. 2013, Drukovanui et al. 2014).

For such a machine, the following essential properties are characteristic, such as continuity of operation, presence of a load-carrying body, and others. With the development of the current method of organising production, the following properties of conveyors take on weight: the proportionality, dynamism and direction of the process are realised with the help of them. In modern conveyor production, a number of new defining faces arise, associated with the spread of the application areas of this equipment. This is the synchronism and proportionality between the main and auxiliary operations, the absence of a contradiction between transport and technological movements, the universality of the machine and the possibility of implementing a complex technological action in the composition of automated production lines.

An essential sign of a conveyor vibrating technological machine determines the possibility of its existence as such is the presence of a vibration technological action with the current method of organising the machine operation. The main essential properties of conveyor vibrating machines are shown in Figure 21.1, where the dynamics of changes in these properties are visually illustrated as technological improvement of machines from transporting to transport-technological and technological processes (Bandura et al. 2014, 2015, Kolyanovska et al. 2019).

Thus, the movement acts accordingly as a single, basic and then as an auxiliary technological movement. Vibration action is the main dynamic factor of the machine's operation and is aimed at implementing a pure transport movement in vibrating delivery conveyors, technological movement or a combination of both recent operations – in vibrating technological machines. The way of realisation of transport and technological movements, that is, the main components of the production process, basically determines the variety of the design of conveyor technological machines.

The proportionality between the transport and technological movement serves as a classification feature, determines both the perfection and the design features of

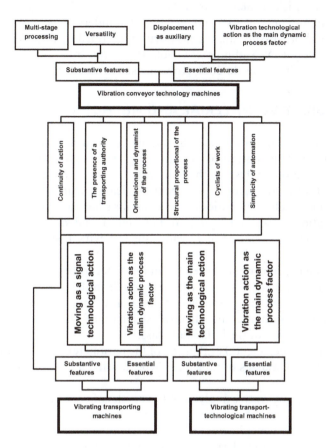

Figure 21.1 The main features of conveyor vibrating machines.

the process equipment. On this basis, cars with separate, concomitant and compatible performance of transport and technological action are distinguished.

Existing vibrating conveyors are used mainly for moving loads in different directions. Transport-technological machines are also being developed, they are efficiently carried out in the process of transportation and processing of the moving bulk product: drying, extraction, crystallisation and other heat-mass-exchange processes. Such technological systems, as a rule, are noted for the complexity of the design of their elements, and the need for special transporting devices often negates the advantages of the current processing regime. In addition, significant problems are the provision of mutual functioning of transport and technological links, and the possibility of increasing the level of mechanisation and automation of such systems. Therefore, the search for efficient vibrating conveyor schemes for transport-technological processing of loose agricultural and food raw materials, the analysis of which is presented in Figure 21.1, becomes urgent (Vedmitskyi et al. 2017, Kukharchuk et al. 2016, 2017a,b).

The efficiency of implementation, high potential and prospects for improving these processes are observed precisely from the creative unification of these areas, as evidenced by the work of both known vibration technologists and designers such as

A. Babichev, P. Bernick, I. Blekhman, I. Goncharevich, V. Nadutyi, V. Poturaeva, A. Spivakovsky and A. Chervonenko and recognised founders of technologies and equipment for food and processing silk-pig production, O. Burdo, G. Limonova, V. Popov, P. Rebinder, Z. Rogov, V. Stabnikov, A. Sokolenko and other outstanding scientists (Ogorodnikov et al. 2004, Burdo 2010, Kolyanovska et al. 2019).

21.3 AIM, OBJECTIVE, MATERIAL AND METHODS

This scientific work aims to develop the energy-efficient and reliable technological schemes for the implementation of vibration conveyor heat exchange processing of bulk products based on analysis of the main features, structural components and development trends of vibrating transport-technological machines; determination of the kinematic and power characteristics of the non-vibrating nature of the vibrating systems; substantiation of rational regimes of raw material advancement in the processing zone and for the implementation of which the following tasks are set:

- to analyse the relationship between the main structural elements of conveyor and vibration machines and establish the main trends in their development;
- to determine the main parameters of the non-vibrant nature of the vibrating systems and substantiate the effective schemes of the vibrational transport-technological motion of the bulk load mass;
- to carry out an experimental evaluation of the parameters of the vibrational transportation of loose masses of rape and soybean using a developed vibration infrared dryer;
- to carry out a comparative analysis of the results of experimental and theoretical studies.

Theoretical research uses modern concepts of physical and mathematical modelling of granular media, the theory of similarity, the theory of vibrational processing of freely granulated masses, the theory of vibrational and wave transport of loose masses or the methods of Lagrange and Cauchy in the mathematical analysis of the equations of motion of the machine's executive organs. MathCAD software is used to process the data.

During the implementation of the experimental studies, the German Robotron equipment, modern complexes for estimating the kinematic, power and energy characteristics of the vibration-wave transport of the rapeseed and soybean masses are used.

21.4 RESULTS AND DISCUSSION

The equipment for drying products in the field of high-frequency jets includes a carrier conveyor, electrodes with a power supply system and devices for transforming the jet of industrial frequency into high-frequency and voltage current. The developed installation contains a flexible load-carrying body, on which a running or standing wave is created at the work of mechanical vibroscopes 2 and 3 (Figure 21.2) (Palamarchuk, Bandura & Palamarchuk 2012, Sukhenko et al. 2018).

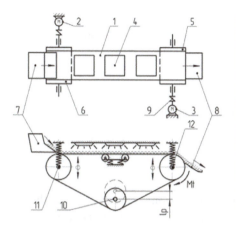 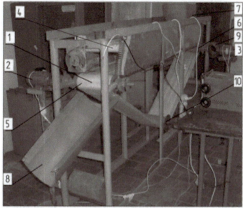

Figure 21.2 Scheme of the conveyor vibrowave infrared dryer: 1, belt; 2 and 3, exciter engines; 4, infrared emitters; 5 and 6, rollers; 7, feeder; 8, receiving hopper; 9, flexible coupling; 10, tensioning roller; and 11 and 12, unbalance vibration exciter.

Such a wave promotes both the transportation of products coming from the feeder (7) and its intense mixing. This reduces the thermal intensity of the surface layer while maintaining a sufficiently high flow rate. At the same time, it is enough to provide oscillations of the rollers (5, 6) to maintain high kinetics of the investigated process. This significantly reduces the power consumption of the drive compared to the vibration conveyor installation. On the basis of the present scheme, the experimental-industrial model was developed of infrared vibration-wave drying (Burdo 2008, Palamarchuk et al. 2015).

To create a mathematical model of the process, the main factors influencing the process, their experimental values and the dependences of their changes, which were obtained with the help of the developed experimental model, were analysed.

The mass transfer coefficient β is determined from the formula:

$$\beta = \frac{\Pi_V}{\Delta X \cdot S_g}, \text{m/s}, \tag{21.1}$$

where S_g is the surface of the irradiated grain: $S_g = 0.1413\,\text{m}^2$;

$$\Delta X = \frac{m_a}{m_h} = X_p - X_i,$$

where m_a is the mass of air coolant and m_h is the mass of lost humidity.

Then we calculate the mass exchange criteria for the process. The Pecle criterion is defined as:

$$Pe = \frac{v_C \cdot d}{a} \tag{21.2}$$

where $a = 12{,}6 \cdot 10^{-8}\,\mathrm{m^2/s}$ (a coefficient of temperature conductivity of the grain) and d is the diameter of the grain.

The Stanton number could be found by using the following formula:

$$St = \frac{\beta}{v_C} \qquad (21.3)$$

The Burdo (2010) criterion is calculated by using the following formula:

$$Bu = \frac{E_I}{Q_B} \qquad (21.4)$$

where $E_I = N_I \cdot \tau$ is energy released by infrared irradiation; $N_I = N_{IR} \cdot N_P$ is the total energy consumption; $N_P = 300$ W is the power of the drive mechanism; N_{IR} is the power when the radiator unit is operated; $Q_B = m_B \cdot r$ is the amount of heat needed to evaporate the moisture mass m_B; and r is the specific heat of steam generation, $r = 2.3 \cdot 10^6$ J/kg (m²/s²).

So, the Burdo criterion looks as follows:

$$Bu = \frac{N_I \tau}{m_B r}$$

For further calculations, we use the value of the volume performance of the dryer for the removed moisture, which is defined as $\Pi_V = V_m \cdot \Pi_B$, where Π_B is the mass productivity of the dryer for the removed moisture, $\Pi_B = m_B/\tau$; $V_m = 1/\rho V = 0.849\,\mathrm{m^3/kg}$ is the volume of unit mass of removed moisture.

Since

$$\Pi_V = \frac{m_B}{\rho_B \tau} \cdot \frac{N_I}{N_I r} = \frac{N_I}{\rho_B \tau} \cdot \frac{1}{Bu} \qquad (21.5)$$

then

$$Bu = \frac{N_I}{\Pi_V \rho_B \tau} \qquad (21.6)$$

The basis of this method lies in the second theorem of the similarity of Federmann–Buckingham, which enables the processing of experimental data in the form of a general criterion equation (Bandura & Palamarchuk 2012, Dragobetskii et al. 2015, Vorobyov et al. 2017).

The intensity of the drying process studied is determined by the coefficient of mass transfer β, the magnitude of which is significantly influenced by such parameters as the specific mass of products of P_S, the speed of the conveyor belt v_c, the size of the raw material particles d_g, the moisture density ρ_B, the coefficient of temperature conductivity a, the volume performance of the process for moisture P_V and specific vaporisation temperature r. In addition, the course of the study process under the action of microwave irradiation is determined by the power of irradiation N_{on} in relation to the energy Q_B for vapour formation and, accordingly, the value of r.

Thus, the desired functional dependence can be represented in the form:

$$\beta = f\left(\Pi_V, N_I, P_S, v_C, a, \rho_B, r, d_g\right) \quad (21.7)$$

Obviously, for a given factor space, the number of variables 9 is equal to the quantity of dimensions 3. Therefore, by the π-theorem of the number of dimensionless complexes 9−3 = 6.

In accordance with the principle of dimensional analysis, dependence (21.7) is represented as a power series.

$$\beta = \Pi_V^k, N_I^l, P_S^m, v_C^n, a^\delta, \rho_B^p, r^\varepsilon, d_g^\eta \quad (21.8)$$

To implement the mathematical analysis of equation (21.8), we construct a matrix based on which the following system of algebraic equations is formed (Polishchuk et al. 2010, Polishchuk, Bilyy & Kharchenko 2016, Polishchuk & Kozlov 2018).

$$\begin{cases} l + m + p = 0 \\ 3k + 2l - 2m + n + 2\delta - 3p + 2\varepsilon + \eta = 1, \text{ so} \\ -k - 3l - n - \delta - 2\varepsilon = -1 \end{cases}$$

$$l = -m - p; 2k - l - 2m + \delta - 3p + \eta = 0; \delta = 1 - k - 3l - n - 2\varepsilon;$$
$$m = 2k - 2p + \delta + \eta; \eta = m - 2k + 2p - \delta \quad (21.9)$$

Taking into account expressions (21.9), dependence (21.8) can presented as follows:

$$\beta = \Pi_V^k, N_I^l, P_S^m, v_C^n, a^{(1-k-3l-n-2\varepsilon)}, \rho_B^{(-l-m)}, r^\varepsilon, d_g^{(m-2k+2p-\delta)} \text{ or}$$

$$\beta = \left(\frac{\Pi_V}{ad_g^2}\right)^k \cdot \left(\frac{N_I}{a^3 \rho}\right)^l \cdot \left(\frac{P_S \cdot d_g}{\rho}\right)^m \cdot \left(\frac{v_C}{a}\right)^n \cdot \left(\frac{r}{a^2}\right)^\varepsilon \cdot a \cdot d_g^{(2p-\delta)} \quad (21.10)$$

We rewrite (21.10) as follows:

$$\frac{\beta}{v_C} = \frac{a}{v_C} \left(\frac{v_C d_g}{a}\right)^k \cdot \left(\frac{N_I}{\Pi_V \rho r}\right)^l \cdot \left(\frac{\Pi_V \cdot r}{a^3}\right)^l \cdot \left(\frac{P_S \cdot d_g}{\rho}\right)^m \cdot \left(\frac{r}{a^2}\right)^\varepsilon \cdot d_g^{(2p-\delta-m)} \quad (21.11)$$

Taking into account the above-mentioned dependences for the similarity criteria, we transform (21.11) into the following

$$St = A \cdot Pe^n \cdot Bu^l \cdot k,$$

where $k = \frac{1}{v_C}\left(\frac{\Pi_V \cdot r}{a^3}\right)^l \cdot \left(\frac{P_S \cdot d_g}{\rho_B}\right)^m$ is a function determined by the technological parameters that determine the productivity of the investigated process and A is the constant value determined by the parameters of the object of processing:

$$A = a\left(\frac{r}{a^2}\right)^{\varepsilon} \cdot d_g^{(2p-\delta-m)}$$

As a result

$$St = \left(\frac{r_C d_g^2}{a^2}\right)^{\varepsilon} \cdot Pe^n \cdot Bu^l \left(\frac{a}{v_C d_g}\right) \cdot \left(\frac{\Pi_V \cdot r \cdot d_g}{a^3}\right)^l \cdot \left(\frac{P_S \cdot d_g}{\rho \cdot d_g}\right)^m \quad (21.12)$$

The indices n, l and A were derived from the results of the graph-analytical analysis using the following method (Bandura & Palamarchuk 2012).

According to experimental data, a graph of function $St = f(Pe)$ was constructed, from which we find the power index $n = tg(\alpha)$. Similarly, the second indicator of power $l = tg(\gamma)$ was determined from the function graph $\dfrac{St}{Pe^n} = f(Bu)$.

From the graph of function $\dfrac{St}{Pe^n Bu^l} = f(k)$, we find the value of the parameter A. Using the obtained database (Derevenko 2004, Palamarchuk, Bandura & Palamarchuk 2013, Bandura, Zozuliak & Palamarchuk 2014), we find the equation of the investigated process of mass transfer in the criterion form.

$$St = A \cdot Pe^n \cdot Bu^l \cdot \frac{1}{v} \cdot \left(\frac{\Pi_V \cdot r}{a^3}\right)^l \cdot \left(\frac{P_S \cdot d}{\rho}\right)^m \quad (21.13)$$

As a result of the conducted graph-analytical analysis (Figures 21.2–21.4), we determined that $n = 1.08$; $l = 1.2$; $m = 1.56$.

Using the compound equation (21.13) and the developed programme (Palamarchuk, Bandura & Palamarchuk 2012, Bandura, Turcan & Palamarchuk 2015, Kolyanovska et al. 2019), we find the recommended range of parameters of the working regime for infrared drying of rape in the moving layer of products (Figure 21.4).

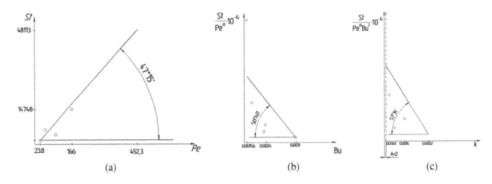

Figure 21.3 Graphic dependences for finding power coefficients to the equation of similarity.

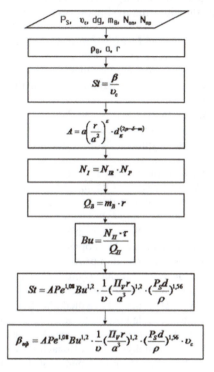

Figure 21.4 Block diagram of the calculation of the intensity of mass transfer in the infrared drying of rape crops.

21.5 CONCLUSION

Intensification of processes of drying of friable lubricants of raw materials in modern technological processes is realised by applying radiation and inductive methods of heat transfer, when combined with convection of energy carrier and when creating pseudo-liquefied layers of products, in particular, under the action of vibrational impulses. According to the experimental data, using the method of "dimensional analysis" and the Feldman–Buckingham theorem, a criterion equation for the process of mass transfer under infrared drying in a moving product layer, which is determined by the criteria of Pekle, Stanton and Burdo, is a function that characterises the productivity of the investigated process and allows to form a recommended set of working mode parameters.

REFERENCES

Bandura, V.M. & Palamarchuk, V.I. 2012. Experimental studies of kinetics of rape and soya drying in a stationary layer in the infrared field. *Scientific Papers of Odessa National Academy of Food Technologies* 41(2): 110–113.

Bandura, V.M., Turcan, O. & Palamarchuk, V. 2015. Experimental study of technological parameters of the process of infrared drying of a moving ball of oilseed crops. *MOTROL Commission of Motorization and Energetics in Agriculture* 17(4): 211–214.

Bandura, V.M., Zozuliak, I. & Palamarchuk, V. 2014. Description of the heat exchange in the similarity of the theory of the vibrating drying process of sunflower. *Ukrainian Journal of Food Science* 2(2): 305–311.

Burdo, O.G. 2008. *Applied Modeling of Transfer Processes in Technological Systems*. Odessa: Druk.

Burdo, O.G. 2010. *The Evolution of Drying Plants*. Odessa: Polygraph.

Derevenko, V. 2004. The main technological regularities of the thermoprocessing of oil material to oil extraction. *Russian School on Science and Technology, devoted to the 80th anniversary of Academician VP Makeev*. Sat tr - Miass: 144–146.

Dragobetskii, V., Shapoval, A., Mos'pan, D., Trotsko, O. & Lotous, V. 2015. Excavator bucket teeth strengthening using a plastic explosive deformation. *Metallurgical and Mining Industry* 4(1): 363–368.

Drukovanui, M., Bandura, V., Kolyanovskaya, L. & Palamarchuk, V. 2014. *Improvement of Heat Technology in the Production of Oil and Biodiesel Fuel*. Vinnytsya: RVB VNAU.

Khir, R.Z. & Pan, A.S. 2006. Drying rates of thin layer roughing drying using infrared radiation. *ASABE Annual International Meeting* 4(2): 17–26.

Kolyanovska, L.M., Palamarchuk, I.P., Sukhenko, Y.G., Mushtruk, M.M., Sukhenko, V.Y., Vasuliev, V.P., Semko, T.V., Tyshchenko, L.M., Popiel, P., Mussabekova, A. & Bissarinov, B. 2019. Mathematical modeling of the extraction process of oil-containing raw materials with pulsed intensification of heat of mass transfer. *Optical Fibers and Their Applications 2018. International Society for Optics and Photonics* 11045: 143–151.

Kukharchuk, V.V., Bogachuk, V.V., Hraniak V.F., Wójcik, W. & Suleimenov, B. 2017a. Simulation of dynamic fracture of the borehole bottom taking into consideration stress concentrator. *Eastern-European Journal of Enterprise Technologies* 3/1(87): 53–62.

Kukharchuk, V.V., Hraniak, V.F., Vedmitskyi, Y.G., & Bogachuk, V.V. 2016. Noncontact method of temperature measurement based on the phenomenon of the luminophor temperature decreasing, *Photonics Applications in Astronomy, Communications, Industry, and High-Energy Physics Experiments 2016* 100312F: 523–533.

Kukharchuk, V.V., Kazyv, S.S., & Bykovsky, S.A., 2017b. Discrete wavelet transformation in spectral analysis of vibration processes at hydropower units. *Przeglad Elektrotechniczny* 93(5): 65–68.

Ogorodnikov, V.A., Savchinskij, I.G. & Nakhajchuk, O.V. 2004. Stressed-strained state during forming the internal slot section by mandrel reduction. *Tyazheloe Mashinostroenie* 12(1): 31–33.

Palamarchuk, I.P., Bandura, V.M. & Palamarchuk, V.I. 2012. Industrial design and technological circuits of confectionary drying machines. *Vibration in Technology and Technology* 2(66): 116–125.

Palamarchuk, I.P., Bandura, V.M. & Palamarchuk, V.I. 2013. Analysis of dynamics of vibroconveyor technological system with kinematic combined vibroexcitation. *MOTROL Commission of Motorization and Energetics in Agriculture* 15(4): 314–323.

Palamarchuk, I.P., Turkan, O.V. & Palamarchuk, V.I. 2015. Justification of the design and technological scheme of the infra-red vibrating conveyor dryer for post-harvest processing of loose agricultural products. *Collection of Scientific Works of Vinnitsa National Agrarian University. Technical Sciences* 1(89): 117–123.

Polishchuk, L., Bilyy, O. & Kharchenko, Y. 2016. Prediction of the propagation of crack-like defects in profile elements of the boom of stack discharge conveyor. *Eastern-European Journal of Enterprise Technologies* 6(1): 44–52.

Polishchuk, L., Kharchenko, Y., Piontkevych, O. & Koval, O. 2010. The research of the dynamic processes of control system of hydraulic drive of belt conveyors with variable cargo flows. *Eastern-European Journal of Enterprise Technologies* 2(8): 22–29.

Polishchuk, L.K. & Kozlov, L.G. 2018. Study of the dynamic stability of the conveyor belt adaptive drive. *Proceedings of SPIE 10808, Photonics Applications in Astronomy, Communications, Industry, and High-Energy Physics Experiments 2018* 1080862: 211–214.

Sorochinsky, V.F. 2002. Estimation of the uniformity of a fluidized bed of a grain by the change in the local density of the heat flow on a vertical heat exchange surface. *M. SETT* 4: 72–75.

Sukhenko, Y., Sukhenko, V., Mushtruk, M. & Litvinenko, A. 2018. Mathematical model of corrosive-mechanic wear materials in technological medium of food industry. *Advances in Design, Simulation and Manufacturing Proc. Intern. Conference on Design, Simulation, Manufacturing: The Innovation Exchange*, DSMIE-2018, June 12–15, Sumy, Ukraine 1(1): 498–507.

Sukhenko, Yu., Sukhenko, V., Mushtruk, M., Vasuliv, V. & Boyko, Y. 2017. Changing the quality of ground meat for sausage products in the process of grinding. *Eastern European Journal of Enterprise Technologies* 11(88): 56–63.

Vedmitskyi, Y.G., Kukharchuk, V.V. & Hraniak, V.F. 2017. New non-system physical quantities for vibration monitoring of transient processes at hydropower facilities, integral vibratory accelerations. *Przeglad Elektrotechniczny* 93(3): 69–72.

Vorobyov, V., Pomazan, M., Vorobyova, L. & Shlyk, S. 2017. Simulation of dynamic fracture of the borehole bottom taking into consideration stress concentrator. *Eastern-European Journal of Enterprise Technologies* 3/1(87): 53–62.

Chapter 22

Energy efficiency of gear differentials in devices for speed change control through a carrier

*Volodymyr O. Malashchenko, Oleh R. Strilets,
Volodymyr M. Strilets, Vladyslav L. Lutsyk, Andrzej Smolarz,
Paweł Droździel, A. Torgesizova, and Aigul Shortanbayeva*

CONTENTS

22.1 Introduction..255
22.2 Analysis of the general method of PE determination of gear differentials......256
22.3 Materials and methods for determining of the energy efficiency of gear differentials with closed circuit hydrosystems of speed control devices through PE evaluation..257
22.4 Discussion of the results of the determination of the PE of gear differentials of speed change devices through a carrier with two internal or external engagements..263
22.5 Conclusion...263
References...264

22.1 INTRODUCTION

In the drives of hoisting-transport, building, road, land reclamation, and other machines, there is a need to control changes in speed by the magnitude and direction of their executive mechanisms. For this purpose, devices with discrete and stepless speed control are often used, with the help of stepped and stepless speed boxes, which have simple and complex gear transmissions, or chain, pass, and friction variators. The main disadvantages of the existing stepped speed control are the emergence of dynamic loads during transitions from one speed to another and traditional stepless – the intense wear of the drive's parts due to the use of friction in the belt, block, or disk brakes and locking friction clutches. This significantly affects the reduction of durability and reliability of parts of the drives and the machines in general. Therefore, the actual scientific and technical task is the development of new combined devices for the continuously variable control of the process of speed change in the form of gear differentials with closed circuit hydrosystems and their research.

In modern technical literature, for example, Strilets (2015), Malashchenko et al. (2016), and others, a new method of stepless speed changes controlled with the help of single-stage and multi-stage gear differentials with a closed circuit hydrosystem has been proposed. Reports at scientific conferences, symposiums and other scientific discussions often draw attention to energy efficiency, which can be estimated by the coefficient of performance efficiency (PE) of such devices developed at the level of

DOI: 10.1201/9781003224136-22

patents of Ukraine that require further theoretical studies of their kinematic, energy, power, and geometric parameters. The PE of the mechanisms is very widely described in the well-known classical technical literature on the theory of mechanisms and machines, for example, Babu and Srithar (2010), Bhandari (2007), and Kinytskyi (2002), but this does not apply to the cases of the work of specific mechanisms. This disadvantage was partially eliminated in Malashchenko et al. (2017), where the PE of one of the possible designs of the proposed speed control devices in the form of a differential gear with a closed hydrosystem is considered. In Pawar and Kulkarni (2015), the theoretical effectiveness formulas for the two-stage differential gear transmission were obtained and verified by experimental studies. In Chen and Chen (2015), the PE of complex gear transmissions was investigated on the basis of graphic and screw theories, which allow approximate values to be obtained. In Xie et al. (2015), the reduction in the construction cost of many high-speed planetary gear trains based on system synthesis is justified taking into account such requirements as the gear ratio, PE, the planarity of the mechanism, and one shifting for heavy goods vehicles. In Pennestri et al. (2018), attention is drawn to the full understanding of the basic mechanics of planetary transmissions and the evaluation of their mechanical efficiency, and it is concluded that for the same input and output units, power loss has a peculiar mathematical expression for each actual sequence of angular velocities. In Dankov (2018), the reasons that hinder the application of smoothly controlled planetary transmission, due to the structural complexity of the ratio control device and variants to simplify it, are described. In Mahesh (2018), attention was drawn to the fact that planetary transmissions are used in industry due to their many benefits, which include high torque, increased efficiency, and a very compact drive, and as they are made of gears, the failure of one link affects the whole transmission, so you need to know the reasons of the failure (Kozlov et al. 2019, Polishchuk et al. 2018, 2019).

Thus, the problem of determining the energy efficiency of differential gears is relevant and requires further research.

22.2 ANALYSIS OF THE GENERAL METHOD OF PE DETERMINATION OF GEAR DIFFERENTIALS

The new suggested method of energy efficiency estimation includes determining the coefficients of PE of the gear differential with two internal and two external gears meshing with a closed circuit hydrosystem in devices for controlling changes in the speed of rotation through the carrier. This is done for the case where the driving link is the sun gear and the driven is the ring gear or vice versa. To solve the problem, it is necessary and sufficient to obtain the analytical dependencies of the PE between the driving and the driven links and establish their number in the investigated gear differential. Traditionally, the coefficient of PE of gear differentials is determined based on the frictional losses in each pair of gear wheels. These losses are proportional to the product of the circular force on the teeth and the velocity of the point on the initial circle of the differential's planetary relative to its carries or the product of the torque of this force at the angular velocity. These products are called potential power and are used in determining the PE of a gear differential transmission. Further, knowing the necessary analytical expressions and using computer simulation, we will be able to obtain graphical dependencies of the PE coefficients from the gear ratio and angular velocities of the driving and control links, and by analyzing them, it is possible to draw appropriate conclusions. This method is effective in terms of time spent on research and its accuracy (Mashkov et al. 2014, Ogorodnikov et al. 2018, Ogorodnikov & Zyska 2018, Titov et al. 2017, Tymchyk et al. 2018).

22.3 MATERIALS AND METHODS FOR DETERMINING OF THE ENERGY EFFICIENCY OF GEAR DIFFERENTIALS WITH CLOSED CIRCUIT HYDROSYSTEMS OF SPEED CONTROL DEVICES THROUGH PE EVALUATION

Energy efficiency estimation is a proper way to evaluate the perfection of a machine or mechanism (when the PE is higher). The PE lies within the limits $0 \leq \eta < 1$ and is a value determined by the ratio of the work of useful forces A_{uf} to full work A_f:

$$\eta = A_{uf} / A_f \tag{22.1}$$

Full work A_f in turn consists of the work of useful forces A_{uf} and the work of harmful forces A_{hf}:

$$A_f = A_{uf} + A_{hf} \tag{22.2}$$

The abovementioned general definition of PE can be specified for individual cases, and importantly, it is possible to obtain formulas for determining the PE through other parameters of the mechanisms.

In practice, the method of determining the PE is mainly used, for example (Malashchenko et al. 2016), where it is assumed that the frictional losses in each pair of gear wheels are proportional to the product of the circular force on the teeth and the velocity of the point of the initial circle of the differential planetary relative to the carrier, or the product of the torque on the angular velocity. This product is often called "potential power."

The PE expression for a device for speed change control by using a gear differential with a closed circuit hydrosystem through a carrier, where the driving link is the sun gear and the driven is the ring gear, can be written in the form:

$$\eta = \eta_{13} \eta_6 \eta_7 \tag{22.3}$$

where η_{13} is the PE of the gear differential transmission; η_6 is the PE of the closed circuit hydrosystem drive (gear transmission); and η_7 is the PE of the closed circuit hydrosystem (power losses for the pump operation when pumping a liquid in a closed circuit hydrosystem).

A block diagram of possible power losses in the device for speed change control using a gear differential with a closed circuit hydrosystem 7 through the carrier 4, where the driving link is the sun gear 1 and the driven ring gear 3, is shown in Figure 22.1.

In determining the PE, it is necessary to know the circular forces acting in each tooth mesh of the gear differential.

To determine the PE, let us consider the scheme of forces acting on the mesh of the sun gear and of the differential's planetary $\bar{F}_{12}$, of the planetary and the ring gear $\bar{F}_{23}$

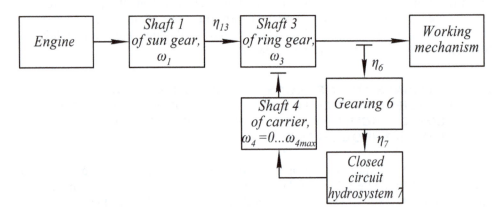

Figure 22.1 Block diagram of power losses in a device for speed change control when the driving link is the sun gear, driven is the ring gear, and control is carried out through a carrier.

and of the planetary and the carrier $\bar{F}_{24}$, and record the equilibrium condition of the planetary (Figure 22.2) in the form:

$$\bar{F}_{12} + \bar{F}_{23} + \bar{F}_{24} = 0 \tag{22.4}$$

Figure 22.2 shows the following: 1, sun gear; 2, planetary gear; 3, ring gear; 4, carrier; 5, frame; 6, pump drive; 7, closed hydrosystem; and z_i, number of teeth of the corresponding gear wheel.

In addition, the sum of the moments of forces acting on the planetary relative to the axis of its rotation is equal to zero:

$$\bar{F}_{12}r_2 + \bar{F}_{23}r_2' = 0 \tag{22.5}$$

where r_2 and r_2' are the radii of the initial circles of planetaries, respectively, with the number of teeth z_2 and z_2'.

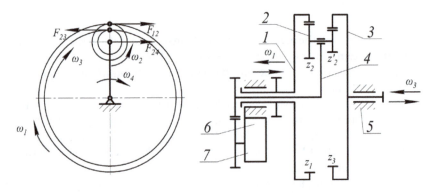

Figure 22.2 Scheme of forces in the gear differential with internal gearing.

From the expressions (22.4) and (22.5), the forces acting on the links of the gear differential transmission will be:

$$\bar{F}_{12} = -\bar{F}_{23}\frac{r_2'}{r_2};\qquad(22.6)$$

$$\bar{F}_{24} = -\bar{F}_{23}\left(1 - \frac{r_2'}{r_2}\right)\qquad(22.7)$$

If we have a given driving torque T_1, then we can record

$$F_{12} = T_1/r_1 = 0 \qquad(22.8)$$

where r_1 is the radius of the initial circle of sun gear z_1.

The torque on a gear wheel z_3 in a transmission with a fixed carrier is:

$$T_3 = F_{23}r_3 = -\frac{r_3}{r_1}T_1 = -u_{13}^{(4)}T_1, \qquad(22.9)$$

that is, the torques T_1 and T_3 without taking into account the friction relate, as in the gear transmission with fixed axles. Given the friction on the teeth meshing surface, the connection between the torques takes the form:

$$T_3 = -T_1 u_{13}^{(4)} \eta_{13}^k \qquad(22.10)$$

where η_{13} is the PE of transmission with fixed axes, defined as for successive engagement; $k = +1$ when the power transmission is carried out from the gear wheel z_1 to the gear wheel z_3 and $k = -1$ when the transmission of power is carried out from the gear wheel z_3 to the gear wheel z_1; and $u_{13}^{(4)} = \frac{z_2 z_3}{z_1 z_2'}$ is the fixed carrier gear ratio of the gear differential.

The connection between the torques on the carrier 4 and the gear wheel z_1 can be presented as:

$$T_4 = -T_1(1 - u_{13}\eta_{13}^k) \qquad(22.11)$$

The connection between the torques acting on the links of the gear differential transmission can be established by considering the condition of the equilibrium of the whole transmission as a system. In this case, we will receive:

$$T_1 + T_3 + T_4 = 0 \qquad(22.12)$$

If the torque T_1 is driving, T_3 is driven and T_4 is controlling, then the PE at the driving sun gear z_1 can be expressed by the ratio of useful power to full power:

$$\eta_{13} = -\frac{T_3\omega_3}{T_1\omega_1 + T_4\omega_4} \qquad(22.13)$$

We substitute the value of T_1, T_3, and T_4 in (22.13) and express ω_3 through ω_1, using formula (22.1) (Strilets 2015), after simple transformations, we obtain:

$$\eta_{13} = \frac{[\omega_1 - \omega_4(1-u_{13}^{(4)})]\eta_{13}^{(4)}}{\omega_1 - \omega_4(1-u_{13}^{(4)}\eta_{13}^{(4)})} \qquad (22.14)$$

To more clearly show the nature of the change in the PE of the gear differential transmission with the device in the form of a closed circuit hydrosystem, when the driving link is the sun gear and the driven is the ring gear, from the transmission parameters, the following has been done. Formula (22.14) was calculated and the computer graphic dependences $\eta_{13} = f\left(\omega_1, \omega_4, u_{13}^{(4)}\right)$ were obtained for the gear ratios, which vary within the limits $u_{13}^{(4)} = 1...10$, and the angular velocity of the driving link was $\omega_1 = 100$ rad/s. Graphic dependencies $\eta_{13} = f\left(\omega_1, \omega_4, u_{13}^{(4)}\right)$ obtained for $\eta_{13}^{(4)} = 0.97$; $\omega_1 = 100$ rad/s; and $\omega_4 = 0...40$ rad/s are shown in Figure 22.3.

The total coefficient of PE of the device for speed change control using a gear differential with a closed circuit hydrosystem through a carrier, where the driving link is the ring gear and the driven is sun gear, is of the form:

$$\eta = \eta_{31}\eta_6\eta_7 \qquad (22.15)$$

where η_{31} is the PE of gear differential transmission; η_6 is the PE of the closed circuit hydrosystem drive (gear transmission); and η_7 is the PE of the closed circuit

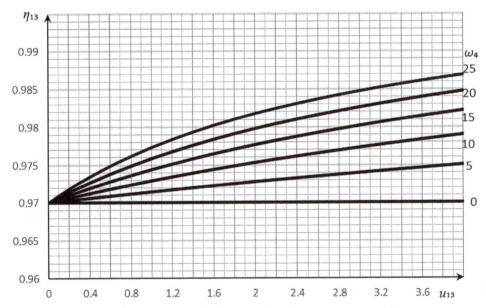

Figure 22.3 Graphs of the change of PE $\eta_{13} = f\left(\omega_1, \omega_4, u_{13}^{(4)}\right)$ in the gear differential, when the driving link is the sun gear and the driven is the ring gear, depending on the gear ratio $u_{13}^{(4)}$ and the angular velocity ω_4 of the control link, the carrier.

hydrosystem (power losses for the pump operation when pumping a liquid in a closed circuit hydrosystem).

A block diagram of possible power losses in the device for speed change control using a gear differential with a closed circuit hydrosystem 7 through the carrier 4, where the driving link is the ring gear 3 and the driven is the sun gear 1, is shown in Figure 22.4.

If the torque T_1 is the moment of resistance of the working machine with the driven sun gear, T_3 is the torque of the driving ring gear, and T_4 is the torque of the control link, the carrier, then the PE of the gear differential with driven sun gear z_1 can be expressed by the ratio of useful power to the full power of the system:

$$\eta_{31} = -\frac{T_1 \omega_1}{T_3 \omega_3 + T_4 \omega_4} \tag{22.16}$$

If we substitute the values T_1, T_3, and T_4 in formula (22.16) from formulas (22.10) and (22.11) and replace ω_1 by ω_3 using formula (22.1) from (Bhandari 2007), then after carrying out simple transformations, we obtain an expression for the PE in the case of a driving gear wheel z_1 in the form:

$$\eta_{31} = \frac{[\omega_3 u_{13}^{(4)} - \omega_4(1 + u_{13}^{(4)})]\eta_{13}^{(4)}}{\omega_3 u_{13}^{(4)} - \omega_4(\eta_{13} + u_{13}^{(4)})} \tag{22.17}$$

Similar to the previous computer modeling, the graphic images of the expression (22.17) shown in Figure 22.5 are for $\eta_{13}^{(4)} = 0.97$; $\omega_3 = 100\,\text{rad/s}$; $\omega_4 = 0...40\,\text{rad/s}$; and $u_{13}^{(4)} = 1...10$.

Figure 22.6 shows the scheme of a gear differential with an external meshing of the gear wheels.

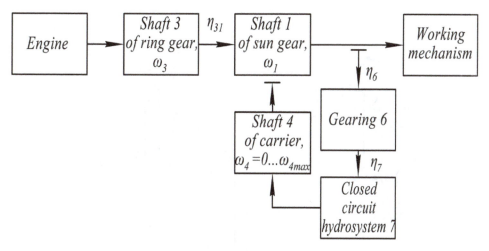

Figure 22.4 Block diagram of power losses in a device for speed change control when the driving link is the ring gear, driven is the sun gear, and control is carried out through a carrier.

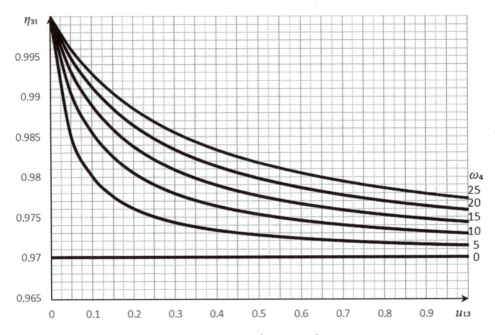

Figure 22.5 Graphs of the change of PE $\eta_{31} = f\left(\omega_3, \omega_4, u_{13}^{(4)}\right)$ in a gear differential, when the driving link is the ring gear and the driven is the sun gear, depending on the gear ratio $u_{13}^{(4)}$ and the angular velocity ω_4 of the control link, the carrier.

For the gear differential shown in Figure 22.6, the analytical expressions for determining the PE, graphic dependencies, and conclusion will be identical to the transmission with internal meshing of the gears shown in Figure 22.2.

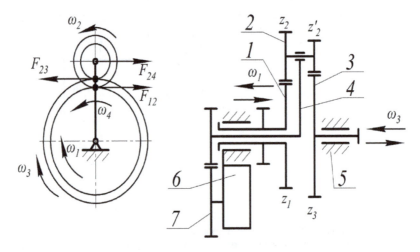

Figure 22.6 Scheme of a gear differential with an external meshing of the gear wheels.

22.4 DISCUSSION OF THE RESULTS OF THE DETERMINATION OF THE PE OF GEAR DIFFERENTIALS OF SPEED CHANGE DEVICES THROUGH A CARRIER WITH TWO INTERNAL OR EXTERNAL ENGAGEMENTS

The analytical expressions (22.14) and (22.17) have been obtained to determine the PE of gear differential transmissions of devices for controlling the change of speed through a carrier with two internal or external engagements. With these expressions, it is possible to more accurately determine the PE of gear differentials with known angular velocity of the driving link, the angular velocity of the control link, and the gear ratio. The obtained expressions for PE, with some minor transformations, can be applied in the case of any number of stages of the type of gear differentials under consideration. The calculation of PE for gear differentials with a greater number of stages with the use of the developed effective method can be simplified by the use of computer technology. In this chapter, graphic dependences for the PE of the gear differentials shown in Figures 22.3 and 22.5 have been obtained by computer simulation. On the basis of these dependencies, the possibility and expediency of application of such gear differentials in the machinery has been proven. Using this technique, you can determine the PE of two- or more-stage gear transmissions and from the point of view of energy consumption and self-braking, draw conclusions about their application in technology. It has been proven that the results of computer modeling are convenient to use, but particular numerical parameters of one-stage and multi-stage gear differentials should be known: angular the velocity of the driving link, the angular velocity of the control link, and the gear ratios. Theoretical research is the basis for further power calculations and experimental research of gear differential transmissions of speed change devices for the case when the driving link is the first-stage sun gear and the driven link is the ring gear of the last stage or vice versa while the control links are the carriers.

22.5 CONCLUSION

1. The analytical and graphic dependences of the PE between the driving and driven links (sun gear and ring gear, or vice versa) have been obtained for gear differential transmissions with two internal or two external meshings of the gear wheels with control by the carrier and the closed circuit hydrosystem, using the computer simulation tools shown in Figures 22.3 and 22.5. These data allow you to make sure about the value change of the PE depending on the gear ratio and the angular velocity of the control link, the carrier, and evaluate it from the point of view of self-braking.
2. From Figure 22.3, it is convincingly seen that in a gear differential with two internal or two external meshings of gear wheels with speed control through a carrier and a closed circuit hydrosystem, where the driving link is the sun gear and the driven the ring gear, the PE is higher than in a simple gear transmission, when $\omega_4 = 0$, and further increases with an increase in the angular speed of the control link ω_4 for the transmissions with a gear ratio $\mu_{13}^{(4)}$ in the range from 0 to 4. The lower limit of the PE is limited to the value $\eta_{13}^{(4)}$, that is, self-braking is excluded.

3. From Figure 22.5, it is evident that in a gear differential transmission with two internal or two external meshings of gear wheels with a control through a carrier and a closed circuit hydrosystem, where the driving link is the ring gear and the driven is the sun gear, the PE is higher than in a simple gear transmission, when $\omega_4 = 0$, and significantly decreases with an increase in gear ratio $u_{13}^{(4)}$ in the range from 0 to 0.3, but higher for transmissions with higher speed of the control links ω_4, and the PE is limited to magnitude $\eta_{13}^{(4)}$, that is, self-braking is eliminated.

REFERENCES

Babu, G.K. & Srithar K. 2010. *Design of Machine Elements*. 2nd Edition. New Delhi: Tata McGrew-Hill Education.

Bhandari, V.B. 2007. *Design of Machine Elements*. New Delhi: Tata McGrew-Hill Education.

Chen, Ch. & Chen, J. 2015. Efficiency analysis of two degrees of freedom epicyclic gear transmission and experimental validation. *Mechanism and Machine Theory* 87: 115–130.

Dankov, A.M. 2018. Planetary continuously adjustable gear train with force closure of planet gear and central gear: from idea to design. *Science & Technique* 17(3): 228–237.

Kinytskyi, Y.T. 2002. *Theory of Machines and Mechanisms*. Kyiv: Naukova Dumka.

Kozlov, L.G., Polishchuk, L.K., Piontkevych, O.V., Korinenko, M.P., Horbatiuk, R.M., Komada, P., Orazalieva, S. & Ussatova, O. 2019. Experimental research characteristics of counter balance valve for hydraulic drive control system of mobile machine. *Przeglad Elektrotechniczny* 95(4): 104–109.

Mahesh, D. 2018. Design of compound planetary gear train. *International Journal for Research in Applied Science and Engineering Technology* 6(4): 3179–3184.

Malashchenko, V., Strilets, O. & Strilets V. 2016. Method and device for speed change by the epicyclic gear train with stepped-planet gear set. *Research Works of Air Force Institute of Technology* 38: 13–19.

Malashchenko, V., Strilets, O. & Strilets, V. 2017. Determining performance efficiency of the differential in a device for speed change through epicycle. *Eastern-European Journal of Enterprise Technologies* 6(7): 51–57.

Mashkov, V., Smolarz, A., Lytvynenko, V. & Gromaszek, K. 2014. The problem of system fault-tolerance. *Informatyka, Automatyka, Pomiary w Gospodarce i Ochronie Środowiska* 4: 41–44.

Ogorodnikov, V.A., Dereven'ko, I.A. & Sivak, R.I. 2018. On the influence of curvature of the trajectories of deformation of a volume of the material by pressing on its plasticity under the conditions of complex loading. *Materials Science* 54(3): 326–332.

Ogorodnikov, V.A., Zyska, T. & Sundetov, S. 2018. The physical model of motor vehicle destruction under shock loading for analysis of road traffic accident. *Proceedings of SPIE* 108086C: 1–5.

Pawar, P.V. & Kulkarni, P.R. 2015. Design of two stage planetary gear train for high reduction ratio. *International Journal of Research in Engineering and Technology* 4(6): 150–157.

Pennestri, E., Esmail, E.L. & Hussein A. 2018. Power losses in two-degrees-of-freedom planetary gear trains: A critical analysis of Radzimovsky's formulas. *Mechanism and Machine Theory* 128: 191–204.

Polishchuk, L., Gromaszek, K., Kozlov, L.G. & Piontkevych, O.V. 2019. Study of the dynamic stability of the belt conveyor adaptive drive. *Przeglad Elektrotechniczny* 95(4): 98–103.

Polishchuk, L., Kozlov, L.G., Piontkevych, O.V., Gromaszek, K. & Mussabekova, A. 2018. Study of the dynamic stability of the conveyor belt adaptive drive. *Proceedings of SPIE* 1791–1800.

Strilets, O.R. 2015. Speed change management by differential gear through carrier. *Bulletin of Kremenchutsk National University by Mykhailo Ostrohradskyi* 6(95): 87–92.

Titov, A.V., Mykhalevych, V.M., Popiel, P. & Mussabekov, K. 2017. Statement and solution of new problems of deformability theory. *Proceedings of SPIE* 1611–1617.

Tymchyk, S.V., Skytsiouk, V.I., Klotchko, T.R., Ławicki, T. & Denisova, N. 2018. Distortion of geometric elements in the transition from the imaginary to the real coordinate system of technological equipment. *Proceedings of SPIE* 108085C: 1595–1604.

Xie, T., Hu, J., Peng, Z. & Liu, Ch. 2015. Synthesis of seven-speed planetary gear trains for heavy-duty commercial vehicle. *Mechanism and Machine Theory* 90: 230–239.

Chapter 23

Influence of electrohydraulic controller parameters on the dynamic characteristics of a hydrosystem with adjustable pump

*Volodymyr V. Bogachuk, Leonid H. Kozlov,
Artem O. Tovkach, Valerii M. Badakh, Taras V. Tarasenko,
Yevhenii O. Kobylianskyi, Zbigniew Omiotek,
Gauhar Borankulova, and Aigul Tungatarova*

CONTENTS

23.1 Introduction.. 267
23.2 Problem setting... 267
23.3 Study object and research methodology...................................... 268
23.4 Conclusion.. 277
References... 277

23.1 INTRODUCTION

In the last decade, the range of mobile working machines with variable working elements has considerably expanded. Such machines can be quite easily reconfigured to perform various operations and can be used almost all year round. The hydraulic system of these machines should provide work of the working elements across a wide range of speed and load modes. At the same time, their high efficiency should be ensured (Kozlov et al. 2015, Dreher 2015, Polishchuk et al. 2016).

The widespread use of adjustable pumps is one of the most effective ways of improving the hydraulic system of mobile machines (Sveshnikov 2005). A hydrosystem based on adjustable pumps provides the possibility of adjusting the movement parameters of the working elements. This increases the productivity of mobile machines as well as their energy efficiency along with improvement of the operator's working conditions (Andreyev et al. 2012, 2013, Minav 2011).

23.2 PROBLEM SETTING

The adjustable pumps should provide the work of the hydrosystem of mobile working machines in a variety of modes: constant pressure, constant power, load sensitivity, and in combinations of the abovementioned (Avrunin et al. 2009). These functions are provided most of all by adjustable pumps equipped with multifunctional electrohydraulic controllers.

DOI: 10.1201/9781003224136-23

The transition to proportional electrohydraulic control in hydraulic drives of mobile working machines enables the wide use of controllers to provide optimal modes of operation and automation of work cycles performing various operations. Controller usage properties improving the efficiency of hydraulic drives on the basis of adjustable pumps are considered in a number of papers (Burennikov et al. 2012, Minav 2011, Kozlov 2013).

Questions of the schemes and design optimization of adjustable pumps with electrohydraulic control are considered (Kozlov et al. 2018). However, it should be noted that at present, a number of issues related to the study of dynamic processes in hydraulic drives with electrohydraulic control on the basis of adjustable pumps require further research.

The task of choosing the most reasonable pump control scheme to prove the proposed control algorithms and the parameters which provide the necessary static, dynamic, and power characteristics of the hydraulic drives is not fully completed. There is still a need to develop and improve the nonlinear mathematical models that describe the working processes in complex systems which combine a hydraulic drive and digital control system (Kozlov 2013, Kukharchuk 2016, Kukharchuk et al. 2017, Titov et al. 2017).

23.3 STUDY OBJECT AND RESEARCH METHODOLOGY

A hydraulic system based on an adjustable pump is shown in Figure 23.1. The hydraulic system includes an adjustable pump 1, with serial adjustable throttles 2 and 18 at the output. The adjustable pump 1 is equipped with an electrohydraulic controller that consists of a slide 8 and a servo valve 7. A pressure sensor 3 is connected to the hydroline and connects the adjustable throttles 2 and 18.

The signal from the pressure sensor 3 via the controller 4 and the amplifier 5 is fed to a proportional electromagnet 6 which interacts with the servo valve 7. A photo of the electrohydraulic controller is shown in Figure 23.2. The electromagnet 1, the unit 2 with the servo valve 3, and the unit 4 with the slide 5 are the main components of the controller. The hydraulic system (Figure 23.1) operates in three modes: idle, feed regulation, and overload protection.

In idle mode, in the absence of a signal from the controller 4, the working fluid from the pump will enter into the feed line 10 and then to the slide 8 and the servo valve 7. In this case, the electromagnet 6 does not engage the servo valve. The working fluid will enter the tank under low pressure p_x and the pump 1 outlet will be at pressure p_n, determined by the spring 9 (Kukharchuk et al. 2017).

The pressure p_n in this case will be small (about 1.5 MPa). The working fluid through the slide opening 8 will come through the throttle 14 into the tank and through the throttle 15 to the servo valve 11 creating pressure p_e. Under the pressure p_n, p_e, and spring 19, the face plate 1 will be in a position at which the angle γ will be small and the pump 1 will compensate only the flow through the slide 8 and servo valve 7.

In feed adjustment mode, with the partial overlapping of the throttle 18 and the open throttle 2, the pressure p_c, which will generate a signal to the controller 4 via the sensor 3, will be formed. The amplified signal from the controller is sent to the electromagnet 6, which proportionally engages the servo valve 7 to the pressure p_c. The servo valve 7 forms the pressure p_x, the value of which is also proportional to the pressure p_c, and the slide 8 will provide a pressure value p_n in the feed line 10. The pressure value p_n

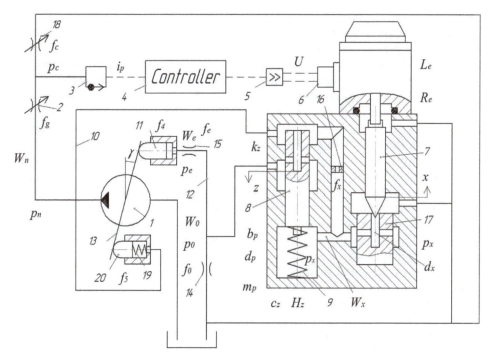

Figure 23.1 Diagram of a hydraulic system based on an adjustable pump.

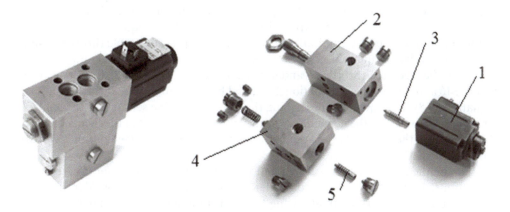

Figure 23.2 Photo of electrohydraulic controller.

will exceed the pressure p_c for the constant difference Δp_g, the value of which is determined by the force of the spring 9 and the slide end area 8. Thus, the pump controller will create and maintain a constant pressure drop Δp_g at the adjustable throttle 2. This drop will actually depend on the pressure p_c.

This enables the working opening of the throttle 2 and the pump feed value to change. The feed value of the pump Q_n will be constantly maintained regardless of the pressure p_c, and the change in the size of the working opening area f_g of the adjustable throttle 2 provides the pump feed value Q_n settings.

In overload protection mode, in the case of complete overlapping of the adjustable throttle 18, the pressure p_c will increase abruptly, and in accordance with the signal from the sensor 3, it will continue increasing. The controller signal i_n is restricted with a program, henceforth the signal U which is sent by the controller 4 to the electromagnet 6 will not increase (Vasilevskyi et al. 2016, 2018, Fatyga et al. 2018).

In this case, the force will be restricted due to the electromagnet 6 engaging with the servo valve 7, and the pressure p_x will be limited. This will restrict the pressure rise p_n in the hydroline 10. The working fluid under the pressure p_n will come through the working opening of the slide 8 and will generate the pressures p_0 and p_e. The face plate of the pump 1 will be moved to the position close to zero angle γ due to the pressure of p_e, p_n, and the spring 19. The pump feed 1 will be reduced and compensate only minor losses of the working fluid through the servo valve working opening 7 and the one of the slide 8. The pressure p_n at the pump 1 output will be set at a high level. The value of the pressure p_n will be determined by the maximum value of the signal U, which is sent by the controller to the electromagnet 6 and by the resistance of the spool spring 9. Therefore, the pump feed 1 will be low. Thus, the pump overload protection will be secured. Certain dynamic characteristics must be ensured when designing adjustable pumps with electrohydraulic controllers for working machines. At the design stage, dynamic characteristics can be estimated on the basis of adjustable pumps studies, which were conducted with the application of mathematical models.

The mathematical model of the hydraulic circuit containing the adjustable pump with the controller consists of equations: equation of the moments applied to the face plate 13 of the pump 1 (equation 23.1), equation of force applied to the slide 8 (equation 23.2), and the equation of forces applied to the servo valve 7 (equation 23.3), as well as the equation of the flow continuity between the pump 1, adjustable throttle 2, slide 8, and servo valve 7 (equation 23.4); between the slide 8 and throttles 14 and 15 (equation 23.5); between the throttle 15 and servo cylinder 11 (equation 23.6); between the throttle 16 and servo valve 7 (equation 23.7); and between adjustable throttles 2 and 18 (equation 23.8). The equations describing the operation of the sensor and the controller (equation 23.9), and the dependence of the resistance moment at the face plate 13 of the pump 1 on the pressure at the pump outlet and the pump feed (equation 23.10) are parts of the mathematical model.

The mathematical model is developed with the following assumptions and simplifications. Lumped pump and controller parameters are under consideration: the temperature of the working fluid during the transition period is unchangeable; wave processes in pipelines are not taken into account; throttle and servo valve flow factors are constant; pump operation mode is non-cavitation; pressure losses in hydrolines are not taken into account; the pressure sensor and amplifier operation are simulated by proportional links; and the force of dry friction in the spool and servo valve is not considered.

$$I\frac{d^2\gamma}{dt^2} = p_n f_5 l - p_e f_4 l - b_\gamma \frac{d\gamma}{dt} - M_c \qquad (23.1)$$

$$p_n \frac{\pi d_p^2}{4} - p_x \frac{\pi d_p^2}{4} - c_z(z + H_z) - b_p \frac{dz}{dt} = 0 \tag{23.2}$$

$$p_x \frac{\pi d_x^2}{4} = \left(L_e \frac{di}{dt} + iR_e \right) \cdot k_e - b_x \frac{dx}{dt} \tag{23.3}$$

$$F_7 d_8 k_1 n_n tg\gamma = \mu f_x \sqrt{\frac{2|p_n - p_x|}{\rho}} \sin(p_n - p_x) + \mu f_g \sqrt{\frac{2|p_n - p_c|}{\rho}}$$
$$\times \sin(p_n - p_c) + \mu k_{zz} \sqrt{\frac{2|p_n - p_0|}{\rho}} \sin(p_n - p_0) + \beta_n W_n \frac{dp_n}{dt} \tag{23.4}$$

$$\mu k_{zz} \sqrt{\frac{2|p_n - p_0|}{\rho}} \sin(p_n - p_0) = \mu f_e \sqrt{\frac{2|p_0 - p_e|}{\rho}} \sin(p_0 - p_e)$$
$$+ \mu f_0 \sqrt{\frac{2 p_0}{\rho}} + \beta_p W_0 \frac{dp_0}{dt} \tag{23.5}$$

$$\mu f_e \sqrt{\frac{2|p_0 - p_e|}{\rho}} \sin(p_0 - p_e) = \beta_p W_e \frac{dp_e}{dt} - f_4 l \frac{d\gamma}{dt} \cdot \cos\gamma \tag{23.6}$$

$$\mu f_x \sqrt{\frac{2|p_n - p_x|}{\rho}} \sin(p_n - p_x) = \mu \pi d_x x \cdot \sin\frac{2x}{2} \sqrt{\frac{2 p_x}{\rho}} + \beta_p W_x \frac{dp_x}{dt} \tag{23.7}$$

$$\mu f_g \sqrt{\frac{2|p_n - p_c|}{\rho}} \operatorname{sign}(p_n - p_c) = \mu f_c \sqrt{\frac{2 p_c}{\rho}} + \beta_n W_c \frac{dp_c}{dt} \tag{23.8}$$

$$p_c k_4 F_k(i_p) = L_e \frac{di}{dt} + iR_e \tag{23.9}$$

$$M_c = m_0 + m_1 Q_n + m_2 p_n + m_3 Q_n^2 + m_4 p_n^2 + m_5 p_n Q_n \tag{23.10}$$

where p_n, p_c, p_e, p_0, p_x are the pressure at the pump outlet 1 and throttle input 18 of the pump control system in the hydraulic cylinder 11 and at the inlet of the servo valve 7; z, x are the coordinate positions of spool 8, servo valve; γ is the rotation angle of the pump 1 face plate; f_0, f_e, f_g, f_c are the area of throttles 14 and 15, working opening area of adjustable throttles 2 and 18; d_p, d_8, d_x are the slide 8 diameter, pump piston and pump face plate contact area diameter, and seat diameter 17, respectively; i, i_p are the electromagnet coils current at the pressure sensor output; k_e, k_u, k_1 are the force ratio of the electromagnet, pressure sensor, and the number of the pistons of the pump 1, respectively; L_e, R_e are the inductance and active resistance of electromagnet coils; c_z is the stiffness of slide spring 8; μ is the throttle and slide flow ratio; ρ is the working fluid density; l is the servo cylinder arm of the pump 1; I is the pump face plate moment of inertia; W_n, W_0, W_c, W_e, W_x are the pump outlet hydroline volume, line volume between the spool 8 and throttle 14, line volume at the throttle 8 inlet, line volume between the throttle 15 and the servo cylinder 11 of the pump 1, and line volume between the throttle 16 and servo valve 7, respectively; n_n is the number of pump shaft revolutions; $F_k(i_p)$ is the transfer function of the controller 4; H_z is the preliminary compression of the spring 8; $m_0, m_1, m_2, m_3, m_4, m_5$ are the relation indexes of the resistance moment at the face plate of the pump 1 to the feed and pressure value; M_c is the resistance moment at the face plate of the pump 1; β_p is the compliance factor of the gas–liquid mixture; β_n is the rubber–metal pipeline and gas–liquid mixture factor; and b_p, b_x are the spool 8 and servo valve 7 damping factors.

Transient processes in a hydrosystem equipped with an adjustable pump are defined according to the Rosenbrock method with the help of mathematical model equations using the MATLAB-Simulink software.

A flowchart of the mathematical equation solution, which was created with MATLAB-Simulink software, is presented in Figure 23.3.

The flowchart includes nine subsystems, where each of them solves one mathematical model equation. The FACEPLATE subsystem solves equation (23.1), moments ongoing to a face plate, determining the angle of rotation γ and the angular velocity of the pump 1 (Figure 23.1). The SLIDE subsystem solves equation (23.2) of the forces applicable to slide 8 and determines the value of the Z axial coordinate of the slide. The SERVOVALVE subsystem solves equation (23.3) of the forces applicable to the servo valve 7 and determines the value of the X axial coordinate of the servo valve 7. The PUMP subsystem solves equation (23.4) of the flow continuity at the pump outlet 1 and determines the pressure value p_n at the pump output 1. The SLIDE FLOW subsystem solves equation (23.5) of the flow continuity for the hydroline 12 between the slide 8 and the throttles 14 and 15 and determines the pressure value p_0. The SERVOCYLINDER subsystem solves equation (23.6) of the flow of continuity for the chamber between

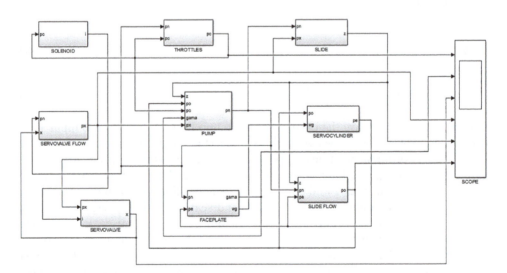

Figure 23.3 Flowchart of mathematical model equation solution.

the throttle 15 and the servo cylinder 11 and determines the value of the pressure p_e. The SERVOVALVE FLOW subsystem solves equation (23.7) of the flow continuity between the throttle 16 and servo valve 7 and determines the value of the pressure p_x. The THROTTLES subsystem solves equation (23.8) of the flow continuity for the hydroline between throttles 2 and 18 and determines the value of the pressure p_c. The SOLENOID subsystem solves equation (23.9) of the voltage drop of the electromagnet coils 6 and determines the current I value. The SCOPE oscilloscope provides registration and visualization of the variables which are the components of mathematical model equations.

The structure of the SLIDE subsystem is shown in Figure 23.4. The subsystem contains units for entering pressure values p_n and p_x acting on the slide (equation (23.2) of the mathematical model). The product2 and product4 units determine the forces that are formed by the pressure p_n and p_x at the slide end 8 (Figure 23.1), and the product3 unit generates the pressure of spring 9. The product1 unit defines the displacement derivative Z of time $\frac{dz}{dt}$. The integrator unit defines the slide 8 axial coordinate–time relation. The saturation unit simulates the structural restriction of the slide 8 in the pump regulator housing.

The time relation of the variables is the result of solving mathematical model system equations 1–9: $i=f(t)$, $p_c=f(t)$, $z=f(t)$, $p_x=f(t)$, $p_n=f(t)$, $p_e=f(t)$, $x=f(t)$, $y=f(t)$, and $p_0=f(t)$, which describe the condition of the hydrosystem when adjustable throttle area f_c is stepwise changed from $6 \cdot 10^{-6}$ to $3 \cdot 10^{-6} m^2$. The systems of equations were solved using the following initial variable values: $i(0)=4 \cdot 10^{-4}$ A, $z(0)=1.5 \cdot 10^{-3}$m, $y(0)=0.11$ rad, $p_c(0)=8 \cdot 10^5 N/m^2$, $p_n(0)=22.6 \cdot 10^5 N/m^2$, $p_0(0)=16 \cdot 10^5 N/m^2$, $x(0)=14 \cdot 10^{-4}$m, $p_e(0)=16 \cdot 10^5 N/m^2$, $p_x(0)=8 \cdot 10^5 N/m^2$. The main parameters of the hydrosystem were: $d_p=10 \cdot 10^{-3}$m, $L_e=50 \cdot 10^{-6}$ Gn, $R_e=20$ Ω, $c_z=1.0 \cdot 10^4 N/m$, $\mu=0.67$, $\rho=900$ kg/m^2, $\beta_p=0.\ 6 \cdot 10^{-9} N/m^2$,

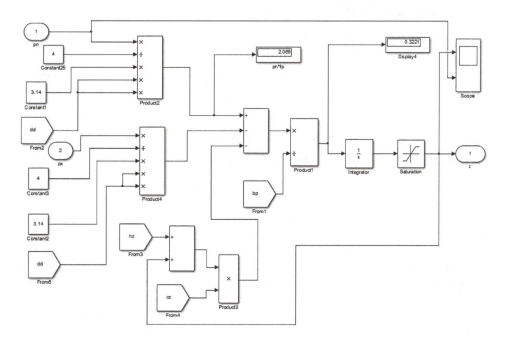

Figure 23.4 Structure of the SLIDE subsystem.

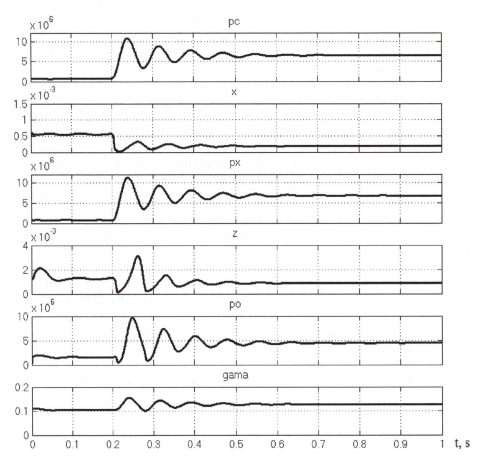

Figure 23.5 Calculation of time dependencies of variables included in the mathematical model equation.

$lW_n = 0.5 \cdot 10^{-3} \text{m}^3$, $W_c = 0.5 \cdot 10^{-3} \text{m}^3$, $W_c = 0.05 \cdot 10^{-3} \text{m}^3$, $W_0 = 0.05 \cdot 10^{-3} \text{m}^3$, and $I = 0.02$ kg·m². The transfer function of the controller was simulated by the link $F_k(s) = \dfrac{135(s+15)}{(s+100)(s+20)}$. The maximum pump flow $Q_n = 1 \cdot 10^{-3} \text{m}^3/\text{s}$, and the maximum pressure in the hydraulic system $p_n = 16$ MPa.

A sample of the calculated transition process of a hydrosystem with stepwise working opening change of the adjustable throttle 18 is shown in Figure 23.5. A pressure increase and pump feed decrease occur when the value of f_c is changed. The servo valve 7 opening reduces, and as a result, the pressure p_x increases. The pressure p_x increases and the slide 8 moves, reducing its working opening, resulting in the pressure p_0 dropping in the 12. The pressure p_e applying to servo cylinder 11 decreases. Under the pressure of p_n, p_e, and the spring 19, the face plate of the pump 1 turns in a way that increases the rotation angle γ, and the feed value Q_n increases. As a result of the process of fluctuation taking place in the pump control system, which is the

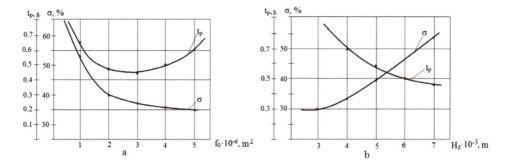

Figure 23.6 Influence of f_0 and H_z on adjustment and readjustment time.

outcome of a stepwise area decrease f_c of the throttle 18, the set feed value Q_n will be restored. The time of the transition process and resetting of the hydrosystem will depend, first of all, on design parameter ratio of the pump and its regulator, on the controller, and the dynamic characteristics of feed and control lines. The influence of the design parameters of the slide, servo valve, and throttle in the pump control system on the dynamic characteristics of the hydraulic system is considered in this chapter. At the blueprint stage of the pump controller, it is important to have the following information.

The influence of the area values f_0 of the throttle 14 on the dynamic characteristics is shown in Figure 23.6a. The throttle 14 (Figure 23.1) area value has an ambiguous effect on the adjustment time in the hydrosystem. For small values of $f_0 < 1.0 \cdot 10^{-6} \text{m}^2$, the adjustment time increases abruptly and the best range is $f_0 = (2 \ldots 4) \cdot 10^{-6} \text{m}^2$. When choosing a bigger value of f_0, the adjustment time increases once again. Considering the value of f_0, we should take into account the fact that when the value of f_0 increases, it causes a parasitic feed rise which is not directed to the consumer but is sent to a tank via throttle 14. This increases the pump capacity loss. Increasing the value of f_0 unambiguously decreases the pressure adjustment value p_n in the hydrosystem.

The influence of the previously compressed H_z spring 9 on dynamic characteristics is shown in Figure 23.6b. More dynamic operation of the pump is ensured by choosing higher values of H_z from the proposed range. At the same time, the response speed and adjustment time increase. To improve the dynamic characteristics, we recommend higher H_z values from the abovementioned range. However, in this case, we have to take into account that increasing the value of H_z will lead to a parasitic pump feed rise, which is not directed to a consumer but is sent to a tank through the throttle 16 and the working opening of the servo valve 7. This slightly increases hydrosystem capacity loss.

The relation of the k_z dynamic characteristics to the hydrosystem is shown in Figure 23.7a. Increasing k_z to the maximum values from the proposed range simultaneously increases pump operation speed and decreases the pressure adjustment value p_n. However, increasing the value of k_z is complicated by the simultaneous expansion of the dimensions of the slide. The slide 8 damping factor b_p is shown in Figure 23.7b. The damping of the slide should be optimal and the adjustment time significantly

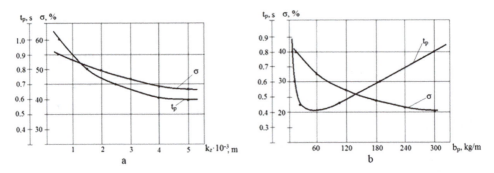

Figure 23.7 Influence of k_z and b_p on adjustment and readjustment time.

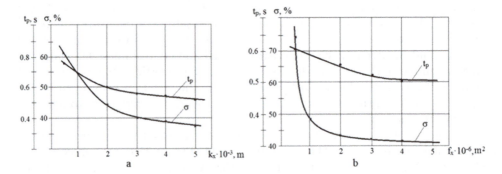

Figure 23.8 Influence of k_x and f_x on adjustment and readjustment time.

decreases when choosing a ratio value from the range $b_p = (15...180)$ kg/s. When the value of b_p is increased up to 300 kg/s, the readjustment value drops up to 20% and stabilization of σ takes place.

The influence of the value of gain factor k_x of the working opening of the servo valve 7 on the dynamic characteristics of the hydrosystem is shown in Figure 23.8a.

Choosing higher values of k_x from the abovementioned range increases the hydrosystem operation speed and decreases the readjustment value σ regarding the pressure p_n. It should, however, be kept in mind that the expansion of the dimensions of the servo valve 7 determines the seat size of the servo valve 17. Moreover, increasing the gain factor k_x value is restricted with the maximum force of the electromagnet. The relation between the throttle 16 area f_x (Figure 23.1) and the dynamic characteristics of the hydrosystem is shown in Figure 23.8b. Increasing the value of the throttle area f_x increases the operating speed of the hydrosystem. Readjustment decreases when increasing the value of f_x from $0.5 \cdot 10^{-6}$ to $2 \cdot 10^{-6}$ m^2, in particular. Further increasing area f_x is followed by stabilization of the value of σ at 40%. When choosing the value of the throttle area f_x, we should keep in mind that increasing this value leads to a parasitic feed rise,, which is redirected into a tank through the throttle 16 and the working opening of the servo valve 7, thus increasing pump capacity loss.

23.4 CONCLUSION

1. The specifications of controller significantly influence the dynamic characteristics of a hydrosystem with an adjustable pump. When designing such a system, the reasonable values of these parameters should be used.
2. If it is necessary to ensure high operating speed of the system, choose a value of f_0 from the range $f_0 = (2...4) \cdot 10^{-6} m^2$, the damping factor is recommended to be $b_p = (15...180)$ kg/s, as well as higher values of H_z from the range $H_z = (2...7) \cdot 10^{-3} m$, k_z from the range $k_z = (2...7) \cdot 10^{-3} m$, k_x from the range $k_x = (2...7) \cdot 10^{-3} m$, and a value of throttle area f_x from the range $f_x = (1...5) \cdot 10^{-6} m^2$.
3. If it is necessary to minimize the adjustment values, higher values of the throttle area f_0 from the range $f_0 = (1...5) \cdot 10^{-6} m^2$, k_z from the range $k_z = (1...5) \cdot 10^{-3} m$, slide damping factor $f_p = (15...300)$ kg/s, index k_x from the range $k_x = (1...5) \cdot 10^{-3} m$, and throttle area f_x from the range $f_x = (1...5) \cdot 10^{-6} m^2$ and smaller values of previously compressed spring H_z from the range $H_z = (2...7) \cdot 10^{-3} m$ are recommended.

REFERENCES

Andreyev, M.A. et al. 2012. Correction of dynamic characteristics of the axial-piston pump regulator with electro-hydraulic proportional control. *Science and Education* 12: 47–54.

Andreyev, M.A. et al. 2013. Dynamics of the electro-hydraulic pressure regulator. *Science and Education* 1: 39–47.

Avrunin, G.A., Nazarov, L.V., Maznichenko, V.A. 2009. Modern Controllers for Changing the Working Volume of Hydromashines. *Collection of Scientific Works* 23: 72–83.

Burennikov, Yu.A. Kozlov, L.G. Repinsky, S.V. Polischuk, O.V. 2012. Optimization of design parameters of the combined regulator of supply of axial-piston-controlled pump. *Industrial Hydraulics and Pneumatics* 1(35): 73–77.

Dreher, T. 2015. *Energie ef fizienz von Konstantdruck systemen sekundargeregelten Antrieben beim Einsatz in mobilen Arbeitsmaschinen.* Karlsruhe: KIT Scientific Publishing.

Fatyga, K. et al. 2018. A comparison study of the features of DC/DC systems with Si IGBT and SiC MOSFET transistors. *Informatyka, Automatyka, Pomiary w Gospodarce i Ochronie Środowiska* 8(2): 68–71.

Kozlov, L. 2013. Digital PD controller for dynamic correction of the differential component coefficient for a mechatronic hydraulic system. *Tehnomus Journal: Proceedings of the XVIIth International Conference "New Technologies and Products in Machine Manufacturing Technologies"*, Suceava, Romania, 120–125.

Kozlov, L.G. 2015. *Scientific Foundations for Designing the Systems of Manipulator Hydraulic Drives with an Adaptive Neural Network-Based Controllers for Mobile Working Machines.* Kyiv: National Technical University of Ukraine Kyiv Polytechnic Institute.

Kozlov, L.G. et al. 2018. Determining of the optimal parameters for a mechatronic hydraulic drive. *Proceedings of SPIE* 10808: 1080861.

Kukharchuk, V.V. 2016. Noncontact method of temperature measurement based on the phenomenon of the luminophor temperature decreasing. *Proceedings of SPIE* 10031: 100312F.

Minav, T. 2011. *Electric-Drive-Based Control and Electric Energy Regeneration in a Hydraulic System.* Thesis for the degree of Doctor of Science. Lappeenranta Finland.

Polishchuk, L.K. et al. 2016. Investigation of dynamic processes in the hydraulic drive control system of belt conveyors with variable load flows. *European Journal of Advanced Technologies* 8(80): 22–29.

Sveshnokiv, V.K. 2005. Axial-piston pumps in modern hydraulic drives. *Hydraulics and Pneumatics: Informational and Technical Journal* 18: 8–12.

Titov, A.V., Mykhalevych, V.M., Popiel, P. & Mussabekov, K. 2017. Statement and solution of new problems of deformability theory. *Proceedings of SPIE* 108085E: 1611–1617.

Vasilevskyi, O.M. et al. 2016. The method of translation additive and multiplicative error in the instrumental component of the measurement uncertainty. *Proceedings of SPIE* 10031: 1003127.

Vasilevskyi, O.M. et al. 2018. Method of evaluating the level of confidence based on metrological risks for determining the coverage factor in the concept of uncertainty. *Proceedings of SPIE* 10808: 108082C.

Chapter 24

High-precision ultrasonic method for determining the distance between garbage truck and waste bin

*Oleh V. Bereziuk, Mykhailo S. Lemeshev,
Volodymyr V. Bogachuk, Piotr Kisala, Aigul Tungatarova, and
Bakhyt Yeraliyeva*

CONTENTS

24.1 Introduction 279
24.2 Methods 280
24.3 Main results of the research 283
24.4 Results of experiments 287
24.5 Conclusion 288
References 289

24.1 INTRODUCTION

One of the items of agreement between the European Union and Ukraine determines the relevance of a municipal solid wastes (MSW) system study. The annual volume of MSW amounted to more than 54 mln m^3 in 2018 in Ukraine. Of this, 93.8% was buried in landfills, 2% was burned in incinerators, and 4.2% of MSW went to procurement points of secondary raw materials and recycling plants. The collection and transportation of MSW for burial and incineration is carried out by 4,000 garbage trucks, which are distinguished by the following methods of loading MSW into the body: rear (70%), lateral (25%), and front (5%) (Bereziuk 2015). For the garbage truck to stop in the correct position to align the manipulator with the waste bin of MSW, the driver operator needs to know the exact distance between the truck and the waste bin. Therefore, in the garbage truck, it is advisable, in our opinion, to use a high-precision method to determine the distance with the possibility of indicating the distance in the cab for the driver operator. This is especially true for the rear and side methods of loading MSW, which are the most common.

According to the resolution of the Cabinet of Ministers of Ukraine No 265, ensuring the application of modern high-efficiency garbage trucks in the country's communal economy, as the main link in the structure of machines for the collecting and primary processing of MSW, is a current scientific and technical problem. Improving the accuracy of the method to determine the distance between the garbage truck and the waste bin of MSW with taking into account the environmental parameters is one of the important tasks for solving this problem.

DOI: 10.1201/9781003224136-24

The aim of the research is to determine the distance between the garbage truck and the waste bin of MSW, on the basis of a study of the existing methods of signal processing and measurement of environmental parameters.

During an analysis of the relevant literary sources, it was found that many studies focus on the application of sensors and data logging systems in many fields of science and technology (Osadchuk et al. 2013, 2016, 2017). In recent years, the Arduino microcontroller has become increasingly popular for the development of digital measuring instruments (Hadaichuk & Krekoten 2018). It, alongside many other well-known brands such as BeagleBone and RaspberryPi (Hadaichuk & Krekoten 2019), belongs to the class of small, inexpensive single-board computers, with programming and application development supported by a large community of developers and users who provide various open-source libraries, sample solutions, and forums spanning different thematic and additional aspects (Cvjetkovic & Stankovic 2017). The advantages of Arduino microcontroller boards are low price of the project, autonomy, relationship with expansion boards and sensors to perform more complex tasks, a low level of energy consumption, small size, and simple installation requirements. Work, based on Arduino, has been performed by Kumar et al. (2016) dedicated to the development of an intelligent alert system for filled waste bins of MSW to empty them with a garbage truck in a timely manner. The work (Yerraboina et al. 2018), based on an ultrasonic sensor and an Arduino Uno microcontroller, developed a smart waste bin for MSW, which tracks their quantity by sending information about the status of waste bin filling. In a monograph by Bohachuk and Mokin (2008), methods have been developed and, based on these methods, means have been created for the continuous control of the humidity of powdered materials under the conditions of the technological process of their manufacture. A peculiarity of a device for measuring the specific electroconductivity of milk in a milk-receiving chamber is that it takes into account the temperature correction via a temperature sensor, which increases the accuracy of measurement (Kucheruk et al. 2017). Bereziuk et al. (2018) proposed a method that allows express analysis to be conducted for the measurement of the relative humidity during experimental research on the dewatering of MSW. Balhabaev et al. (2016) and Kaliuzhnyi et al. (2016) describe an ultrasonic rangefinder and ultrasonic meter distances, respectively, based on a HC-SR04 sensor, which, when calculating distances, does not take into account the dependence of the speed of sound on environmental parameters and which leads to significant measurement errors. In the article by Bereziuk et al. (2019), an ultrasound method is proposed to measure distances between a garbage truck and a waste bin, which takes into account the correction of the temperature and relative humidity by means of a low-precision DHT11 sensor but ignores changes in atmospheric pressure.

24.2 METHODS

The following methods are used for research and analysis: analog and digital signal processing techniques, contactless resistive and piezoresistive methods of measuring physical quantities, analysis and synthesis, computer processing of information, and experimenting.

To convert the value of an ultrasonic sensor's digital signal to a distance value, the following conversion equation (Bereziuk et al. 2019) is used:

$$l = \frac{\Delta\tau}{2\cdot 10^4}\sqrt{\frac{\gamma R(273.15+t)}{M_d - (M_d - M_v)\dfrac{\phi}{p_a} 6.112 e^{\frac{17.62t}{243.12+t}}}} \text{[cm]}, \quad (24.1)$$

where $\Delta\tau$ is the width pulse, μs; γ is the adiabatic index (for air $\gamma = 7/5$); $R = 8.31441$ is the universal gas constant, J/(mol·K); t is the air temperature, °C; $M_v = 0.018$ is the molar mass of water vapor, kg/mol; $M_d = 0.029$ is the molar mass of dry air, kg/mol; φ is the relative air humidity, %; and pa is the atmospheric pressure, Pa.

To convert the digital signal value of a DHT11 relative humidity sensor into the relative humidity value, you can use the following conversion equation (Bereziuk et al. 2019):

$$\phi = \frac{100}{\sigma_H - \sigma_D}\left[\frac{I(2^n - 1)}{U_b N_\phi l_t} - \sigma_D\right] [\%], \quad (24.2)$$

where $N\varphi$ is the value of the digital signal of air relative humidity; n is the digit in analog-digital converter (ADC), bit; I is the current strength, A; U_b is the base voltage, V; σ_H is the specific electrical conductivity of humidity in the air, Sm/m; σ_D is the dielectric specific electrical conductivity, Sm/m; and l_t is the thickness of the dielectric layer, m.

Since the digital signal of a DHT22 high-precision air humidity sensor is two bytes, this value can be calculated with the formula:

$$N_\varphi = \frac{256 N_{\varphi 0} + N_{\varphi 1}}{10} \quad (24.3)$$

where $N\varphi_0$ and $N\varphi_1$ are the values of the most significant (MSB) and least significant bytes of the digital signal of the relative humidity reading, respectively.

After substitution of equation (24.3) into equation (24.2), we get the equation for the conversion of the digital signal value of a DHT22 high-precision relative humidity sensor:

$$\varphi = \frac{100}{\sigma_H - \sigma_D}\left[\frac{10 I(2^n - 1)}{U_b (256 N_{\varphi 0} + N_{\varphi 1}) l_t} - \sigma_D\right] [\%] \quad (24.4)$$

To convert the values of the digital signals of the air temperature and atmospheric pressure sensor into air temperature and atmospheric pressure values, the following transformation equations can be used:

$$t = 12.8 t' + 0.005 [°C]; \quad (24.5)$$

$$t' = 2^{-14}\left(2^{-8} N_t - K_{t1}\right)\left[K_{t2} - 2^{-12}(N_t - K_{t1}) K_{t3}\right]; \quad (24.6)$$

$$p_a = 2^{-16} p'_a \left(2^{-51} K_{p9} p'_a + 2^{-19} K_{p8} + 1\right) + 2^{-4} K_{p7} [\text{pa}]; \quad (24.7)$$

$$p'_a = \frac{2^{51}\left(1-2^{-14}N_p\right)-3125\left\{(t'-128,000)\left[K_{p6}(t'-128,000)+2^{17}K_{p5}\right]+2^{35}K_{p4}\right\}}{2^{-21}K_{p1}\left\{2^{35}+(t'-128,000)\left[2^{-20}K_{p3}(t'-128,000)+K_{p2}\right]\right\}},$$

(24.8)

where N_t and N_p are the values of digital air temperature and atmospheric pressure signals, respectively; K_{t1}, K_{t2} and K_{t3} are the gauge air temperature coefficients; K_{p1}, K_{p2}, ..., K_{p9} are the gauge atmospheric pressure coefficients; and t' is the intermediate results of air temperature and atmospheric pressure calculations.

After substitution of (24.4), (24.5), and (24.7) into equation (24.1), we get the complete equation of conversion values of ultrasonic sensor distance digital signals, atmospheric pressure, temperature, and humidity into distance values:

$$l = \frac{\Delta\tau}{2\cdot 10^4} \times \sqrt{\frac{\gamma R(273.155+12.8t')}{M_d - \frac{611.2(M_d-M_v)\left[10I(2^n-1)/(U_b(256N_{\phi 0}+N_{\phi 1})l_t)-\sigma_D\right]}{\left[2^{-16}p'_a(2^{-51}K_{p9}p'_a+2^{-19}K_{p8}+1)+2^{-4}K_{p7}\right](\sigma_H-\sigma_D)} e^{\frac{2560t'+1}{145.3t'+2760}}}} \text{ [cm]},$$

(24.9)

The total error of reproduction of the real value of the distance is composed of systematic and random errors in the elements of the measuring channel and can be determined by the quadratic dependence:

$$\delta_\Sigma = \sqrt{\delta_{in}^2 + \delta_{dn}^2 + \delta_{qe}^2 + \delta_{ce}^2 + \delta_{ze}^2}\, [\%],$$

(24.10)

where δ_{in} is the integral nonlinearity, %; δ_{dn} is the differential nonlinearity, %; δ_{qe} is the quantization error, %; δ_{ce} is the conversion factor error, %; and δ_{ze} is the zero bias error, %.

The documentation for the microcontroller gives values of individual accuracy parameters. In the article by Bereziuk et al. (2019), the following values of these errors are defined, which will be true for a high-precision ultrasonic rangefinder: $\delta_{in} = 0.05\%$; $\delta_{dn} = 0.025\%$; $\delta_{ce} = 0.2\%$; and $\delta_{ze} = 0.2\%$.

After the mathematical transformations of expression (24.1), the measurement of the quantization of the high-precision ultrasonic device to determine the distance, taking into account the environmental parameters, is:

$$\delta_{K\beta} = \frac{100\%}{\Delta\tau_{\min}} =$$

$$= \frac{1}{200 l_{\min}} \sqrt{\frac{\gamma R(273.155+12.8t')}{M_d - \frac{611.2(M_d-M_v)\left[10I(2^n-1)/(U_b(256N_{\phi 0}+N_{\phi 1})l_t)-\sigma_D\right]}{\left[2^{-16}p'_a(2^{-51}K_{p9}p'_a+2^{-19}K_{p8}+1)+2^{-4}K_{p7}\right](\sigma_H-\sigma_D)} e^{\frac{2560t'+1}{145.3t'+2760}}}}\,[\%].$$

(24.11)

Using expression (24.11), we determine that the maximum error quantization of a high-precision ultrasonic device to determine the distance, taking into account the environmental parameters, does not exceed 0.4%.

Substituting known values into expression (24.10), we determine that the total accuracy of the high-precision ultrasonic device to determine the distance, taking into account the environmental parameters, is 0.493%.

24.3 MAIN RESULTS OF THE RESEARCH

Figure 24.1 shows a block diagram of a high-precision device to determine the distance based on the environmental parameters, consisting of an ultrasonic distance sensor (USD), atmospheric pressure and temperature sensor, temperature and air humidity sensor, microcontroller unit (MCU), and displaying and monitoring module using LCD keypad shield. The schema also shows the object of measurement. To ensure the operation of the microcontroller, a clock pulse G and a source of base voltage are included in the scheme.

The microcontroller ports are configured as follows: port 1, indication of the results of measurement and control of display parameters; port 2, providing communication with a personal computer via the USB interface; port 3, channel of output ultrasonic signal at the beginning of the distance measurement; port 4, channel of the input ultrasonic signal at the end of the distance measurement; port 5, channel of measurement of relative humidity; and port 6, channel for measuring atmospheric pressure and air temperature.

To determine the distance, a HC-SR04 ultrasonic distance sensor is used, described in detail by Bereziuk et al. (2019). To take into account the correction of the air velocity in air to its relative humidity to increase the accuracy of the measurement of the relative humidity of the air, instead of the DHT11 sensor described in detail by Bereziuk et al. (2019), it is suggested to use a DHT22 relative air humidity sensor, which has increased accuracy, and the differences in the technical characteristics are given in Table 24.1.

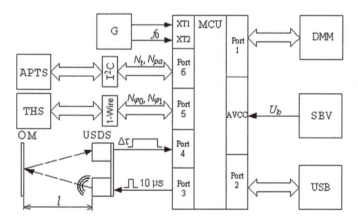

Figure 24.1 Block diagram of high-precision device for determining the distance taking into account the environmental parameters.

Table 24.1 Comparison of relative air humidity sensors

Parameter	DHT11	DHT22
Range of measurement of relative humidity, %	20...80	0...100
Error of measurement of relative humidity, %	5	2
Range of measurement of air temperature, °C	0...+50	−40...+125
Error of measurement of air temperature, %	2	0.5
Measurement frequency, not more than, Hz	1	0.5

To account for the correction of the speed of sound in the air by the amount of atmospheric pressure, a high-precision GY-BMP280-3.3 digital module is used, which measures the current values of the atmospheric pressure and ambient temperature. Another application of this module is the definition of the altitude based on the air pressure and is calculated by the international barometric formula. The module board is designed based on the improved BMP280 chip and other auxiliary components. The accuracy of the sensor allows you to fix a change in height from 20 cm. So, it allows the use of a module for building a home weather station, as well as for the creation of such devices as barometer and altimeter. Also, this module is often used in self-made planes, quadrocopters, etc. to determine the flight altitude.

The pressure sensor is based on the BMP280 chip (Bosch Module Pressure), equipped with a level converter on the basis of the PCA9306 chip and a linear RT9193 chip stabilizer. The BMP280 chip is equipped with a piezoresistance sensor, a temperature sensor, an ADC, EEPROM and RAM, and a microcontroller that supports cyclic computing of measurements (when the request is received, the module immediately returns the answer without spending time calculating). The PCA9306 chip allows data to be transmitted by the I2C bus with levels from 3.3 to 5.5 V. The RT9193 chip allows you to connect the BMP280 module to a power supply from 3.3 to 5.5 VDC.

The atmospheric pressure module supports serial interfaces of I2C and SPI (3–4 wires), which makes it easy to connect the sensor to the Arduino platform and other microcontroller devices.

The BMP280 sensor has an ADC resolution of up to 20 bits for temperature and pressure.

Characteristics of the GY-BMP280-3.3 module [20]:

- Power supply: 1.7...3.6V DC. The module contains a stabilizer and alignment of logic levels.
- Consumption current: up to 2 mA during measurement (depends on the mode of accuracy) and up to 0.2 mA in standby mode.
- Measured pressure: from 30 to 110 kPa (resolution 0.16 Pa at 0...65°C).
- Measured temperature: −40...+85°C (resolution 0.01°C).
- Bus frequency: up to 3.4 MHz for I2C and up to 10 MHz for SPI.
- Module address on I2C bus: 0x77 or 0x76.
- Logic level "1" on the I2C bus: from 0.7 to VCC (where VCC is the voltage of the module).
- Preparation for the first start-up after feeding: not less than 2 ms.
- Dimensions: 21 × 18 mm.

Because the BMP280 sensor's error of temperature measurement is much smaller than in the DHT22 sensor, it is used in the high-precision device for the determination of distance and to account for the temperature correction during calculation of the speed of sound in the air data obtained from the sensor.

The BMP280 sensor has three operating modes: sleep mode, forced mode, and normal mode.

Sleep – low power or sleep mode (measurements are not performed).

Forced is a mode where the sensor measures the required parameters by command, and the sensor enters sleep mode afterwards.

Normal – the mode in which the sensor cycles independently and measures the parameters at predetermined times. That is, it independently, at a certain time which can be set, wakes up from sleep mode and performs measurements, stores the data in memory registers, and returns to sleep mode again at a certain time.

Assigning the outputs of the module board:

- VCC – 3.3 voltage supply;
- GND – "Ground";
- SCL – serial bus clock;
- SDA – data line;
- CSB – output device;
- SDO – output of sequential data.

Connecting to the I2C bus (SCL and SDA contacts):

- VCC to the Arduino 3.3 V contact;
- GND to the Arduino GND contact;
- SCK to the Arduino SCL;
- SDA to the Arduino SDA contact.

In case of connection to the SPI bus:

- The VCC module output is connected to the Arduino 3.3 V contact;
- GND module output – to the Arduino GND contact;
- The SCL of the module – to the SCK (Arduino Uno pin 13) output of the SPI;
- The SDO module output – to contact with the SPI MISO bus (Arduino Uno output pin 12);
- The SDA module output – to the MOSI contact (Aarduino Uno output pin 11);
- The CSB (CS) module output – to any digital output of the Arduino Uno.

For the BMP280 sensor to work, the Adafruit_BMP280_Library control program library must be installed.

To perform a measurement, the BMP280 sensor should be read from the sensor gauge coefficients that are unique to each instance. Then the ADC indicators for temperature and pressure should be read and the real temperature and pressure values calculated by formulas (24.5)–(24.8).

To read data from the BMP280 sensor, you do not need to wait for the measurement to be complete. In addition, the sensor has a filter that can be adjusted to your tasks. The filter works based on the previous value. Thus, for specific problems, you can configure the programmatic filter and the accuracy of measurements using the following parameters: OVERSAMPLING for temperature (16…20 bits), OVERSAMPLING for pressure (16…20 bits), TSB – time of the sensor waiting between measurements (0.5; 62.5; 125; 250; 500; 1,000; 2,000; and 4,000 ms), and FILTER_COEFFICIENT – coefficient of the filter.

The control of the sensors from the microcontroller board of the Arduino or from another controller of a microprocessor device is carried out using a special program.

The Arduino Uno R3 was chosen to be the microcontroller board – the most common version of Arduino microcontroller boards – which is described in detail by Bereziuk et al. (2018).

The controller is programmed from the Arduino integrated development environment. A block diagram of the algorithm is shown in Figure 24.2.

For distance detection results, an LCD keypad shield is used – one of the most popular expansion cards for Arduino – as described in detail by Bereziuk et al. (2018).

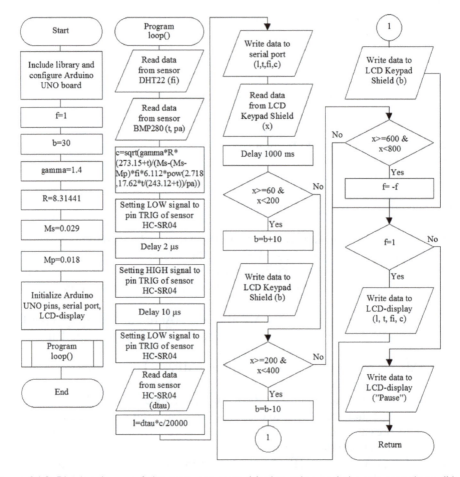

Figure 24.2 Block scheme of the main program (a) algorithm and the program loop (b).

Table 24.2 Key features of high-precision distance detection method tailored to environmental parameters

Characteristics	Value
Range of measurements, cm	2...500
Resolution, cm	0.3
The measurement error, %, no more	0.5
Effective viewing angle, °	15
Working angle of observation, °	30
Power supply, V: from USB	5
Power supply, V: from an external power supply	6...20
Consumption current, mA, no more	187
Interfaces	USB, LCD+Keypad

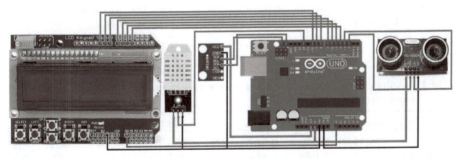

Figure 24.3 Connection of the indication and control module sensors to the Arduino Uno R3 to determine the distance based on the environmental parameters.

In Table 24.2, the main characteristics of the high-precision distance detection method, taking into account the environmental parameters, are shown.

Connection of the indication and control module sensors to the Arduino Uno R3 card to determine the distance based on the environmental parameters is shown in Figure 24.3.

The SELECT button is used to implement the pause/start function to fix the current value of the distance. The UP and DOWN buttons of the LCD keypad module are used to increase and decrease the brightness of the screen, respectively, to improve the visibility of measurement results under different lighting conditions.

The schemes of measurement of distance between the waste bin of MSW and the garbage truck with different types of loading (Karhyn 2011) are shown in Figure 24.4.

24.4 RESULTS OF EXPERIMENTS

With the help of a high-precision ultrasonic range finder, an experimental determination of the distance l in the range 25–500 cm was performed using BMP280, DHT22 (l_1) sensors, and DHT11 sensor (l_2), the results of which are shown in Table 24.3. In Table 24.3, the absolute values of relative errors $|\delta_1|$ and $|\delta_2|$ when using different sensors are also shown.

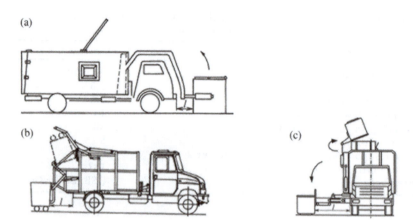

Figure 24.4 Schemes of determining the distance between the waste bin of MSW and the garbage truck with different types of loading: rear (a), lateral (b), and front (c).

Table 24.3 Results of the experiment on distance determination by high-precision and basic ultrasonic range finders

l, cm	l1, cm	l2, cm	$|\delta_1|$, %	$|\delta_2|$, %	l, cm	l1, cm	l2, cm	$|\delta_1|$, %	$|\delta_2|$, %
25	25.12	25.22	0.498	0.887	275	275.17	276.24	0.063	0.450
50	50.04	50.23	0.079	0.466	300	299.50	300.66	0.166	0.220
75	74.80	75.09	0.260	0.125	325	325.73	326.99	0.223	0.611
100	99.94	100.33	0.060	0.326	350	351.62	352.98	0.462	0.850
125	124.84	125.33	0.124	0.262	375	375.27	376.72	0.072	0.459
150	149.49	150.06	0.343	0.043	400	401.69	403.25	0.423	0.812
175	174.82	175.50	0.100	0.286	425	427.14	428.80	0.505	0.893
200	199.02	198.25	0.490	0.873	450	452.24	453.98	0.497	0.885
225	223.95	223.09	0.466	0.849	475	477.06	478.90	0.433	0.821
250	249.40	248.44	0.240	0.624	500	500.64	498.72	0.129	0.257

Figure 24.5 shows the graphical dependence of the absolute values of relative errors $|\delta_1|$ and $|\delta_2|$ from the measured distance l.

Consequently, the experimental research on determining the distance by a high-precision ultrasonic rangefinder showed that taking into account such a setting of the environment as atmospheric pressure and increasing the accuracy of temperature measurements and relative air humidity can significantly reduce the error of measurement, and the high-precision method to determine the distance in view of the environmental parameters is suitable for determining the distance between the garbage truck and the waste bin of MSW.

24.5 CONCLUSION

1. Based on the analysis of existing methods of signal processing and measurement of physical quantities, the accuracy of the vehicle is increased, which allows the distance to obstacles to be determined for the development of highly effective

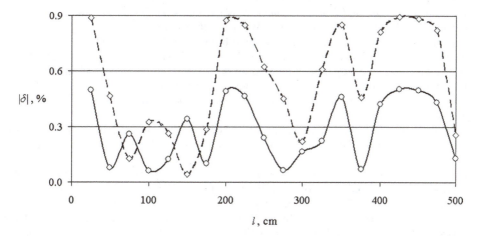

Figure 24.5 Dependence of the absolute values of the relative error $|\delta|$ on the measured distance *l*: using the BMP280, DHT22 sensors – ○; and using the DHT11 sensor – ◊.

garbage trucks, as the main link in the structure of machines for the collecting and primary processing of MSW.
2. A structural diagram of the method and a flowchart diagram of the algorithm of the program, which controls the operation of a high-precision device to determine distance based on environmental parameters, is proposed.
3. Experimental studies to determine the distance using a high-precision ultrasonic rangefinder showed that taking into account such a setting of the environment as atmospheric pressure and increasing the accuracy of temperature measurement and relative air humidity allows you to significantly reduce the error of measurement, and the highly accurate method for measuring distances taking into account the environmental parameters is suitable for determining the distance between the garbage truck and the waste bin of MSW.

REFERENCES

Balhabaev, N. N., Karlykhanov, O. K. & Stulnev, V. Y. 2016. Ultrazvukovoi urovnemer. *Patent Respublyky Kazakhstan G01F 23/28, G01F 23/68, G01F 23/296. No. 1779 MPK(2016.01)*.

Bereziuk, O., Lemeshev, M., Bogachuk, V., Wójcik, W., Nurseitova, K. & Bugubayeva, A. 2019. Ultrasonic microcontroller device for distance measuring between dustcart and container of municipal solid wastes. *Przeglad Elektrotechniczny* 4(2019): 146–150.

Bereziuk, O. V. 2015. Ohliad konstruktsii mashyn dlia zbyrannia ta pervynnoi pererobky tverdykh pobutovykh vidkhodiv. *Visnyk mashynobuduvannia ta transportu* 1: 3–8.

Bereziuk, O. V., Lemeshev, M. S., Bohachuk, V. V. & Duk, M. 2018. Means for measuring relative humidity of municipal solid wastes based on the microcontroller Arduino UNO R3. *Proc. SPIE, Photonics Applications in Astronomy, Communications, Industry, and High Energy Physics Experiments 2018* 108083G.

Bohachuk, V. V. & Mokin, B. I. 2008. *Metody ta zasoby vymiriuvalnoho kontroliu volohosti poroshkopodibnykh materialiv*. Vinnytsia: UNIVERSUM.

Cvjetkovic, V. M. & Stankovic, U. 2017. Arduino Based Physics and Engineering Remote Laboratory. *International Journal of Online Engineering* 13(1): 87–105.

Hadaichuk, N. M. & Krekoten, Ye. H. 2018. What exactly is Arduino?. XLVII reh. nauk.-tekhn. konf. profesorsko-vykladatskoho skladu, spivrobitnykiv ta studentiv VNTU.

Hadaichuk, N. M. & Krekoten, Ye. H. 2019. What exactly is a Raspberry Pi?. XLVIII reh. nauk.-tekhn. konf. profesorsko-vykladatskoho skladu, spivrobitnykiv ta studentiv VNTU.

Kaliuzhnyi, V. O. & Suslov, Ye. F. 2016. Ultrazvukovyi vymiriuvach vidstani na bazi peretvoriuvacha HC-SR04. *Mizhnar. nauk.-prakt. konf. molodykh uchenykh i studentiv: Polit. Suchasni problemy nauky.* Kyiv: Ukraina.

Karhyn, R. V. 2011. Klassyfykatsyia mashyn dlia sbora y vyvoza tverdykh bytovykh otkhodov. *Yzvestyia vysshykh uchebnykh zavedenyi. Severo-Kavkazskyi rehyon. Tekhnycheskye nauky* 2: 69–74.

Kucheruk, V. Yu., Kulakov, P. I. & Mostovyi, D. V. 2017. Prystrii vymiriuvannia pytomoi elektroprovidnosti moloka u molokopryimalnii kameri z temperaturnoiu kompensatsiieiu. *Patent Ukrainy G01N 27/00, G01N 33/04, G01R 27/00. No.121665 MPK(2017.01)*.

Kumar, N. S., Vuayalakshmi, B., Prarthana, R. J. & Shankar, A. 2016. IOT based smart garbage alert system using Arduino UNO. *2016 IEEE Region 10 Conference (TENCON), Singapore*: 1028–1034.

Osadchuk, A. V., Semenov, A. A., Baraban, S. V., Semenova, S. V. & Koval, K. O. 2013. Non-contact infrared thermometer based on a self-oscillating lambda type system for measuring the human body's temperature. *Proceedings of the 2013 23rd International Crimean Conference "Microwave & Telecommunication Technology" (CriMiCo)*, 8–14 Sept. 2013: 1069–1070.

Osadchuk, O., Koval, K., Prytula, M. & Semenov, A. 2016. Comparative analysis of radiomeasuring frequency converters of the magnetic field. *Proceedings of the XIII International Conference : Modern problems of radio engineering, telecommunications, and computer science*, Lviv-Slavsko, Ukraine, February 23–26, 2016: 275–278.

Osadchuk, O., Semenov, A., Zviahin, O., & Savytskyi, A. 2017. Numerical method for processing frequency measuring signals from microelectronic sensors based on transistor structures with negative resistance, *2017 IEEE First Ukraine Conference on Electrical and Computer Engineering (UKRCON), Conference Proceedings*, May 29 – June 2, 2017, Kyiv, Ukraine: 721–725.

Yerraboina, S., Kumar, N. M., Parimala, K. S. & Jyothi, N. A. 2018. Monitoring the smart garbage bin filling status: An IoT application towards waste management. *Technology* 9(6): 373–381.